# International Press Publications

## Mathematical Physics

*Quantum Groups: From Coalgebras to Drinfeld Algebras*
  Steven Schinider and Shlomo Sternberg
*75 Years of Radon Transform*
  edited by Simon Gindikin and Peter Michor
*Perspectives in Mathematical Physics*
  edited by Robert Penner and S.-T. Yau
*Essays On Mirror Manifolds*
  edited by S. T. Yau
*Mirror Symmetry II*
  edited by Brian Greene

## Number Theory

*Fermat's Last Theorem: Elliptic Curves and Modular Forms*
  edited by John Coates and Shing Tung Yau

## Geometry and Topology

$L^2$ *Moduli Spaces with 4-Manifolds with Cylindrical Ends*
  by Clifford Henry Taubes
*The $L^2$ Moduli Space and a Vanishing Theorem for Donaldson Polynomial Invariants*
  by J. Morgan, T. Mrowka, and D. Ruberman
*Algebraic Geometry and Related Topics*
  edited by J.-H. Yang, Y. Namikawa, and K. Ueno
*Lectures on Harmonic Maps*
  by R. Schoen and S.-T. Yau
*Lectures on Differential Geometry*
  by R. Schoen and S.-T. Yau
*Geometry, Topology and Physics for Raoul Bott*
  edited by S.-T. Yau
*Lectures on Low-Dimensional Topology*
  edited by K. Johannson
*Chern, A Great Geometer*
  edited by S.-T. Yau
*Surveys in Differential Geometry*
  edited by C.C. Hsiung and S.-T. Yau

**Analysis**

*Proceedings of the Conference on Complex Analysis*
    edited by Lo Yang
*Integrals of Cauchy Type on the Ball*
    by S. Gong
*Proceedings of the Conference in honor of Robert Finn*
    edited by P. Concus and K. Lancaster

**Physics**

*Physics of the Electron Solid*
    edited by S.-T. Chui
*Proceedings of the International Conference on Computational Physics*
    edited by D.H. Feng and T.-Y. Zhang
*Chen Ning Yang, A Great Physicist of the Twentieth Century*
    edited by S.-T. Yau

**Collected and Selected Works**

*The Collected Works of Freeman Dyson*
*The Collected Works of C. B. Morrey*
*The Collected Works of P. Griffiths*
*V. S. Varadarajan*

# Elliptic Curves, Modular Forms, & Fermat's Last Theorem

Edited by
John Coates
University of Cambridge

S. T. Yau
Harvard University

International Press

Editorial Board

Barry Mazur
Harvard University
1 Oxford Street
Cambridge, MA 02138
U.S.A.

International Press Incorporated, Boston
P.O. Box 2872
Cambridge, MA 02238-2892

Copyright ©1995 by International Press Incorporated

All rights are reserved. No part of this work can be reproduced in any form, electronic or mechanical, recording, or by any information storage and data retrieval system, without specific authorization from the publisher. Reproduction for classroom or personal use will, in most cases, be granted without charge.

Library of Congress, Card Catalog Number: 95-079521

Fermat's Last Theorem: Elliptic curves and modular forms
Edited by John Coates, and S.-T. Yau

ISBN 1-57146-026-8

Typeset using AMS-LaTeX
Printed on acid free paper, in the United States of America

## Table of Contents

**Forward by John Coates and Shing-Tung Yau** ............................ 1

**On The Symmetric Square Of A Modular Elliptic Curve**
J. Coates, A. Sydenham ................................................. 2

**The Refined Conjecture Of Serre**
Fred Diamond ......................................................... 22

**Wiles Minus Epsilon Implies Fermat**
Noam Elkies .......................................................... 38

**Geometric Galois Representations**
J.-M. Fontaine and B. Mazur .......................................... 41

**On Elliptic Curves**
Gerhard Frey ......................................................... 79

**Complete Intersections And Gorenstein Rings**
H.W. Lenstra, Jr. ..................................................... 99

**Homologie Des Courbes Modulaires**
L. Merel ............................................................. 110

**Irreducible Galois Representation**
Kenneth A. Ribet .................................................... 131

**Mod $p$ Representations Of Elliptic Curves**
K. Rubin, A. Silverberg .............................................. 148

**A Review Of Non-Archimedean Elliptic Functions**
John Tate ........................................................... 162

**On Galois Representations Associated to Hilbert Modular Forms II**
Richard Taylor ...................................................... 185

Conference on Elliptic Curves and Modular Forms
Hong Kong, December 18-21, 1993
Copyright ©1995 International Press

# **Forward**

A conference, on the general theme of "Elliptic curves and modular forms", was held in the Mathematics Department of the Chinese University of Hong Kong from December 18–21, 1993. The impetus for organizing the conference arose from Andrew Wiles' deep and spectacular work on the celebrated conjecture that every elliptic curve over Q is modular, although only some of the lectures at the conference were specifically related to this theme. At the time of the conference, the difficulties in the last hurdle in Wiles' work (the proof of the conjectural upper bound for the order of the Selmer group attached to the symmetric square of a modular form) had still not been overcome. However, the optimism shared by all at the conference that is was only a matter of time until the proof would become complete has happily been borne out by subsequent events. It is now history that Wiles himself, assisted by R. Taylor, found a beautiful proof of the desired upper bound. As a result, we now know today the remarkable fact that every semi-stable elliptic curve over Q is modular. Not only is this result revolutionary in its own right for the study of the arithmetic of these elliptic curves, but it has the added bonus that it provides at last a proof of Fermat's last theorem, thanks to the earlier work of Frey, Ribet and others. During the conference, lectures were given by John Coates, Noam Elkies, Matthias Flach, Jean-Marc Fontaine, Gerhard Frey, Dick Gross, Victor Kolyvagin, Ken Ribet, Karl Rubin, Jean-Pierre Serre, John Tate, Richard Taylor, and Don Zagier. The present short volume is a mixture of the texts of some of these lectures, together with a number of recent articles related to the general theme of the conference. Finally, the organizers of the conference wish to express their warmest thanks to Professor Charles Gao, Vice-Chancellor, and to Professor K.W. Leung, Chairman of the Mathematics Department, of the Chinese University of Hong Kong for their support and assistance throughout the preparation and running of the conference.

John Coates, Cambridge University<br>
Shing-Tung Yau, Harvard University

Conference on Elliptic Curves and Modular Forms
Hong Kong, December 18-21, 1993
Copyright ©1995 International Press

# On the symmetric square of a modular elliptic curve

J. COATES
A. SYDENHAM
DEPT. OF PURE MATHEMATICS AND MATHEMATICAL STATISTICS
UNIVERSITY OF CAMBRIDGE
16 MILL LANE, CAMBRIDGE CB2 1SB ENGLAND
*E-mail address*: JHC13@DPMMS.CAM.AC.UK

**1 Introduction.** Let $E$ be a modular elliptic curve over $\mathbb{Q}$, and let $Sym^2(E)$ denote the 3-dimensional motive whose $l$-adic realizations are the Galois representations $Sym^2(H^1_l(E))$, where $H^1_l(E)$ denotes the $l$-adic cohomology of $E$ (see §2). The work of Bloch and Kato [1] enables us to define analogues for $Sym^2(E)$ of the classical Selmer group of $E$. It is conjectured that these Selmer groups of $Sym^2(E)$ are finite for every $E$ and every prime $p$, and this has been proven for all but a finite number of primes $p$ and every $E$ by the deep work of $M$. Flach [5], [6]. There is also a conjectural exact formula for the orders of these Selmer groups due to Bloch and Kato [1], and a proof of this exact formula has become one of the major goals of number theory because of Andrew Wiles' momentous discovery that it is deeply related to the conjecture that every elliptic curve over $\mathbb{Q}$ is modular. The present note is not concerned with this exact formula per se, but rather with proving some basic facts about the Galois cohomology of $Sym^2(E)$, which may eventually be useful in establishing the exact formula. In fact, we carry out for $Sym^2(E)$ arguments which are entirely parallel to those given for $E$ itself (subject to the hypothesis that the analytic rank of $E$ is $\leqslant 1$) in the paper [4]. Indeed, B. Perrin-Riou pointed out to one of us in an unpublished letter some years ago that the proofs in [4] can be generalized to motives, and she has subsequently included a discussion of these questions for $Sym^2(E)$ in [11]. Thus the only real novelty in the present paper lies in the detailed arguments, as well as one or two finer results about the relevant Iwasawa modules (see Theorem 4.2). It seemed to us worthwhile to give proofs as simply and as fully as possible, both because of the great interest in $Sym^2(E)$ created by Wiles' work, and also because they provide a beautiful illustration of some aspects of the motivic theory. For simplicity, we have basically only considered $E$ without complex multiplication. For curves $E$ with complex multiplication and primes $p$ which do not divide the number of roots of unity in the field of complex multiplication, it is well known (see [2], [5], [12]) that one can establish full analogues of the results of this paper, together with exact formula for the order of the Selmer groups of $Sym^2(E)$, by using Rubin's proof of the main conjectures of Iwasawa theory for such $E$.

## 2 The critical twists of $Sym^2(E)$ and their local behaviour.

Let $E$ be an elliptic curve defined over $\mathbb{Q}$. We begin by briefly recalling the definition of the motive $Sym^2(E)$ and its two Tate twists which are critical. We then derive a number of basic local properties of the two critical twists.

Let $l$ be a prime number. For each $n \geqslant 1$, let $E_{l^n}$ be the group of $l^n$-division points, and, as usual, put

$$T_l(E) = \varprojlim E_{l^n}, \quad V_l(E) = T_l(E) \otimes \mathbb{Q}_l.$$

Of course, these spaces are endowed with their canonical action of the Galois group $G_{\mathbb{Q}} = G(\overline{\mathbb{Q}}/\mathbb{Q})$. Let $H^1_l(E) = Hom(V_l(E), \mathbb{Q}_l)$, so that we can view the $H^1_l(E)$ as the $l$-adic realizations of the motive $H^1(E)$. For each $n \geqslant 1$, let $\mu_{l^n}$ be the group of $l^n$-th roots of unity, and put $\mathbb{Q}_l(1) = (\varprojlim \mu_{l^n}) \otimes_{\mathbb{Z}_l} \mathbb{Q}_l$. The Weil pairing defines a canonical isomorphism of $G_{\mathbb{Q}}$-modules

(2.1) $$V_l(E) \xrightarrow{\sim} Hom(V_l(E), \mathbb{Q}_l(1)).$$

Let $Sym^2(E)$ denote the motive whose $l$-adic realizations are given by $Sym^2(H^1_l(E))$. This is a motive of dimension 3, and we refer to [2] for a more detailed discussion of its realizations and the Euler factors of its $L$-series. In particular, it has just two Tate twists which are critical, namely

(2.2) $$A = (Sym^2((E)))(1), \quad B = (Sym^2(E))(2).$$

Using (2.1), it is easy to see that the $l$-adic realizations of these motives are given by

(2.3) $$H_l(A) = Sym^2(V_l(E))(-1), \quad H_l(B) = Sym^2(V_l(E)).$$

Let us also note that (2.1) implies that $\check{A} = A$, where $\check{A}$ denotes the dual motive (i.e. the motive whose realizations are the dual vector spaces of the realizations of $A$). Hence $B = \check{A}(1)$, so that $A$ and $B$ correspond under the functional equation. Let $d^{\pm}(A)$ denote the dimensions of the subspaces of $H_l(A)$ on which complex conjugation acts like $\pm 1$, respectively, and let $d^{\pm}(B)$ be defined similarly for $B$. We have

(2.4) $$d^+(A) = d^-(B) = 1, \quad d^-(A) = d^+(B) = 2.$$

The vector spaces $H_l(A)$ and $H_l(B)$ contain respectively the $G_{\mathbb{Q}}$-invariant lattices

(2.5) $$T_l(A) = (Sym^2(T_l(E)))(-1), \quad T_l(B) = Sym^2(T_l(E)).$$

We can then form the quotients

(2.6) $$\mathfrak{A}_{l^\infty} = H_l(A)/T_l(A), \quad \mathfrak{B}_{l^\infty} = H_l(B)/T_l(B),$$

which are discrete, divisible, torsion $G_{\mathbb{Q}}$-modules. The Weil pairing (2.1) then gives rise to a canonical isomorphism of $G_{\mathbb{Q}}$-modules

(2.7) $$\mathfrak{A}_{l^n} \xrightarrow{\sim} Hom(\mathfrak{B}_{l^n}, \mu_{l^n}),$$

where $n$ is any integer $\geq 1$, and $\mathfrak{A}_{l^n}$ (resp. $\mathfrak{B}_{l^n}$) denotes the kernel of multiplication by $l^n$ in $\mathfrak{A}_{l^\infty}$ (resp. $\mathfrak{B}_{l^\infty}$).

For each finite prime $v$ of $\mathbb{Q}$, let $L_v(Sym^2(E), s)$ be the Euler factor at $v$ of the $L$-function of the motive $Sym^2(E)$. We refer the reader to [2], where $L_v(Sym^2(E), s)$ is calculated explicitly for all $v$. Of course, the Euler factors for $A$ and $B$ are then given by

$$L_v(A, s) = L_v(Sym^2(E), s+1), \quad L_v(B, s) = L_v(Sym^2(E), s+2),$$

and the next lemma is then plain from the explicit formulae of [2].

**Lemma 2.1.**
(i) $L_v(A, s)$ has a pole at $s = 0$ for all but a finite number of $v$, and this pole is simple whenever it occurs; in particular, $L_v(A, s)$ has a simple pole at $s = 0$ whenever $v$ is a prime where $Sym^2(E)$ has good reduction;

(ii) $L_v(B, s)$ is homomorphic at $s = 0$ for all finite primes $v$.

We now fix a prime number $p$. Let $v$ be any finite prime of $\mathbb{Q}$ different from $p$. We write $I_v$ for the inertial subgroup of $G_v = G(\overline{\mathbb{Q}}_v/\mathbb{Q}_v)$, and $\text{Frob}_v$ for the Frobenius automorphism in $G_v/I_v$. Let $M$ denote either the motive $A$ or the motive $B$. Following Bloch-Kato, we then define

$$(2.8) \quad H^1_f(\mathbb{Q}_v, H_p(M)) = H^1(G_v/I_v, H_p(M)^{I_v}) \quad (v \neq p).$$

**Lemma 2.2.** Assume $v \neq p$. Then
(i) $H^1_f(\mathbb{Q}_v, H_p(B)) = 0$;
(ii) $H^1_f(\mathbb{Q}_v, H_p(A))$ has dimension 1 or 0 over $\mathbb{Q}_p$, according as the Euler factor $L_v(A, s)$ does or does not have a pole at $s = 0$.

*Proof.* Since $G_v/I_v$ is topologically generated by $\text{Frob}_v^{-1}$, we have

$$H^1(G_v/I_v, H_p(M)^{I_v}) = H_p(M)^{I_v}/(1 - \text{Frob}_v^{-1})H_p(M)^{I_v}.$$

On the other hand, since $v \neq p$, the definition of $L_v(M, s)$ is

$$L_v(M, s) = \det(1 - (Frob_v)^{-1}.v^{-s} \mid H_p(M)^{I_v})^{-1}.$$

Hence Lemma 2.2 is an immediate consequence of Lemma 2.1. $\square$

We next consider the case when $v = p$. we recall that, in this case, Bloch and Kato define

$$H^1_f(\mathbb{Q}_p, H_p(M)) = Ker(H^1(\mathbb{Q}_p, H_p(M))$$
$$\to H^1(\mathbb{Q}_p, H_p(M) \underset{\mathbb{Q}_p}{\otimes} B_{\text{crys}})),$$

where $B_{\text{crys}}$ is the ring defined by Fontaine (see [1]). Prior to calculating the dimension of $H^1_f(\mathbb{Q}_p, H_p(M))$, we need the following basic lemma.

**Lemma 2.3.**
(i) $H^0(\mathbb{Q}_p, H_p(B)) = 0$;

*(ii)  If $E$ does not admit complex multiplication, then $H^0(\mathbb{Q}_p, H_p(A)) = 0$.*

*Proof.* It will be convenient to introduce the following notation. Let $\chi : G_p \to \mathbb{Z}_p^\times$ be the character giving the action of $G_p$ on $\mathbb{Q}_p(1)$. By the irreducibility of the cyclotomic equation over $\mathbb{Q}_p$, $\chi$ is surjective. Let

$$\rho : G_p \to GL(V_p(E)), \quad \varphi : G_p \to GL(Sym^2(V_p(E)))$$

be the homomorphisms arising from the action of $G_p$ on $V_p(E)$ and $Sym^2(V_p(E))$. By the Weil pairing, we know that $\det(\rho) = \chi$. For each $\sigma \in G_p$, we write $\lambda(\sigma)$, $\mu(\sigma)$ for the eigenvalues in $\overline{\mathbb{Q}}_p$ of $\rho(\sigma)$. Recall that if $\{e_1, e_2\}$ is a $\mathbb{Q}_p$-basis of $V_p(E)$, then a $\mathbb{Q}_p$-basis of $Sym^2(V_p(E))$ is given by the set $\{e_1 \otimes e_1, e_2 \otimes e_2, e_1 \otimes e_2 + e_2 \otimes e_1\}$.

We first consider case (i). Suppose that, on the contrary, there does exist $z \neq 0$ in $Sym^2(V_p(E))$ such that $\varphi(\sigma)(z) = z$ for all $\sigma \in G_p$. This means that 1 must be an eigenvalue of $\varphi(\sigma)$ for every $\sigma$ in $G_p$. But linear algebra tells us that the eigenvalues with multiplicity of $\varphi(\sigma)$ are $\lambda(\sigma)^2$, $\mu(\sigma)^2$, $\chi(\sigma) = \lambda(\sigma)\mu(\sigma)$. Now choose $\sigma_0$ in $G_p$ such that $\chi(\sigma_0)$ is not a root of unity. This is clearly possible because $\chi$ is surjective. Since $\chi(\sigma_0) \neq 1$, we must have one of the other eigenvalues of $\varphi(\sigma_0)$ equal to 1, say $\lambda(\sigma_0)^2 = 1$. Taking $\tau = \sigma_0^2$, we conclude that we must have one of the eigenvalues of $\rho(\tau)$ equal to 1, say $\lambda(\tau) = 1$. Since $\chi(\tau) \neq 1$, we can then choose a basis $\{e_1, e_2\}$ of $V_p(E)$ consisting of eigenvectors for the eigenvalues $1, \chi(\tau)$, respectively. Writing $z$ as a linear combination of the basis elements $e_1 \otimes e_1, e_2 \otimes e_2, e_1 \otimes e_2 + e_2 \otimes e_1$, and using the fact $\chi^2(\tau) \neq 1$, we deduce immediately from the equation $\varphi(\tau)(z) = z$, that we can take $z = e_1 \otimes e_1$. Now take any $\sigma$ in $G_p$, and let

$$(2.9) \qquad \rho(\sigma) = \begin{pmatrix} a(\sigma) & b(\sigma) \\ c(\sigma) & d(\sigma) \end{pmatrix}$$

be the matrix of $\rho(\sigma)$ relative to the basis $\{e_1, e_2\}$. Then the equation $\varphi(\sigma)(z) = z$ yields immediately $a(\sigma)^2 = 1$ and $c(\sigma) = 0$ for all $\sigma \in G_p$. Then $\sigma \mapsto a(\sigma)$ defines a quadratic character of $G_p$, and we let $L$ be the fixed field of the kernel of this character. It is then clear that $G(\overline{\mathbb{Q}}/L)$ fixes the 1-dimensional subspace of $V_p(E)$ which is generated by $e_1$. But this is impossible, since the $p$-primary subgroup of $E(L)$ is finite. This completes the proof of (i).

We now turn to (ii). The argument breaks up into cases, depending on the nature of the reduction of $E$.

Case (1). Assume that $E$ has potential supersingular reduction. This means that there exists a finite extension $L$ of $\mathbb{Q}_p$ such that $E$ has good supersingular reduction over $L$. Enlarging $L$ if necessary, we can assume it is Galois over $\mathbb{Q}_p$. Let $O_L$ be the ring of integers of $L$. Let $\hat{E}$ be the height 2 formal group over $O_L$ giving the kernel of reduction on $E$ over $L$. We claim that we have the natural isomorphism

$$(2.10) \qquad End_{O_L}(\hat{E}) \otimes \mathbb{Q}_p \xrightarrow{\sim} H_p(A)^{G_L} \oplus \mathbb{Q}_p,$$

where $G_L = G(\overline{\mathbb{Q}}_p/L)$. To see this, we note that $V_p(E) = V_p(\hat{E})$, and recall that we can identify $\hat{E}$ with a connected $p$-divisible group of height 2 over $O_L$.

Hence, by Corollary 1 of Theorem 4 of [14], we have

(2.11) $$End_{O_L}(\hat{E}) \otimes \mathbb{Q}_p \xrightarrow{\sim} Hom(V_p(E), V_p(E))^{G_L}.$$

The assertion (2.10) now follows on noting that the Weil pairing gives an isomorphism of $G_p$-modules

$$Hom(V_p(E), V_p(E)) \xrightarrow{\sim} H_p(A) \oplus \mathbb{Q}_p.$$

But, since $E$ is defined over $\mathbb{Q}_p$, $G(L/\mathbb{Q}_p)$ operates on $End_{O_L}(\hat{E})$ via its action on the coefficients of the formal power series over $O_L$ defining an endomorphism. With this action, (2.10) is an isomorphism of $G(L/\mathbb{Q}_p)$-modules. Suppose now that $H^0(G_p, H_p(A)) \neq 0$. It follows from (2.10) that $H^0(G(L/\mathbb{Q}_p), End_{O_L}(\hat{E}))$ has $\mathbb{Z}_p$-rank $\geq 2$. But this is impossible, because Lubin [9] has proven that we obtain an injective homomorphism $c : End_{O_L}(\hat{E}) \hookrightarrow O_L$ by defining $c(\alpha)$ to be the coefficient of $t$ in the formal power series in a variable $t$ which defines $\alpha$. Hence $H^0(G(L/\mathbb{Q}_p), End_{O_L}(\hat{E}))$ has $\mathbb{Z}_p$-corank at most 1, which contradicts the above. Thus $H^0(G_p, H_p(A)) = 0$ in this case.

We now make several remarks in linear algebra prior to discussing the two remaining cases. Suppose that we can find $\sigma_0$ in $G_p$ such that the eigenvalues $\lambda(\sigma_0)$ and $\mu(\sigma_0)$ are distinct elements of $\mathbb{Z}_p$ (this will evidently be true in the remaining cases). Let $\{e_1, e_2\}$ be a basis of $V_p(E)$ consisting of eigenvector for $\lambda(\sigma_0)$ and $\mu(\sigma_0)$. Let $W_i = \mathbb{Q}_p e_i$ ($i = 1, 2$). We claim that the assumption that $H^0(G_p, H_p(A)) \neq 0$ implies that $W_1$ and $W_2$ are stable under the action of $G_p$, and that $V_p(E) = W_1 \oplus W_2$. Indeed, suppose that $H^0(G_p, H_p(A)) \neq 0$. This means that there exists $z \neq 0$ in $Sym^2(V_p(E))$ such that

(2.12) $$\varphi(\sigma)(z) = \chi(\sigma) z \quad \text{for all} \quad \sigma \in G_p.$$

Writing $z$ as a $\mathbb{Q}_p$-linear combination of $e_1 \otimes e_1$, $e_2 \otimes e_2$, $e_1 \otimes e_2 + e_2 \otimes e_1$, and applying (2.12) with $\sigma = \sigma_0$, we conclude that $z$ is necessarily a scalar multiple of $e_1 \otimes e_2 + e_2 \otimes e_1$, and hence we can suppose $z = e_1 \otimes e_2 + e_2 \otimes e_1$. Now take $\sigma$ to be any element of $G_p$, and let (2.9) be the matrix of $\rho(\sigma)$ relative to the basis $\{e_1, e_2\}$. Using the equation (2.12), we deduce after a short calculation that $b(\sigma) = c(\sigma) = 0$. Hence $V_p(E)$ is a direct sum of the $G_p$-spaces $W_1$ and $W_2$, as asserted.

Case (2). Assume that $E$ has multiplicative reduction at $p$. This means that there is a quadratic extension $L$ of $\mathbb{Q}_p$ such that $E$ is isomorphic over $L$ to the Tate curve $E_q$ for a suitable $q \in \mathbb{Q}_p^\times$ (see [13], Appendix A.1). But $Sym^2(V_p(E))$ does not change if we replace $E$ by a quadratic twist, and so we can assume that $E = E_q$. But then we have the exact sequence of $G_p$-modules

(2.13) $$0 \to \mathbb{Q}_p(1) \to V_p(E_q) \to \mathbb{Q}_p \to 0$$

of $G_p$-modules. Hence, for all $\sigma$ in $G_p$, the eigenvalues of $\rho(\sigma)$ are equal to $\chi(\sigma)$ and 1. We can therefore choose $\sigma_0$ above to be any element of $G_p$ such that $\chi(\sigma_0) \neq 1$, and we can take $e_1$ to be a basis of $\mathbb{Q}_p(1)$. Then the above argument shows that the sequence (2.13) splits as an exact sequence of $G_p$-modules if $H^0(G_p, H_p(A)) \neq 0$. But it is well known (see [13], Appendix A.1) that (2.13) does not split as an exact sequence of $G_p$-modules. Hence $H^0(G_p, H_p(A)) = 0$ in this case.

Case (3). Suppose next that $E$ has potential good ordinary reduction at $E$. This means that there exists a finite extension $L$ of $\mathbb{Q}_p$ such that $E$ has good ordinary reduction over $L$. Let $\hat{E}$ be the formal group and $\widetilde{E}$ the reduction of $E$ over $L$. Put $G_L = G(\overline{\mathbb{Q}}_p/L)$. Then we have the exact sequence of $G_L$-modules

(2.14) $$0 \to X \to V_p(E) \to Y \to 0,$$

where $X$ comes from the Tate module of $\hat{E}$, and $Y$ from the Tate module of $\widetilde{E}$. In particular, $Y$ is an unramified $G_L$-module. Hence we can choose $\sigma_0$ to be any element of the inertial subgroup of $G_L$ such that $\chi(\sigma_0) \neq 1$, and take $e_1$ to be a basis of $X$. Then the above argument shows that (2.14) necessarily splits as an exact sequence of $G_p$-modules if $H^0(G_p, H_p(A)) \neq 0$. But it is shown in Appendix A.2 of [13] (see the theorem of A.2.4) that the splitting of (2.14) as a sequence of $G_L$-modules implies that $E$ has complex multiplication. Since we assume in (ii) that $E$ does not admit complex multiplication, this completes the proof of Lemma 2.3. □

**Lemma 2.4.** *Assume $E$ does not admit complex multiplication*
(i) $\dim_{\mathbb{Q}_p}(H^1_f(\mathbb{Q}_p, H_p(B))) = 2$,
(ii) $\dim_{\mathbb{Q}_p}(H^1_f(\mathbb{Q}_p, H_p(A))) = 1$.

*Proof.* Let $M$ denote either the motive $A$ or the motive $B$. Let $B_{DR}$ be Fontaine's field of $p$-adic periods, and, as usual, we define

$$DR_p(M) = (H_p(M) \underset{\mathbb{Q}_p}{\otimes} B_{DR})^{G_p}.$$

We obtain a $\mathbb{Q}_p$-vector space, which is endowed with the filtration $Fil^k DR_p(M)$ ($k \in \mathbb{Z}$) coming from the canonical filtration on $B_{DR}$. Now it is well known that $H_p(M)$ is a de Rham representation for every prime $p$, i.e. $DR_p(M)$ has dimension 3 over $\mathbb{Q}_p$. Hence, in view of Lemma 2.3, the formula of Bloch-Kato (see [1], p. 355) gives

(2.15) $$\dim_{\mathbb{Q}_p}(H^1_f(\mathbb{Q}_p, H_p(M))) = \dim_{\mathbb{Q}_p}(DR_p(M)/Fil^0 DR_p(M)).$$

Hence it remains to compute the dimension on the right of this formula. This is done most simply by using the fact that Fontaine's stronger de Rham conjecture is true for $M$ and every prime $p$. Let $H_{DR}(M)$ be the de Rham realization of $M$. We recall that $H_{DR}(M)$ is a 3-dimensional vector space over $\mathbb{Q}$, which is endowed with the Hodge filtration $Fil^k H_{DR}(M)$ ($k \in \mathbb{Z}$). Then there is an isomorphism of $\mathbb{Q}_p$-vector spaces

(2.16) $$\theta_p : DR_p(M) \xrightarrow{\sim} H_{DR}(M) \underset{\mathbb{Q}}{\otimes} \mathbb{Q}_p,$$

which preserves the filtrations. Hence we conclude from (2.15) and (21.6) that

(2.17) $$\dim_{\mathbb{Q}_p}(H^1_f(\mathbb{Q}_p, H_p(M))) = \dim_{\mathbb{Q}}(H_{DR}(M)/Fil^0 H_{DR}(M)).$$

Now $M$ is critical at $s = 0$, and it is well known (see [3], p. 156) that this implies that the right hand side of (2.17) is equal to $d^+(M)$. Since $d^+(B) = 2$ and $d^+(A) = 1$, this completes the proof of Lemma 2.4. $\square$

**3 Galois cohomology over $\mathbb{Q}$.** Although it is not essential, we impose the following assumptions for the rest of the paper to simplify the discussion.

Hypothesis on $E$ and $p$. *We assume that $E$ does not admit complex multiplication, and that the prime $p$ is odd.*

Let $S$ denote any finite set of non-archimedean primes of $\mathbb{Q}$ such that $S$ contains both $p$ and all primes $q \neq p$ where the $G_\mathbb{Q}$-representation $H_p(A)$ is ramified (or equivalently the $G_\mathbb{Q}$-representation $H_p(B)$ is ramified). We write $\mathbb{Q}_S$ for the maximal extension of $\mathbb{Q}$ unramified outside $S$ and $\infty$, and put $G_S = G(\mathbb{Q}_S/\mathbb{Q})$. Then $\mathfrak{A}_{p^\infty}$ and $\mathfrak{B}_{p^\infty}$ as defined by (2.6) are discrete $G_S$-modules, and the aim of this section will be to study their $G_S$-cohomology. The principal tool used will be the Poitou-Tate sequence for finite $G_S$-modules (see, for example, [10], p. 70), together with an important modification of this sequence, which was first pointed out by Cassels in the case of elliptic curves. Our arguments are the direct analogue of those used in [4] for modular elliptic curves over $\mathbb{Q}$ with analytic rank $\leqslant 1$.

We begin by stating the basic exact sequences we shall use. As before, let $M$ denote either $A$ or $B$, and let $T_p(M)$ be the $G_S$-lattice in $H_p(M)$ which is defined by (2.5). We also put $\mathfrak{M}_{p^\infty} = H_p(M)/T_p(M)$, so that we have a canonical exact sequence of $G_S$-modules

(3.1) $$0 \longrightarrow T_p(M) \xrightarrow{\rho_M} H_p(M) \xrightarrow{\pi_M} \mathfrak{M}_{p^\infty} \longrightarrow 0.$$

Let $v$ be any finite place of $\mathbb{Q}$, and let

$$Y_v(M) = H^1_f(\mathbb{Q}_v, H_p(M))$$

be the subspace of $H^1(\mathbb{Q}_v, H_p(M))$ which is defined in §2. We then define

(3.2) $$\begin{aligned} H^1_f(\mathbb{Q}_v, T_p(M)) &= \rho_M^{-1}(Y_v(M)), \\ H^1_f(\mathbb{Q}_v, \mathfrak{M}_{p^\infty}) &= \pi_M(Y_v(M)), \end{aligned}$$

where $\rho_M$ and $\pi_M$ now denote the maps on the corresponding $H^1$'s which are induced by (3.1). Motivated by analogy with elliptic curves, we then put

(3.3) $$H^1_f(\mathbb{Q}, T_p(M)) = Ker\left(H^1(G_S, T_p(M)) \to \bigoplus_{v \in S} \frac{H^1(\mathbb{Q}_v, T_p(M))}{H^1_f(\mathbb{Q}_v, T_p(M))}\right)$$

(3.4) $$H^1_f(\mathbb{Q}, \mathfrak{M}_{p^\infty}) = Ker\left(H^1(G_S, \mathfrak{M}_{p^\infty}) \to \bigoplus_{v \in S} \frac{H^1(\mathbb{Q}_v, \mathfrak{M}_{p^\infty})}{H^1_f(\mathbb{Q}_v, \mathfrak{M}_{p^\infty})}\right)$$

If $X$ is a $\mathbb{Z}_p$-module, we put $T_p(X) = \varprojlim (X)_{p^n}$, where $(X)_{p^n}$ denotes the kernel of $p^n$ on $X$. In addition, if $X$ is locally compact, we write $X^\wedge = Hom_{cont}(X, \mathbb{Q}_p/\mathbb{Z}_p)$ for the Pontrjagin dual of $X$.

**Theorem 3.1.** *We have the following two canonical exact sequences*

$$
(3.5) \quad 0 \to H^1_f(\mathbb{Q}, \mathfrak{A}_{p^\infty}) \to H^1(G_S, \mathfrak{A}_{p^\infty}) \to \bigoplus_{v \in S} \frac{H^1(\mathbb{Q}_v, \mathfrak{A}_{p^\infty})}{H^1_f(\mathbb{Q}_v, \mathfrak{A}_{p^\infty})}
$$
$$
\to H^1_f(\mathbb{Q}, T_p(B))^\wedge \to H^2(G_S, \mathfrak{A}_{p^\infty}) \to 0
$$

$$
(3.6) \quad 0 \to \bigoplus_{v \in S} T_p(\mathfrak{A}_{p^\infty}(\mathbb{Q}_v)) \to H^2(G_S, \mathfrak{B}_{p^\infty})^\wedge
$$
$$
\to H^1_f(\mathbb{Q}, T_p(A)) \to \bigoplus_{v \in S} H^1_f(\mathbb{Q}_v, T_p(A)).
$$

*Proof.* Although both sequences are well known to the experts (see [5] or [7]), we shall sketch their derivation from the Poitou-Tate sequence for finite $G_S$-modules, as this is not easy to extract from the literature. For brevity, put

$$M_n = \mathfrak{A}_{p^n}, \quad R_n = \mathfrak{B}_{p^n}.$$

Recalling (2.7), the Poitou-Tate long exact sequence for these pair of $G_S$-modules is (see [10], p. 70)

$$
(3.7) \quad \begin{aligned} 0 \to H^0(\mathbb{Q}, M_n) &\to \bigoplus_{v \in S} H^0(\mathbb{Q}_v, M_n) \to H^2(G_S, R_n)^\wedge \\ \to H^1(G_S, M_n) &\to \bigoplus_{v \in S} H^1(\mathbb{Q}_v, M_n) \to H^1(G_S, R_n)^\wedge \\ \to H^2(G_S, M_n) &\to \bigoplus_{v \in S} H^0(\mathbb{Q}_v, R_n)^\wedge \to H^0(\mathbb{Q}, R_n)^\wedge \to 0 \end{aligned}
$$

Let us also note that the standard properties of the Weil pairing shows that, under the isomorphism (2.7), the dual of the inclusion $R_n \hookrightarrow R_{n+1}$ is given by the map from $M_{n+1}$ to $M_n$ which is multiplication by $p$, and similarly with $M_n$ and $R_n$ interchanged.

We first establish (3.6). Let us pass to the projective limit in (3.7) relative to the maps from $M_{n+1}$ to $M_n$ given by multiplication by $p$. The limit sequence remains exact because all the groups in (3.7) are finite. We note that $H^0(\mathbb{Q}, T_p(A)) = 0$ because $H^0(\mathbb{Q}_p, H_p(A)) = 0$ by Lemma 2.3. Hence, neglecting the latter part of the limit sequence of (3.7) we obtain the exact sequence

$$
(3.8) \quad \begin{aligned} 0 \to \bigoplus_{v \in S} T_p(\mathfrak{A}_{p^\infty}(\mathbb{Q}_v)) &\to H^2(G_S, \mathfrak{B}_{p^\infty})^\wedge \\ \to H^1(G_S, T_p(A)) &\xrightarrow{\mu_A} \bigoplus_{v \in S} H^1(\mathbb{Q}_v, T_p(A)) \end{aligned}
$$

Recalling the definition (3.3), simple diagram chasing shows that we have an exact sequence

$$0 \to Ker\, \mu_A \to H^1_f(\mathbb{Q}, T_p(A)) \to \bigoplus_{v \in S} H^1_f(\mathbb{Q}_v, T_p(A)).$$

Combining this last sequence with (3.8), we have clearly proven (3.6).

We next establish (3.5). We pass to the inductive limit in (3.7) relative to the natural inclusion of $M_n$ in $M_{n+1}$. Let us note that $H^0(\mathbb{Q}_v, T_p(B)) = 0$ for all $v \in S$, because $H^0(\mathbb{Q}_v, H_p(B)) = 0$ for all $v \in S$ (see Lemma 2.1 and 2.3). Hence, neglecting the first part of the limit sequence of (3.7), we obtain the exact sequence

$$(3.9) \quad H^1(G_S, \mathfrak{A}_{p^\infty}) \xrightarrow{\varphi_A} \bigoplus_{v \in S} H^1(\mathbb{Q}_v, \mathfrak{A}_{p^\infty}) \to H^1(G_S, T_p(B))^\wedge \to H^2(G_S, \mathfrak{A}_{p^\infty}) \to 0$$

Let

$$\psi_A : H^1(G_S, \mathfrak{A}_{p^\infty}) \to \bigoplus_{v \in S} \frac{H^1(\mathbb{Q}_v, \mathfrak{A}_{p^\infty})}{H^1_f(\mathbb{Q}_v, \mathfrak{A}_{p^\infty})}$$

be the natural map, so that, by definition, $Ker(\psi_A) = H^1_f(\mathbb{Q}, \mathfrak{A}_{p^\infty})$. Again, simple diagram chasing shows that we have an exact sequence

$$(3.10) \quad \bigoplus_{v \in S} H^1_f(\mathbb{Q}_v, \mathfrak{A}_{p^\infty}) \xrightarrow{\nu_1} Coker(\varphi_A) \to Coker(\psi_A) \to 0.$$

To proceed further, we must now appeal to the fundamental result of Bloch-Kato (see [1], Proposition 3.8), which asserts that, under the canonical dual pairing

$$(3.11) \quad H^1(\mathbb{Q}_v, \mathfrak{A}_{p^\infty}) \times H^1(\mathbb{Q}_v, T_p(B)) \to \mathbb{Q}_p/\mathbb{Z}_p,$$

the exact orthogonal complement of $H^1_f(\mathbb{Q}_v, \mathfrak{A}_{p^\infty})$ is equal to $H^1_f(\mathbb{Q}_v, T_p(B))$. Hence, dualizing the exact sequence derived from (3.3) which gives the definition of $H^1_f(\mathbb{Q}, T_p(B))$, we obtain the exact sequence

$$(3.12) \quad \bigoplus_{v \in S} H^1_f(\mathbb{Q}_v, \mathfrak{A}_{p^\infty}) \xrightarrow{\nu_2} H^1(G_S, T_p(B))^\wedge \to H^1(\mathbb{Q}, T_p(B))^\wedge \to 0.$$

The idea now is to compare (3.10) and (3.12), noting that (3.9) gives rise to a canonical injection

$$\eta_1 : Coker(\varphi_A) \hookrightarrow H^1(G_S, T_p(B))^\wedge.$$

Indeed, it is easy to see that we have a commutative diagram

$$\begin{array}{ccccccc}
\bigoplus_{v \in S} H^1_f(\mathbb{Q}_v, \mathfrak{A}_{p^\infty}) & \xrightarrow{\nu_2} & H^1(G_S, T_p(B))^\wedge & \to & H^1_f(\mathbb{Q}, T_p(B))^\wedge & \to & 0 \\
\| & & \uparrow \eta_1 & & \uparrow \eta_2 & & \\
\bigoplus_{v \in S} H^1_f(\mathbb{Q}_v, \mathfrak{A}_{p^\infty}) & \xrightarrow{\nu_1} & Coker(\varphi_A) & \to & Coker(\psi_A) & \to & 0
\end{array}$$

where $\eta_2$ is the map induced by $\eta_1$. Note that $\eta_2$ is plainly injective. But now simple diagram chasing shows immediately that the induced map from $Coker(\eta_1)$ to $Coker(\eta_2)$ is an isomorphism. But this is precisely what is needed to show that the exact sequence (3.9) gives rise to the exact sequence (3.5). This completes the proof of Theorem 3.1. $\square$

In the next lemma, $M$ again denotes either $A$ or $B$, and we recall that $\mathfrak{M}_{p^\infty}$ is defined by the exact sequence (3.1).

**Lemma 3.2.** *We have the exact sequence*

(3.13)  $\quad 0 \to W(M) \to H^1_f(\mathbb{Q}, T_p(M)) \to T_p(H^1_f(\mathbb{Q}, \mathfrak{M}_{p^\infty})) \to 0,$

*where $W(M)$ is the torsion subgroup of $H^1(G_S, T_p(M))$. Moreover, $W(M) = \mathfrak{M}_{p^\infty}(\mathbb{Q})$.*

*Proof.* For each $n \geq 1$, define $\mathfrak{M}_{p^n}$ by the exactness of

(3.14) $\quad 0 \to \mathfrak{M}_{p^n} \to \mathfrak{M}_{p^\infty} \xrightarrow{p^n} \mathfrak{M}_{p^\infty} \to 0.$

We first take $G_v$-cohomology of (3.14) for any $v$ in $S$, obtaining the exact sequence

$$0 \to \mathfrak{M}_{p^\infty}(\mathbb{Q}_v)/p^n \mathfrak{M}_{p^\infty}(\mathbb{Q}_v) \to H^1(\mathbb{Q}_v, \mathfrak{M}_{p^n}) \to (H^1(\mathbb{Q}_v, \mathfrak{M}_{p^\infty}))_{p^n} \to 0.$$

Passing to the projective limit relative to the maps given by multiplication by $p$, we obtain the exact sequence

(3.15) $\quad 0 \to W_v(M) \to H^1(\mathbb{Q}_v, T_p(M)) \to T_p(H^1(\mathbb{Q}_v, \mathfrak{M}_{p^\infty})) \to 0$

where $W_v(M)$ denotes the quotient of $\mathfrak{M}_{p^\infty}(\mathbb{Q}_v)$ by its maximal divisible subgroup. Using Lemma 2.3, we see easily that the maximal divisible subgroup of $\mathfrak{M}_{p^\infty}(\mathbb{Q}_v)$ is non-zero if and only if the following three conditions hold (i) $M = A$, (ii) $v \neq p$, and (iii) $L_v(A, s)$ has a pole at $s = 0$. Now it follows by a straightforward formal argument using the definitions of the various subspaces $H^1_f$ that (3.15) gives rise also to the exact sequence

(3.16) $\quad 0 \to W_v(M) \to H^1_f(\mathbb{Q}_v, T_p(M)) \to T_p(H^1_f(\mathbb{Q}_v, \mathfrak{M}_{p^\infty})) \to 0.$

Next take $G_S$-cohomology of (3.14), obtaining the exact sequence

$$0 \to \mathfrak{M}_{p^\infty}(\mathbb{Q})/p^n \mathfrak{M}_{p^\infty}(\mathbb{Q}) \to H^1(G_S, \mathfrak{M}_{p^n}) \to (H^1(G_S, \mathfrak{M}_{p^\infty}))_{p^n} \to 0.$$

Note that $\mathfrak{M}_{p^\infty}(\mathbb{Q})$ is finite since it injects into $\mathfrak{M}_{p^\infty}(\mathbb{Q}_p)$, and this latter group is finite by Lemma 2.3. Passing to the projective limit, we obtain the exact sequence

(3.17) $\quad 0 \to \mathfrak{M}_{p^\infty}(\mathbb{Q}) \to H^1(G_S, T_p(M)) \to T_p(H^1(G_S, \mathfrak{M}_{p^\infty})) \to 0.$

The exact sequence (3.13) now follows by simple diagram chasing from (3.3), (3.15), (3.16), and (3.17), the essential point to note being that (3.15) and (3.16) show that the natural map

$$\bigoplus_{v \in S} \frac{H^1(\mathbb{Q}_v, T_p(M))}{H^1_f(\mathbb{Q}_v, T_p(M))} \longrightarrow \bigoplus_{v \in S} \frac{T_p(H^1(\mathbb{Q}_v, \mathfrak{M}_{p^\infty}))}{T_p(H^1_f(\mathbb{Q}_v, \mathfrak{M}_{p^\infty}))}$$

is injective. Note finally that $\mathfrak{M}_{p^\infty}(\mathbb{Q})$ must be equal to the whole torsion subgroup of $H^1_f(G_S, T_p(M))$ because $T_p(H^1_f(\mathbb{Q}, \mathfrak{M}_{p^\infty}))$ is torsion free. This completes the proof of Lemma 3.2. □

REMARK. If $X$ is any cofinitely generated $\mathbb{Z}_p$-module, then $T_p(X)$ is a free $\mathbb{Z}_p$-module of rank equal to the $\mathbb{Z}_p$-corank of $X$. Hence, as $W(M)$ is finite, Lemma 3.2 implies that $H^1_f(\mathbb{Q}, T_p(M))$ is finite if and only if $H^1_f(\mathbb{Q}, \mathfrak{M}_{p^\infty})$ is finite.

**Proposition 3.3.**
(i) We have $\operatorname{corank}_{\mathbb{Z}_p}(H^1(G_S, \mathfrak{A}_{p^\infty})) = 2 + \delta(A)$, where the integer $\delta(A)$ is given by $H^2(G_S, \mathfrak{A}_{p^\infty}) \cong (\mathbb{Q}_p/\mathbb{Z}_p)^{\delta(A)}$;
(ii) We have $\operatorname{corank}_{\mathbb{Z}_p}(H^1(G_S, \mathfrak{B}_{p^\infty})) = 1 + \delta(B)$, where the integer $\delta(B)$ is given by $H^2(G_S, \mathfrak{B}_{p^\infty}) \cong (\mathbb{Q}_p/\mathbb{Z}_p)^{\delta(B)}$.

*Proof.* This is an easy consequence of Tate's global Euler characteristic theorem. Indeed, if $M$ denotes either $A$ or $B$, then $H^0(G_S, \mathfrak{M}_{p^\infty})$ is finite because $H^0(\mathbb{Q}_p, \mathfrak{M}_{p^\infty})$ is finite. A standard argument with the global Euler characteristic theorem then shows that

$$\operatorname{corank}_{\mathbb{Z}_p}(H^1(G_S, \mathfrak{M}_{p^\infty})) = d^-(M) + \delta(M), \tag{3.18}$$

where $\delta(M) = \operatorname{corank}_{\mathbb{Z}_p}(H^2(G_S, \mathfrak{M}_{p^\infty}))$. Also $H^2(G_S, \mathfrak{M}_{p^\infty})$ is divisible because $G_S$ has $p$-cohomological dimension equal to 2 since $p > 2$. This completes the proof of the proposition. □

**Proposition 3.4.** *The group $H^1_f(\mathbb{Q}, \mathfrak{A}_{p^\infty})$ is finite if and only if $H^1_f(\mathbb{Q}, \mathfrak{B}_{p^\infty})$ is finite.*

*Proof.* We shall use the exact sequence (3.5), and we begin by claiming that the $\mathbb{Z}_p$-corank of

$$\frac{H^1(\mathbb{Q}_v, \mathfrak{A}_{p^\infty})}{H^1_f(\mathbb{Q}_v, \mathfrak{A}_{p^\infty})} \tag{3.19}$$

is equal to 0 or 2, according as $v \neq p$ or $v = p$. Indeed, (3.19) is dual to $H^1_f(\mathbb{Q}_v, T_p(B))$, and the previous assertion follows immediately from Lemmas 2.2 and 2.4. Hence the middle term in the exact sequence (3.5) has $\mathbb{Z}_p$-corank equal to 2.

Suppose first that $H^1_f(\mathbb{Q}, \mathfrak{A}_{p^\infty})$ is finite. It follows from (3.5) that $\operatorname{corank}_{\mathbb{Z}_p}(H^1(G_S, \mathfrak{A}_{p^\infty}))$ is at most equal to 2. Hence Proposition 3.3 shows that $H^2(G_S, \mathfrak{A}_{p^\infty}) = 0$. It is then clear from (3.5) that $H^1_f(\mathbb{Q}, T_p(B))$ is finite, and so $H^1_f(\mathbb{Q}, \mathfrak{B}_{p^\infty})$ is finite. Conversely, assume that $H^1_f(\mathbb{Q}, \mathfrak{B}_{p^\infty})$ is finite, whence also $H^1_f(\mathbb{Q}, T_p(B))$ is also finite. By Lemma 3.2, we conclude that $H^1_f(\mathbb{Q}, T_p(B))$ is equal to $W(B) = \mathfrak{B}_{p^\infty}(\mathbb{Q})$. On the other hand, by (3.16), we have the exact sequence

$$0 \to W_p(B) \to H^1_f(\mathbb{Q}_p, T_p(B)) \to T_p(H^1(\mathbb{Q}_p, \mathfrak{B}_{p^\infty})) \to 0,$$

where $W_p(B) = \mathfrak{A}_{p^\infty}(\mathbb{Q}_p)$, which is finite by Lemma 2.3. Hence the restriction map from $H^1_f(\mathbb{Q}, T_p(B))$ to $H^1_f(\mathbb{Q}_p, T_p(B))$ is injective because it can be identified with the natural inclusion of $W(B)$ into $W_p(B)$. It now follows from

(3.5) that $H^2(G_S, \mathfrak{A}_{p^\infty}) = 0$. Hence, by Proposition 3.3, the $\mathbb{Z}_p$-corank of $H^1(G_S, \mathfrak{A}_{p^\infty})$ is equal to 2, and so necessarily $H^1_f(\mathbb{Q}, \mathfrak{A}_{p^\infty})$ is finite. This completes the proof. □

It is conjectured that $H^1_f(\mathbb{Q}, \mathfrak{A}_{p^\infty})$ (or equivalently $H^1_f(\mathbb{Q}, \mathfrak{B}_{p^\infty})$) is finite for all elliptic curves $E$ over $\mathbb{Q}$ and all primes $p$. Deep progress on this conjecture has been made by M. Flach [6], who has proven it for all modular elliptic curves $E$ over $\mathbb{Q}$ and all primes $p$ satisfying (i) $p \geq 5$, (ii) $E$ has good reduction at $p$, and (iii) the representation

$$\tilde{\rho}_p : G_\mathbb{Q} \to \mathrm{Aut}(E_p) = GL_2(\mathbb{F}_p)$$

giving the action of $G_\mathbb{Q}$ on the group of $p$-division points on $E$, is surjective. I also understand that A. Wiles has found an alternative proof of this finiteness for all modular elliptic curves $E$ over $\mathbb{Q}$ and all primes $p$ such that (i) $p > 2$, (ii) $E$ has either good or multiplicative reduction at $p$, and (iii) $Sym^2(\tilde{\rho}_p)$ is an absolutely irreducible $G_\mathbb{Q}$-representation. Although we are not concerned with elliptic curves with complex multiplication here, let us also note that the finiteness of $H^1_f(\mathbb{Q}, \mathfrak{A}_{p^\infty})$, for $E$ with complex multiplication and all primes $p$ which are relatively prime to the number of roots of unity in $K = \mathrm{End}(E) \otimes \mathbb{Q}$, can be shown to be a consequence of Rubin's proof of the main conjectures of Iwasawa theory for such $E$. In any case, because the above conjecture still has not been proven in complete generality, we shall specifically indicate when our subsequent results depend on it.

Let us put $S' = S \setminus \{p\}$. We then define

(3.20) $$r(S) = \#\{v \in S' : L_v(A, s) \text{ has a pole at } s = 0\}.$$

**Proposition 3.5.** *Assume that $H^1_f(\mathbb{Q}, \mathfrak{A}_{p^\infty})$ is finite. Then*
(i) $\mathrm{corank}_{\mathbb{Z}_p}(H^1(G_S, \mathfrak{A}_{p^\infty})) = 2$, $H^2(G_S, \mathfrak{A}_{p^\infty}) = 0$;
(ii) $\mathrm{corank}_{\mathbb{Z}_p}(H^1(G_S, \mathfrak{B}_{p^\infty})) = 1 + r(S)$, $H^2(G_S, \mathfrak{B}_{p^\infty}) \xrightarrow{\sim} (\mathbb{Q}_p/\mathbb{Z}_p)^{r(S)}$.

*Proof.* The assertion (i) follows from (i) of Proposition 3.3, since it was remarked in the proof of Proposition 3.4 that the finiteness of $H^1_f(\mathbb{Q}, \mathfrak{A}_{p^\infty})$ implies that $H^2(G_S, \mathfrak{A}_{p^\infty}) = 0$. We now turn to (ii). We claim that, for each $v \in S'$, $\mathfrak{A}_{p^\infty}(\mathbb{Q}_v)$ has $\mathbb{Z}_p$-corank equal to 1 or 0, according as $L_v(A, s)$ does or does not have a pole at $s = 0$. This follows easily from taking $G_v$-invariants of the exact sequence (3.1) with $M = A$, and noting that $H_p(A)^{G_v}$ has $\mathbb{Q}_p$-dimension equal to 1 or 0, according as $L_v(A, s)$ does or does not have a pole at $s = 0$ (see Lemma 2.1). Also, we know that $\mathfrak{A}_{p^\infty}(\mathbb{Q}_p)$ is finite, because Lemma 2.3 shows that $H_p(A)^{G_p} = 0$. Hence (3.6) shows that the $\mathbb{Z}_p$-corank of $H^2(G_S, \mathfrak{B}_{p^\infty})$ is always $\geq r(S)$; in addition, it implies that equality holds in this last estimate if $H^1_f(\mathbb{Q}, \mathfrak{B}_{p^\infty})$, or equivalently $H^1_f(\mathbb{Q}, \mathfrak{A}_{p^\infty})$ is finite. Hence (ii) follows from (ii) of Proposition 3.3, and the proof is complete. □

Recall that $S' = S \setminus \{p\}$, and let

(3.21) $$\psi'_M : H^1(G_S, \mathfrak{M}_{p^\infty}) \to \bigoplus_{v \in S'} \frac{H^1(\mathbb{Q}_v, \mathfrak{M}_{p^\infty})}{H^1_f(\mathbb{Q}_v, \mathfrak{M}_{p^\infty})}$$

be the natural map.

**Proposition 3.6.** *Assume that $H^1_f(\mathbb{Q}, \mathfrak{A}_{p^\infty})$ is finite. Then both $\psi'_A$ and $\psi'_B$ are surjective.*

*Proof.* Let $M = A$ or $B$. We claim that our finiteness hypothesis implies that the restriction map

$$(3.22) \qquad j_p(M) : H^1_f(\mathbb{Q}, T_p(M)) \to H^1_f(\mathbb{Q}_p, T_p(M))$$

is injective. Indeed, by Proposition 3.4, $H^1_f(\mathbb{Q}, \mathfrak{M}_{p^\infty})$ is finite because $H^1_f(\mathbb{Q}, \mathfrak{A}_{p^\infty})$ is finite. Hence Lemma 3.2 shows that $H^1_f(\mathbb{Q}, T_p(M)) = \mathfrak{M}_{p^\infty}(\mathbb{Q})$. But, in view of (3.16), we see that $j_p(M)$ can be identified with the natural inclusion of $\mathfrak{M}_{p^\infty}(\mathbb{Q})$ into $\mathfrak{M}_{p^\infty}(\mathbb{Q}_p)$, and so is certainly injective.

Let

$$(3.23) \qquad j_S(M) : H^1_f(\mathbb{Q}, T_p(M)) \to \bigoplus_{v \in S} H^1_f(\mathbb{Q}_v, T_p(M))$$

be the map given by restriction at all $v \in S$. We first consider the map $\psi'_A$. Now $H^2(G_S, \mathfrak{A}_{p^\infty}) = 0$ because $H^1_f(\mathbb{Q}, \mathfrak{A}_{p^\infty})$ is finite. An analysis of the proof of Theorem 3.1 then shows that $j_S(B)$ is the dual of the surjective map

$$\bigoplus_{v \in S} \frac{H^1(\mathbb{Q}_v, \mathfrak{A}_{p^\infty})}{H^1_f(\mathbb{Q}_v, \mathfrak{A}_{p^\infty})} \to H^1_f(\mathbb{Q}, T_p(B))^\wedge$$

occurring in the exact sequence (3.5). Simple diagram chasing then shows that we have an exact sequence

$$0 \longrightarrow Coker(\psi'_A)^\wedge \longrightarrow H^1_f(\mathbb{Q}, T_p(B)) \xrightarrow{j_p(B)} H^1_f(\mathbb{Q}_p, T_p(B)).$$

But $j_p(B)$ is injective, and hence it follows that $Coker(\psi'_A) = 0$, as required.

The argument for the map $\psi'_B$ is slightly different. In fact, in analogy with (3.5), we claim we also have the exact sequence

$$(3.24) \qquad H^1(G_S, \mathfrak{B}_{p^\infty}) \xrightarrow{\psi_B} \bigoplus_{v \in S} \frac{H^1(\mathbb{Q}_v, \mathfrak{B}_{p^\infty})}{H^1_f(\mathbb{Q}_v, \mathfrak{B}_{p^\infty})} \xrightarrow{\theta_B} H^1_f(\mathbb{Q}, T_p(A))^\wedge,$$

where $\theta_B$ is the dual of the map $j_S(A)$. We do not give the detailed proof of (3.24), merely pointing out that one uses the Poitou-Tate sequence (3.7) with $\mathfrak{A}_{p^n}$ and $\mathfrak{B}_{p^n}$ interchanged, passes to the inductive limit as $n \to \infty$, and then makes a similar analysis, based on the orthogonality result of Bloch-Kato, to that given in the last paragraph of the proof of Theorem 3.1. Let us also note in passing that one can show that the continuation of (3.24) to the right as an exact sequence is given by dualizing (3.6). The proof of the exactness of (3.24) does not depend on the finiteness of $H^1_f(\mathbb{Q}, \mathfrak{A}_{p^\infty})$. However, assuming now this finiteness, we claim that $\theta_B$ is then surjective. Indeed, $\theta_B$ is dual to $j_S(A)$, and this latter map is injective, because $j_p(A)$ is injective. Knowing that $\theta_B$ is surjective, we again easily derive the exact sequence

$$0 \longrightarrow Coker(\psi'_B)^\wedge \longrightarrow H^1_f(\mathbb{Q}, T_p(A)) \xrightarrow{j_p(A)} H^1_f(\mathbb{Q}_p, T_p(A)),$$

and so $Coker(\psi'_B) = 0$ because $j_p(A)$ is injective. This completes the proof of Proposition 3.6. □

Let $M$ denote either $A$ or $B$. We define

$$\mathfrak{S}(\mathbb{Q}, \mathfrak{M}_{p^\infty}) = Ker(\psi'_M),$$

where $\psi'_M$ is given by the map (3.21) above.

**Proposition 3.7.** *Assume that $H^1_f(\mathbb{Q}, \mathfrak{A}_{p^\infty})$ is finite. Then*

(3.25) $$corank_{\mathbb{Z}_p}(\mathfrak{S}(\mathbb{Q}, \mathfrak{M}_{p^\infty})) = d^-(M),$$

*where the value of $d^-(M)$ is given by (2.4).*

*Proof.* Using Proposition 3.5 and 3.6, we see that (3.25) is equivalent to showing that

(3.26) $$\bigoplus_{v \in S'} \frac{H^1(\mathbb{Q}_v, \mathfrak{M}_{p^\infty})}{H^1_f(\mathbb{Q}_v, \mathfrak{M}_{p^\infty})}$$

has $\mathbb{Z}_p$-corank equal to 0 or $r(S)$, according as $M = A$ or $B$. But, recalling the local duality of Bloch and Kato, this last statement is an immediate consequence of Lemma 2.2. This completes the proof of Proposition 3.7. □

REMARK. If $M = A$, we have just shown that (3.26) is finite, and it may be of interest to note that there is an exact formula for its order in the spirit of Tamagawa numbers. Define, following Bloch and Kato,

$$H^1_f(\mathbb{Q}_v, T_p(B))^0 = Ker(H^1(\mathbb{Q}_v, T_p(B)) \to H^1(I_v, T_p(B))) \quad (v \neq p).$$

Note that we have $H^1_f(\mathbb{Q}_v, T_p(B)) = H^1(\mathbb{Q}_v, T_p(B))$ for $v \neq p$ by Lemma 2.2. If $u$ and $w$ are in $\mathbb{Q}_p^\times$, we write $u \sim w$ if $u/w$ is a $p$-adic unit. By the definition of $L_v(B, s)$, we see easily that

$$\#(H^1_f(\mathbb{Q}_v, T_p(B))^0) \sim L_v(B, 0)^{-1} \quad (v \neq p).$$

Hence, if we define

(3.27) $$c_v^{(p)}(B) = [H^1_f(\mathbb{Q}_v, T_p(B)) : H^1_f(\mathbb{Q}_v, T_p(B))^0] \quad (v \neq p),$$

we conclude that

(3.28) $$\#\left(\bigoplus_{v \in S'} \frac{H^1(\mathbb{Q}_v, \mathfrak{A}_{p^\infty})}{H^1_f(\mathbb{Q}_v, \mathfrak{A}_{p^\infty})}\right) \sim \prod_{v \in S'} (c_v^{(p)}(B) L_v(B, 0)^{-1}).$$

**4 Galois cohomology over the cyclotomic $\mathbb{Z}_p$-extension of $\mathbb{Q}$.** Let $F_\infty$ denote the cyclotomic $\mathbb{Z}_p$-extension of $\mathbb{Q}$, i.e. the unique subfield of $\mathbb{Q}(\mu_{p^\infty})$ whose Galois group over $\mathbb{Q}$ is topologically isomorphic to the additive group of $\mathbb{Z}_p$. Since $p \in S$, we have $F_\infty \subset \mathbb{Q}_S$, and we put

(4.1) $$G_{\infty,S} = G(\mathbb{Q}_S/F_\infty), \qquad \Gamma = G(F_\infty/\mathbb{Q}).$$

We also write $\Lambda = \mathbb{Z}_p[[\Gamma]]$ for the Iwasawa algebra of $\Gamma$. The aim of this section is to study the Galois cohomology groups

$$H^i(G_{\infty,S}, \mathfrak{M}_{p^\infty}) \qquad (i = 0, 1, 2),$$

where $M$ denotes either $A$ or $B$. Of course, $\Gamma$ acts naturally on these groups, and so we can view them as modules over the Iwasawa algebra $\Lambda$. It is somewhat surprising that one can prove quite a lot about these Iwasawa modules, despite the fact that no analogue of Flach's theorem [6] has been proven for finite extensions of $\mathbb{Q}$ contained in $F_\infty$. We again suppose throughout this section that $E$ does not admit complex multiplication, and that $p$ is an odd prime number. In fact, analogues of the principal results below can be proven for $E$ with complex multiplication and $p$ a prime number which is relatively prime to the number of roots of unity in $K = End(E) \otimes \mathbb{Q}$ by using Rubin's proof of the main conjectures of Iwasawa theory for such $E$ and $p$ (see [2], [5]).

We now state our principal results. Again, $M$ denotes either $A$ or $B$, and $\mathfrak{M}_{p^\infty}$ is the divisible $G_S$-module defined by the exact sequence (3.1).

**Theorem 4.1.** *Assume that $H^1_f(\mathbb{Q}, \mathfrak{A}_{p^\infty})$ is finite. Then*
*(i)* $H^2(G_{\infty,S}, \mathfrak{M}_{p^\infty}) = 0$;
*(ii) The Pontrjagin dual of $H^1(G_{\infty,S}, \mathfrak{M}_{p^\infty})$ has $\Lambda$-rank equal to $d^-(M)$;*
*(iii) The Pontrjagin dual of $H^1(G_{\infty,S}, \mathfrak{M}_{p^\infty})$ has no non-zero finite $\Lambda$-submodule.*

In fact, to prove this result, we need to study a second Iwasawa module attached to $\mathfrak{M}_{p^\infty}$. For each place $w$ of $F_\infty$, let $F_{\infty,w}$ be the union of the completions at $w$ of all finite extensions of $\mathbb{Q}$ contained in $F_\infty$. We define $\mathfrak{S}(F_\infty, \mathfrak{M}_{p^\infty})$ by the exactness of the sequence

(4.2) $$0 \longrightarrow \mathfrak{S}(F_\infty, \mathfrak{M}_{p^\infty}) \longrightarrow H^1(G_{\infty,S}, \mathfrak{M}_{p^\infty}) \xrightarrow{\psi'_{\infty,M}} \bigoplus_{w|S'} H^1(F_{\infty,w}, \mathfrak{M}_{p^\infty}),$$

where $\psi'_{\infty,M}$ is given by the restriction maps at all places $w$ of $F_\infty$ which lie above $S' = S \setminus \{p\}$. Put

(4.3) $$Y_{\infty,M} = \mathfrak{S}(F_\infty, \mathfrak{M}_{p^\infty})^\wedge.$$

**Theorem 4.2.** *Assume that $H^1_f(\mathbb{Q}, \mathfrak{A}_{p^\infty})$ is finite. Then*
*(i) $Y_{\infty,M}$ has $\Lambda$-rank equal to $d^-(M)$;*
*(ii) $Y_{\infty,M}$ has no non-zero finite $\Lambda$-submodule;*

*(iii)* $H^1(\Gamma, \mathfrak{S}(F_\infty, \mathfrak{M}_{p^\infty})) = 0$;

*(iv) The map $\psi'_{\infty,M}$ in (4.2) is surjective.*

We begin the proof of these theorems by two purely local lemmas. If $w$ is a place of $F_\infty$, let $I_{\infty,w}$ be the inertial subgroup of $G(\overline{\mathbb{Q}}_w/F_{\infty,w})$. The situation for restriction to the inertial subgroup over $F_{\infty,w}$ when $w \neq p$ is different from that over $\mathbb{Q}_w$ (see Lemma 2.2).

**Lemma 4.3.** *If $w$ is a place of $F_\infty$ not lying above $p$, then the restriction map from $H^1(F_{\infty,w}, H_p(M))$ to $H^1(I_{\infty,w}, H_p(M))$ is always injective.*

*Proof.* The point is that, since $w$ does not divide $p$, $F_{\infty,w}$ is the unique unramified $\mathbb{Z}_p$-extension of $\mathbb{Q}_w$. Hence, if $\mathbb{Q}_w^{nr}$ denotes the maximal unramified extension of $\mathbb{Q}_w$, the Sylow $p$-subgroup of $G(\mathbb{Q}_w^{nr}/F_{\infty,w})$ is trivial. Hence the kernel of the restriction map, which is equal to

$$H^1(G(\mathbb{Q}_w^{nr}/F_{\infty,w}), H_p(M)^{I_{\infty,w}}),$$

is necessarily zero. $\square$

If $v$ is a finite prime of $\mathbb{Q}$, we define

$$(4.4) \qquad Z_{v,\mathbb{Q}}(M) = \frac{H^1(\mathbb{Q}_v, \mathfrak{M}_{p^\infty})}{H^1_f(\mathbb{Q}_v, \mathfrak{M}_{p^\infty})},$$

$$Z_{v,\infty}(M) = \bigoplus_{w/v} H^1(F_{\infty,w}, \mathfrak{M}_{p^\infty}).$$

Of course, the global Galois group $\Gamma = G(F_\infty/\mathbb{Q})$ operates on $Z_{v,\infty}(M)$ in the natural fashion.

**Lemma 4.4.** *Assume that $v \neq p$. Then the Pontrjagin dual of $Z_{v,\infty}(M)$ is $\Lambda$-torsion. Moreover, $H^1_f(\mathbb{Q}, \mathfrak{M}_{p^\infty})$ is in the kernel of the restriction map from $H^1(\mathbb{Q}_v, \mathfrak{M}_{p^\infty})$ to $Z_{v,\infty}(M)$, and hence restriction induces a homomorphism*

$$(4.5) \qquad \eta_{v,M} : Z_{v,\mathbb{Q}}(M) \to (Z_{v,\infty}(M))^\Gamma.$$

*This homomorphism is surjective, and has finite kernel.*

*Proof.* Lemma 4.3 and the definition (3.2) clearly imply that $H^1_f(\mathbb{Q}, \mathfrak{M}_{p^\infty})$ is in the kernel of the restriction map. For each $w$ dividing $v$, the restriction map from $H^1(\mathbb{Q}_v, \mathfrak{M}_{p^\infty})$ to $H^1(F_{\infty,w}, \mathfrak{M}_{p^\infty})^{\Gamma_w}$, where $\Gamma_w = G(F_{\infty,w}/\mathbb{Q}_v)$, is surjective because $\Gamma_w$ is isomorphic to $\mathbb{Z}_p$ and so has cohomological dimension equal to 1. The surjectivity of $\eta_{v,M}$ follows immediately. Next we note that, by the local duality theorem of Bloch and Kato, $Z_{v,\mathbb{Q}}(M)$ is dual to

$$(4.6) \qquad H^1_f(\mathbb{Q}_v, T_p(R)),$$

where $R$ is the motive $\check{M}(1)$ (i.e. $R = B$ if $M = A$ and vice-versa). By Lemmas 2.1 and 2.2, the $\mathbb{Z}_p$-rank of (4.6) is equal to the order of the pole of $L_v(R, s)$ at

$s = 0$. Hence we must compute the $\mathbb{Z}_p$-corank of $Z_{v,\infty}(M)^\Gamma$. In fact, it is easy to see that there is an isomorphism

$$H^1(F_{\infty,w}, \mathfrak{M}_{p^\infty})^{\Gamma_w} \xrightarrow{\sim} Z_{v,\infty}(M)^\Gamma.$$

But the structure of $H^1(F_{\infty,w}, \mathfrak{M}_{p^\infty})$ as a $\Gamma_w$-module is well known (see [8], p. 112). If we put

$$\mathfrak{R}_{p^\infty} = H_p(R)/T_p(R),$$

then $H^1(F_{\infty,w}, \mathfrak{M}_{p^\infty})$ is pseudo-isomorphic to $\mathfrak{R}_{p^\infty}(F_{\infty,w})$. In particular, we conclude that the Pontrjagin dual of $Z_{v,\infty}(M)$ is a finitely generated $\mathbb{Z}_p$-module, and so is certainly $\Lambda$-torsion. Moreover, it follows that $H^1(F_{\infty,w}, \mathfrak{M}_{p^\infty})^{\Gamma_w}$ has the same $\mathbb{Z}_p$-corank as $\mathfrak{R}_{p^\infty}(\mathbb{Q}_v)$. But is clear from the definition of the Euler factor $L_v(R, s)$ that the order of its pole at $s = 0$ is equal to the $\mathbb{Z}_p$-corank of $\mathfrak{R}_{p^\infty}(\mathbb{Q}_v)$. Hence $Z_{v,\mathbb{Q}}(M)$ and $(Z_{v,\infty}(M))^\Gamma$ have the same $\mathbb{Z}_p$-corank, and so the surjectivity of $\eta_{v,M}$ implies that it must have finite kernel. This completes the proof of Lemma 4.4. □

The next lemma gives the fundamental diagram on which the proofs of Theorem 4.1 and 4.2 crucially depend.

**Lemma 4.5.** *Assume that $H^1_f(\mathbb{Q}, \mathfrak{A}_{p^\infty})$ is finite. Then we have a commutative diagram with exact rows*

(4.7)
$$\begin{array}{ccccccccc}
0 & \to & \mathfrak{S}(F_\infty, \mathfrak{M}_{p^\infty})^\Gamma & \to & H^1(G_{\infty,S}, \mathfrak{M}_{p^\infty})^\Gamma & \to & \left(\bigoplus_{v \in S'} Z_{v,\infty}(M)\right)^\Gamma & \to & 0 \\
& & \uparrow \alpha_1 & & \uparrow \alpha_2 & & \uparrow \alpha_3 & & \\
0 & \to & \mathfrak{S}(\mathbb{Q}, \mathfrak{M}_{p^\infty}) & \to & H^1(G_S, \mathfrak{M}_{p^\infty}) & \xrightarrow{\psi'_M} & \bigoplus_{v \in S'} Z_{v,\mathbb{Q}}(M) & \to & 0
\end{array}$$

*where the top row is given by taking $\Gamma$-invariants of (4.2), $\alpha_2$ is the restriction map, and $\alpha_3 = \oplus \eta_{v,M}$.*

*Proof.* By Proposition 3.6, $\psi'_M$ is surjective. The only point which is not clear about (4.7) is the surjectivity of the map in the top right hand corner. But this is plain from the surjectivity of $\psi'_M$ and $\alpha_3$.

As usual, the second basic ingredient in the proofs is a lower bound for the $\Lambda$-corank of $\mathfrak{S}(F_\infty, \mathfrak{M}_{p^\infty})$, which is obtained by applying Tate's global Euler characteristic theorem to all finite extensions of $\mathbb{Q}$ contained in $F_\infty$ (see §4 of [8]). □

**Lemma 4.6.**

(4.8) $\quad corank_\Lambda(\mathfrak{S}(F_\infty, \mathfrak{M}_{p^\infty})) = d^-(M) + corank_\Lambda(H^2(G_{\infty,S}, \mathfrak{M}_{p^\infty})).$

*Proof.* By Lemma 4.4, the term on the extreme right of the exact sequence (4.2) is $\Lambda$-cotorsion, and so it is clear that

$$corank_\Lambda(\mathfrak{S}(F_\infty, \mathfrak{M}_{p^\infty})) = corank_\Lambda(H^1(G_{\infty,S}, \mathfrak{M}_{p^\infty})).$$

Hence (4.8) is just the statement of Proposition 3 of §4 of [8] for the Galois module $\mathfrak{M}_{p^\infty}$. □

We can now prove Theorems 4.1 and 4.2. On the one hand, it is plain from the structure theory of $\Lambda$-modules that

$$corank_{\mathbb{Z}_p}(\mathfrak{S}(F_\infty, \mathfrak{M}_{p^\infty})^\Gamma) \geq corank_\Lambda(\mathfrak{S}(F_\infty, \mathfrak{M}_{p^\infty})).$$

On the other hand, assuming that $H^1_f(\mathbb{Q}, \mathfrak{A}_{p^\infty})$ is finite, it follows from (4.7) that $Coker(\alpha_1)$ is finite because $\alpha_2$ is surjective and $\alpha_3$ has finite kernel. Hence, by Proposition 3.7,

$$corank_{\mathbb{Z}_p}(\mathfrak{S}(F_\infty, \mathfrak{M}_{p^\infty})^\Gamma) \leq corank_{\mathbb{Z}_p}(\mathfrak{S}(\mathbb{Q}, \mathfrak{M}_{p^\infty})) = d^-(M).$$

Combining these last two estimates, and using (4.8), we conclude that

(4.9)
$$\begin{aligned} corank_\Lambda(\mathfrak{S}(F_\infty, \mathfrak{M}_{p^\infty})) &= d^-(M), \\ corank_\Lambda(H^2(G_{\infty,S}, \mathfrak{M}_{p^\infty})) &= 0. \end{aligned}$$

But, since $p \neq 2$, it is known ([8], Proposition 4) that the Pontrjagin dual of $H^2(G_{\infty,S}, \mathfrak{M}_{p^\infty})$ is a free $\Lambda$-module. Hence (4.9) implies, in particular, that

(4.10) $$H^2(G_{\infty,S}, \mathfrak{M}_{p^\infty}) = 0.$$

Moreover, it is known ([8], Proposition 5) that (4.10) implies that the dual of $H^1(G_{\infty,S}, \mathfrak{M}_{p^\infty})$ has no non-zero finite $\Lambda$-submodule. Thus we have established Theorem 4.1 and (i) of Theorem 4.2. Now (iii) of Theorem 4.2 is equivalent to $(Y_{\infty,M})^\Gamma = 0$, and so it is clear that (ii) of Theorem 4.2 follows from (iii). We next establish (iii) of Theorem 4.2. In view of the surjectivity of the right hand map in the top row of (4.7), we have the exact sequence

(4.11)
$$\begin{aligned} 0 &\longrightarrow H^1(\Gamma, \mathfrak{S}(F_\infty, \mathfrak{M}_{p^\infty})) \\ &\longrightarrow H^1(\Gamma, H^1(G_{\infty,S}, \mathfrak{M}_{p^\infty})) \\ &\xrightarrow{\tau_M} \bigoplus_{v \in S'} H^1(\Gamma, Z_{v,\infty}(M)) \longrightarrow 0. \end{aligned}$$

We claim that the map $\tau_M$ is injective. To this end, let us put

(4.12) $$W_{v,\infty}(M) = \bigoplus_{w|v} H^2(F_{\infty,w}, \mathfrak{M}_{p^\infty}).$$

Since $\Gamma$ has cohomological dimension equal to 1, the Hochschild-Serre spectral sequence shows that the rows of the commutative diagram

$$\begin{array}{ccccccccc} 0 & \to & H^1(\Gamma, H^1(G_{\infty,S}, \mathfrak{M}_{p^\infty})) & \to & H^2(G_S, \mathfrak{M}_{p^\infty}) & \to & H^2(G_{\infty,S}, \mathfrak{M}_{p^\infty})^\Gamma = 0 & \to & 0 \\ & & \downarrow \tau_M & & \downarrow \beta_M & & \downarrow & & \\ 0 & \to & \bigoplus_{v \in S'} H^1(\Gamma, Z_{v,\infty}(M)) & \to & \bigoplus_{v \in S'} H^2(\mathbb{Q}_v, \mathfrak{M}_{p^\infty}) & \to & \bigoplus_{v \in S'} W_{v,\infty}(M)^\Gamma & \to & 0, \end{array}$$

where $\beta_M$ is the restriction map, are exact. Hence it suffices to show that $\beta_M$ is injective. When $M = A$, this is clear because Proposition 3.5 shows that

$H^2(G_S, \mathfrak{A}_{p^\infty}) = 0$. When $M = B$, we dualize the exact sequence (3.6) and use Tate local duality, obtaining the exact sequence

$$\bigoplus_{v \in S} H^1_f(\mathbb{Q}_v, T_p(A))^\wedge \stackrel{\theta_B}{\to} H^f(\mathbb{Q}, T_p(A))^\wedge \to H^2(G_S, \mathfrak{B})$$
$$\stackrel{\beta_M}{\to} \bigoplus_{v \in S} H^2(\mathbb{Q}_v, \mathfrak{B}_{p^\infty}) \to 0.$$

Moreover, $H^2(\mathbb{Q}_p, \mathfrak{B}_{p^\infty}) = 0$ because it is dual to $T_p(\mathfrak{A}_{p^\infty}(\mathbb{Q}_p))$, and this latter group is 0 because $\mathfrak{A}_{p^\infty}(\mathbb{Q}_p)$ is finite. Hence the map on the right can be identified with $\beta_M$, as indicated. But it was shown in the proof of Proposition 3.6 that $\theta_B$ is surjective, and so it follows that $\beta_M$ is injective. Hence $\tau_M$ is injective, and assertion (iii) of Theorem 4.2 now follows from the exact sequence (4.11). Finally, we establish (iv) of Theorem 4.2. Put

$$D = Im(\psi'_{\infty, M}), \quad J = Coker(\psi'_{\infty, M}).$$

By the surjectivity of the top row of (4.7), we have

$$D^\Gamma = \bigoplus_{v \in S'} Z_{v, \infty}(M)^\Gamma,$$

and thus we obtain an exact sequence

(4.13) $$0 \to J^\Gamma \to H^1(\Gamma, D) \to \bigoplus_{v \in S'} H^1(\Gamma, Z_{v, \infty}(M)).$$

But, by virtue of (iii) of Theorem 4.2, the natural map induces an isomorphism

$$H^1(\Gamma, H^1(G_{\infty, S}, \mathfrak{M}_{p^\infty})) \stackrel{\sim}{\to} H^1(\Gamma, D).$$

With this last identification, the map on the right of (4.13) is none other than the map $\tau_M$ of (4.11). But we have just proven that $\tau_M$ is injective, and so $J^\Gamma = 0$, whence also $J = 0$ because $J$ is a discrete $p$-primary $\Gamma$-module. This completes the proof of Theorems 4.1 and 4.2. □

*References*

[1] Bloch, S., Kato, K., *L-functions and Tamagawa numbers of motives*, in *Grothendieck Festschrift 1*, Progress in Math. 86 (1990), Birkhauser 333-400.

[2] Coates, J., Schmidt, C.-G., *Iwasawa theory for the symmetric square of an elliptic curve*, Crelle 375/376 (1987), 104-156.

[3] Coates, J., *Motivic p-adic L-functions*, in *L-functions and Arithmetic*, LMS Lecture Notes 153 (1991), CUP, 141-172.

[4] Coates, J., M$^c$Connell, G., *Iwasawa theory of modular elliptic curves of analytic rank at most 1*, J. London Math. Soc. (2) 50 (1994), 243-264.

[5] Flach, M., *Selmer groups for the symmetric square of an elliptic curve*, Ph.D. thesis, Cambridge (1990).

[6] Flach, M., *A finiteness theorem for the symmetric square of an elliptic curve*, Invent. Math. 109 (1992), 307-327.

[7] Fontaine, J-M., Perrin-Riou, B., *Autour des conjectures de Bloch et Kato I. Cohomologie galoisienne*, C.R. Acad. Sci. Paris 313 (1991), 189-196.

[8] Greenberg, R., *Iwasawa theory for p-adic representations*, in *Algebraic Number Theory*, Adv. Studies Pure Math. 17 (1989), Academic Press.

[9] Lubin, J., *One Parameter formal Lie groups over p-adic integer rings*, Ann. of Math. 80 (1964), 464-484.

[10] Milne, J., *Arithmetic Duality Theorems*, Perspectives in Math. 1 (1986), Academic Press.

[11] Perrin-Riou, B., *Fonctions L p-adiques des représentations p-adiques*, to appear.

[12] Rubin, K., Silverberg, A., *Families of Elliptic curves with constant mod p representations*, this volume.

[13] Serre, J.-P., *Abelian l-adic representations and elliptic curves*, 1968, Benjamin.

[14] Tate, J., *p-divisible groups*, in *Proceedings of a Conference on Local Fields*, 1967, Springer, 158-183.

**Note added in proof:** A. Wiles, assisted by R. Taylor, has now proven the desired exact formula for the order of $H^1_f(\mathbb{Q}, \mathfrak{A}_{p^\infty})$ for all modular elliptic curves $E$ over $\mathbb{Q}$ and all primes $p$ such that

(i) $p > 2$,

(ii) $E$ has either good or multiplicative reduction at $p$, and

(iii) $Sym^2(\tilde{\rho}_p)$ is an absolutely irreducible $G_\mathbb{Q}$. See the paper by Wiles, and the joint paper by paper by Taylor and Wiles, to appear in the "*Annals of Mathematics*".

Conference on Elliptic Curves and Modular Forms
Hong Kong, December 18-21, 1993
Copyright ©1995 International Press

# The refined conjecture of Serre

FRED DIAMOND
DEPARTMENT OF MATHEMATICS, COLUMBIA UNIVERSITY, NEW YORK, NY
10027

**1 Introduction.** Let $\ell$ be a prime, and let $\mathbf{F}$ be an algebraic closure of the field $\mathbf{F}_\ell$. Fix a homomorphism $\alpha : \mathcal{O} \to \mathbf{F}$ where $\mathcal{O}$ is the ring of algebraic integers in $\mathbf{C}$. Suppose that

$$\rho : \mathrm{Gal}\,(\overline{\mathbf{Q}}/\mathbf{Q}) \to GL_2(\mathbf{F})$$

is a continuous irreducible representation and that $\det(\rho(\tau)) = -1$ if $\tau$ is a complex conjugation. Serre [16] has conjectured that any such representation is modular, meaning that there are positive integers $k$ and $N$ such that $\rho$ arises from an eigenform in $S_k(\Gamma_1(N))$ (see [15]). Serre also defines integers $k(\rho)$ and $N(\rho)$ (satisfying $k \geq 2$ and $N$ relatively prime to $\ell$) and formulates a refinement which predicts that $\rho$ arises from $S_{k(\rho)}(\Gamma_1(N(\rho)))$. In [15], Ribet discusses the progress on the problem of establishing the equivalence between Serre's conjecture and this refinement. Combining his work with that of others, Ribet concludes that this problem [15], (1.4) will be solved for $\ell > 3$ if the hypothesis $\det(\rho) = \chi_\ell$ is removed from a crucial intermediate result [15], (1.5). The main purpose of this article is to explain how to remove this restriction as well as include the case $\ell = 3$. We find that Ribet's methods yield the following:

**Theorem 1.1.** *Suppose that $\ell$ is odd and that $\rho : \mathrm{Gal}\,(\overline{\mathbf{Q}}/\mathbf{Q}) \to GL_2(\mathbf{F})$ is an irreducible representation arising from $S_k(\Gamma_1(N))$ for some positive integers $k$ and $N$. Then $\rho$ arises from $S_{k(\rho)}(\Gamma_1(N(\rho)))$.*

The proof so closely follows the arguments of [15] and [14] that we make no attempt to give an independent treatment. Instead we refer to these articles throughout Sections 3, 4 and 5 and detail only the changes needed to adapt Ribet's proof to the more general situation.

Before explaining those changes, however, we recall that Serre also defines a Dirichlet character $\chi(\rho) : (\mathbf{Z}/N\mathbf{Z})^\times \to \mathbf{C}^\times$ and predicts that $\rho$ arises from $S_{k(\rho)}(\Gamma_1(N(\rho))^{\chi(\rho)}$. Serre subsequently found counterexamples in which $\ell \leq 3$. On the other hand, a lemma of Carayol discussed in Section 2 yields the following corollary to Theorem 1.1.

**Corollary 1.2.** *Suppose that $\rho$ is irreducible and arises from $S_k(\Gamma_1(N))$ for some positive integers $k$ and $N$. If $\ell = 3$ then suppose also that $\rho$ is not induced from a character of $\mathrm{Gal}\,(\overline{\mathbf{Q}}/\mathbf{Q}(\sqrt{-3}))$. Then $\rho$ arises from $S_{k(\rho)}(\Gamma_1(N(\rho)))^{\chi(\rho)}$.*

In Section 6 we explain how to deduce a variant of these results formulated by Wiles [19]. We prove that if $\rho$ is as in the corollary, then $\rho$ arises from a certain space of cusp forms of weight two.

Finally we remark that Theorem 1.1 allows us to weaken the hypotheses on the weight and determinant in the main result of [6]. Theorem 1 of [6] now applies if $\rho$ arises from $S_k(\Gamma_1(N))$ with $2 \leq k \leq \ell - 1$ and $\ell$ prime to $N$, provided that $\ell > 3$, or that $\ell = 3$ and $\rho$ is not induced from a character of $\text{Gal}(\overline{\mathbf{Q}}/\mathbf{Q}(\sqrt{-3}))$.

The author is grateful to Ken Ribet and Andrew Wiles for helpful comments during the preparation of this article.

**2 Carayol's Lemma.** Suppose that $f = \sum_{n=1}^{\infty} a_n(f) e^{2\pi i n z}$ is an eigenform in $S_k(\Gamma_1(N))$, where by "eigenform" we shall always mean a normalized eigenform for the Hecke operators $T_n$, $n \geq 1$. Let $K_f$ denote the number field generated by the eigenvalues $a_n(f)$, and let $\mathcal{O}_f$ denote the ring of integers of $K_f$. Recall that we have fixed a homomorphism $\alpha : \mathcal{O} \to \mathbf{F}$ where $\mathcal{O}$ is the ring of algebraic integers in $\mathbf{C}$ and $\mathbf{F}$ is an algebraic closure of $\mathbf{F}_\ell$. Let $\lambda = \ker(\alpha|_{\mathcal{O}_f})$ and $R_f = \mathcal{O}_{f,\lambda}$. Recall that the $\lambda$-adic representation $\text{Gal}(\overline{\mathbf{Q}}/\mathbf{Q}) \to GL_2(K_{f,\lambda})$ associated to $f$ is unramified outside $N\ell$, and that if $p$ is a prime not dividing $N\ell$ then $a_p(f)$ is the trace of the image of $\text{Frob}_p$. This representation has a model

$$\rho_f : \text{Gal}(\overline{\mathbf{Q}}/\mathbf{Q}) \to GL_2(R_f),$$

which is unique up to isomorphism if $\rho_f \otimes_{R_f} \mathbf{F}$ is irreducible. If $S$ is a space of cusp forms stable under the action of the Hecke operators, we say that $\rho$ arises from $S$ if there is an eigenform $f$ in $S$ such that $\rho \cong \rho_f \otimes_{R_f} \mathbf{F}$.

For relatively prime positive integers $m$ and $N$, we will write $\Gamma_1(m, N)$ for $\Gamma_1(m) \cap \Gamma_0(N)$. Note that $(\mathbf{Z}/N\mathbf{Z})^\times \cong \Gamma_1(m, N)/\Gamma_1(mN)$ acts naturally on $S_k(\Gamma_1(mN))$. For a Dirichlet character $\chi$ modulo $N$, we write $S_k(\Gamma_1(mN))^\chi$ for the associated isotypic component. Thus $S_k(\Gamma_1(mN))^\chi$ consists of the forms $f$ in $S_k(\Gamma_1(mN))$ satisfying $\langle a \rangle f = \chi(a) f$ for all integers $a \equiv 1 \bmod m$ relatively prime to $N$. Note also that $S_k(\Gamma_1(mN))^\chi$ is stable under the action of the Hecke operators $T_n$.

To deduce Corollary 1.2 from Theorem 1.1, one can apply the following version of Carayol's Lemma.

**Lemma 2.1.** *Suppose that $\chi$ and $\psi$ are Dirichlet characters modulo $N$ satisfying $\alpha \circ \chi = \alpha \circ \psi$. Suppose that one of the following holds:*

1. $m > 3$,
2. $\ell > 3$, or
3. $\ell = 3$ and $\rho$ is not induced from a character of $\text{Gal}(\overline{\mathbf{Q}}/\mathbf{Q}(\sqrt{-3}))$.

*If $\rho$ is irreducible and arises from $S_k(\Gamma_1(mN))^\chi$ with $k \geq 2$ and $(m, N) = 1$, then $\rho$ arises from $S_k(\Gamma_1(mN))^\psi$.*

Note that if $m > 3$, then $\Gamma_1(m, N)$ has no elliptic elements, and $(\mathbf{Z}/N\mathbf{Z})^\times$ acts without fixed points on $\Gamma_1(mN)\backslash \mathcal{Z}$ where $\mathcal{Z}$ is the upper-half plane. So under the first condition, the conclusion is immediate from [3], Lemme 1 (see

[3], Proposition 3). Under the second condition, this is [3], Proposition 3). Under the third condition, one can apply a variant of [3], Lemme 1 explained in [3], (4.4).

If $f$ is an eigenform in $S_k(\Gamma_1(mN))^\chi$ such that $\rho \cong \rho_f \otimes_{R_f} \mathbf{F}$, then under the hypotheses of the lemma, we see that there is an eigenform $g$ in $S_k(\Gamma_1(mN))^\psi$ such that $\alpha(a_p(f)) = \alpha(a_p(g))$ for all primes $p$ not dividing $\ell mN$. We will need a stronger version of the above lemma in the case $k = 2$; we will show that $g$ can be chosen so that $\alpha(a_n(f)) = \alpha(a_n(g))$ for all $n \geq 1$. The argument is essentially the one found in [3] or in [6], Section 3, except that we consider the action of the Hecke operators on group cohomology rather than the Galois action on etale cohomology. Though a similar argument works for $k > 2$, we state, prove and use the result only for $k = 2$.

Before doing so, we need a few definitions. Let $\tilde{\mathbf{T}}$ denote the polynomial ring over $\mathbf{Z}$ generated by the variables $T_n$ for $n \geq 1$. Let us call a maximal ideal of $\tilde{\mathbf{T}}$ *Eisenstein* if it contains $T_r - (r+1)$ for all primes $r \equiv 1 \bmod mN$. We say that a $\tilde{\mathbf{T}}$-module is *Eisenstein* if all maximal ideals in its support are Eisenstein. Let $\mathbf{m}$ denote the kernel of $\alpha \circ \theta_f$ where $\theta_f : \tilde{\mathbf{T}} \to \mathcal{O}$ is the homomorphism defined by $T_n \mapsto a_n(f)$. Note that $\mathbf{m}$ is Eisenstein if and only if $\rho_f \otimes_{R_f} \mathbf{F}$ is reducible; hence we assume that $\mathbf{m}$ is not Eisenstein.

**Lemma 2.2.** *Suppose that $k = 2$ and let $\ell$, $m$, $N$, $\chi$, $\psi$ and $\rho$ be as in Lemma 2.1. Let $f$ be an eigenform in $S_2(\Gamma_1(mN))^\chi$ such that $\rho \cong \rho_f \otimes_{R_f} \mathbf{F}$. Then there is an eigenform $g$ in $S_2(\Gamma_1(mN))^\psi$ such that $\alpha(a_n(f)) = \alpha(a_n(g))$ for all $n \geq 1$.*

*Proof.* Suppose first that $m > 3$. Let $K \subset \mathbf{C}$ be a finite extension of $\mathbf{Q}$ containing the values of $\chi$ and $\psi$. Let $\mathcal{O}'$ be the localization of $\mathcal{O}_K$ at the kernel of $\alpha|_{\mathcal{O}_K}$, $\lambda'$ a uniformizer in $\mathcal{O}'$ and $\mathbf{F}' = \alpha(\mathcal{O}_K) \cong \mathcal{O}'/\lambda'\mathcal{O}'$. For an $\mathcal{O}'$-module $M$, write $\mathcal{L}^i(M, \chi)$ for the cohomology group $H^i(\Gamma_1(m, N), M)$ where $\Gamma_1(m, N)$ acts on $M$ by $\chi^{-1}$. Using for example that $\Gamma_1(m, N)$ is a free group on a finite set of generators, we see that $\mathcal{L}^i(\mathcal{O}', \chi)$ is finitely generated over $\mathcal{O}'$, that $\mathcal{L}^i(M, \chi) = 0$ if $i > 1$, and that $\mathcal{L}^i(M, \chi) \cong \mathcal{L}^i(\mathcal{O}', \chi) \otimes_{\mathcal{O}'} M$ if $M$ is flat over $\mathcal{O}'$. As in [17], Chapter 8, $\mathcal{L}^i(M, \chi)$ is endowed with a natural action of the Hecke operators $T_n$ and there is a $\tilde{\mathbf{T}}$-equivariant inclusion

$$\mathrm{Sh}_\chi : S_2(\Gamma_1(mN))^\chi \oplus S_2(\Gamma_1(mN))^\chi \hookrightarrow \mathcal{L}^1(\mathbf{C}, \chi).$$

We can replace $\chi$ by $\psi$ in the above definitions and assertions.

Since $\mathbf{m}$ contains the annihilator of $S_2(\Gamma_1(mN))^\chi$ and $\mathcal{L}^1(\mathcal{O}', \chi) \otimes_{\mathcal{O}'} \mathbf{C} \cong \mathcal{L}^1(\mathbf{C}, \chi)$, we see that $\mathbf{m}$ contains the annihilator of $\mathcal{O}'$-module $\mathcal{L}^1(\mathcal{O}', \chi)$. Since $\mathcal{L}^1(\mathcal{O}', \chi)_{\mathbf{m}}$ is finitely generated over $\mathcal{O}'$, $\mathbf{m}$ is in the support of

$$\mathcal{L}^1(\mathcal{O}', \chi) \otimes_{\mathcal{O}'} \mathbf{F}' \cong \mathcal{L}^1(\mathbf{F}', \chi) \cong \mathcal{L}^1(\mathbf{F}', \psi),$$

where the first isomorphism is gotten from the vanishing of $\mathcal{L}^2(\mathcal{O}', \chi)$ and the second from the hypothesis that $\alpha \circ \chi = \alpha \circ \psi$. To reverse the implications and conclude that $\mathbf{m}$ contains the annihilator of $S_2(\Gamma_1(mN))^\psi$, it suffices to check that the following $\tilde{\mathbf{T}}$-modules are Eisenstein:
- the torsion subgroup of $\mathcal{L}^1(\mathcal{O}', \psi)$;

- the cokernel of $\text{Sh}_\psi$.

That the former module is Eisenstein follows from the fact that $T_r = r+1$ on $\mathcal{L}^0(\mathbf{F}',\psi)$ for all primes $r$ not dividing $mN$. The latter module is Eisenstein as it is isomorphic to a space of modular forms spanned by Eisenstein series of weight two and level $mN$. (See [10], Section 5 for an explicit description.)

We have now proved that $\mathbf{m}$ contains the annihilator of $S_2(\Gamma_1(mN))^\psi$. It follows that $\alpha \circ \theta_f = \alpha \circ \theta$ for some homomorphism $\theta : \tilde{\mathbf{T}} \to \mathcal{O}$ which factors through the image of $\tilde{\mathbf{T}}$ in $\text{End}\,(S_2(\Gamma_1(mN))^\psi)$. Such a homomorphism is necessarily of the form $\theta_g$ for some eigenform $g$ in $S_2(\Gamma_1(mN))^{\sigma \circ \psi}$ with $\sigma \in \text{Gal}\,(\overline{\mathbf{Q}}/\mathbf{Q})$. Since this implies that $\alpha \circ \sigma \circ \psi = \alpha \circ \psi$, we may assume that $\sigma$ is in the inertia subgroup for $\ker \alpha$ and replace $g$ by $g^{\sigma^{-1}}$ which is in $S_2(\Gamma_1(mN))^\psi$. Then $\alpha(a_n(f)) = \alpha(a_n(g))$ as desired for all $n \geq 1$.

Now suppose that either of the last two conditions in the lemma are satisfied. Choose a prime $q$ as in [6], Lemma 2 so that $q$ does not divide $6mN$. By what we have already proved, there is an eigenform $g'$ in $S_2(\Gamma_1(mqN))^\psi$ so that $\alpha(a_n(f)) = \alpha(a_n(g'))$ for all $n$ relatively prime to $q$. But since $q$ does not divide the conductor of the newform associated to $g'$, there is an eigenform $g$ in $S_2(\Gamma_1(mN))^\psi$ so that $a_n(g) = a_n(g')$ for all $n$ relatively prime to $q$. Considering the trace and determinant of the image of $\text{Frob}_q$ in $\rho_f \otimes_{R_f} \mathbf{F} \cong \rho_g \otimes_{R_g} \mathbf{F}$, we see also that $\alpha(a_{q^i}(f)) = \alpha(a_{q^i}(g))$ for all $i > 0$. So in fact $\alpha(a_n(f)) = \alpha(a_n(g))$ for all $n \geq 1$. $\square$

## 3 Character groups.

A key case of "lowering the level" was treated by Ribet in [15] and [14] under the assumption that $\rho$ arises from an eigenform on $\Gamma_0(N)$. An important ingredient in Ribet's proof is a certain relationship between character groups arising from Jacobians of modular curves and Shimura curves. We now undertake the analysis of character groups in Sections 3 and 4 of [14], but in the context of modular forms on $\Gamma_1(N)$ rather than $\Gamma_0(N)$.

For relatively prime positive integers $N$ and $c$, the modular curve $X_1(N,c)$ over $\mathbf{Q}$ arises from the moduli problem parametrizing triples $(E, P, C)$ where $E$ is an elliptic curve, $P$ is a point of $E$ of order $N$ and $C$ is a cyclic subgroup of $E$ of order $c$. Suppose that $N$ and $d$ are relatively prime and let $q$ be a prime not dividing $Nd$. Applying the results of Raynaud discussed in [14], Section 2 to the model for $X_1(N, dq)$ over $\mathbf{Z}_q$ described by Deligne and Rapoport [5], we consider the connected component $J^0$ of the special fiber of the Neron model over $\mathbf{Z}_q$ of the Jacobian $J_1(N, dq)$ of $X_1(N, dq)$. We thus obtain an exact sequence

$$1 \to T \to J^0 \to J_1(N,d) \times J_1(N,d) \to 0,$$

where $T$ is a torus whose character group $X$ we describe as follows. The analogue of [14], (3.1) is that $X$ is isomorphic to the group of degree-0 divisors on the set $\Sigma_1(N, d)$ of supersingular points of $X_1(N,d)(\overline{\mathbf{F}}_q)$. This set may be identified with the set of isomorphism classes of "enhanced elliptic curves" $(E, P, C)$ over $\overline{\mathbf{F}}_q$, where $E$ is a supersingular elliptic curve, $P$ is a point of $E$ of order $N$ and $C$ is a cyclic subgroup of $E$ of order $d$. Fix such a point $\mathbf{E}_0 = (E_0, P_0, C_0)$, and let $R = \text{End}\,(E_0)$ and $H = R \otimes \mathbf{Q}$. Then $R$ is a maximal order in the quaternion algebra $H$ over of discriminant $q$. (Note that $R$ is used to denote an Eichler order in [**14**].) We again have a natural notion of an "enhanced Tate module" of $\mathbf{E} = (E, P, C)$, which we denote $T(\mathbf{E})$. Let $U$ denote the open compact subgroup of $R_{\mathbf{f}}^{\times}$ consisting of those elements which induce automorphisms of $T(\mathbf{E}_0)$ (where $R_{\mathbf{f}}$ denotes $R \otimes \hat{\mathbf{Z}}$). If $\mathbf{E}$ defines an element of $\Sigma_1(N, d)$, then we may choose an isogeny $\lambda : E \to E_0$ and an element $g$ of $H_{\mathbf{f}}^{\times}$ so that $g\lambda$ induces an isomorphism of enhanced Tate modules. This construction, similar to [**14**], (3.3), defines a bijection between $\Sigma_1(N, d)$ and the double coset space

$$U \backslash H_{\mathbf{f}}^{\times} / H^{\times}.$$

The curve $X_1(N, dq)$ comes with the usual Hecke correspondences $T_n$ for integers $n \geq 1$ and $\langle m \rangle$ for integers $m$ relatively prime to $N$. For example, if $r$ is a prime not dividing $Ndq$, then $T_r$ is described by

$$(E, P, C) \mapsto \sum_B (E/B, P \bmod B, (C+B)/B)$$

where the sum is over subgroups $B$ of $E$ of order $r$. These correspondences are defined over $\mathbf{Q}$, as is the operator $w_q$ defined by

$$(E, P, C) \mapsto (E/C_q, P \bmod C_q, (C + E[q])/C_q).$$

We use the same symbols to denote the endomorphisms of $J_1(N, dq)$, $T$ and $X$ these induce by Pic functoriality, and we find that [**14**], (3.7),(3.8) hold with "involution" replaced by "automorphism", $X_0(N)$ replaced by $X_1(N, dq)$ and $\Sigma(M)$ replaced by $\Sigma_1(N, d)$. (See [**15**], Section 8.)

We let $\mathbf{T} = \mathbf{T}_1(N, dq)$ be the subring of $\mathrm{End}\,(J_1(N, dq))$ generated by the operators $T_n$ for $n \geq 1$. We recall that $\langle m \rangle$ is in $\mathbf{T}$ for $m$ relatively prime to $N$ and that $\mathbf{T}$ also acts faithfully on $S = S_2(\Gamma_1(N, dq))$. Decompose $S = S_0 \oplus S_1$ where $S_0$ is the "$q$-old" subspace of $S$ and $S_1$ is the "$q$-new" subspace. Thus $S_0$ is isomorphic to the direct sum of two copies of $S_2(\Gamma_1(N, d))$ and $S_1$ is its orthogonal complement. The action of $\mathbf{T}$ on $S$ respects the decomposition, and we may consider the $q$-old quotient $\mathbf{T}_0$ of $\mathbf{T}$ and the $q$-new quotient $\mathbf{T}_1$ of $\mathbf{T}$. No changes are then needed for [**14**], (3.10), (3.11). We postpone the discussion of monodromy pairings and component groups until the next section.

For distinct primes $p$ and $q$ not dividing $N$, we consider the two natural degeneracy maps

$$\alpha_p, \beta_p : X_1(N, pq) \to X_1(N, q).$$

The maps induced on the character groups are described explicitly by the corresponding maps $\Sigma_1(N, p) \to \Sigma_1(N, 1)$ defined by $\alpha_p(E, P, C) = (E, P, 0)$ and $\beta_p(E, P, C) = (E/C, P \bmod C, 0)$. These combine to produce a map

$$\delta_p : L_q \to X_q$$

where $L_q$ is the character group $X$ defined with $d = p$ and $X_q$ is the product of two copies of the corresponding character group defined with $d = 1$. Thus $L_q$ replaces the group $L$ in [**14**] and $X_q$ replaces $X \oplus X$. To prove that $\delta_p$ is surjective ([**14**], (3.15)), we appeal directly to the strong approximation theorem rather than the arithmetic of Eichler orders. Fix a supersingular enhanced elliptic curve $\mathbf{E}_0 = (E_0, P_0, 0)$, and let $R = \text{End}(E_0)$ and $H = R \otimes \mathbf{Q}$. Define $U \subset R_{\mathbf{f}}^\times$ as above and note that $\nu(U) = \hat{\mathbf{Z}}$ where $\nu$ is the reduced norm. If $\mathbf{E} = (E, P, 0)$ defines an element of $\Sigma_1(N, 1)$, choose an isogeny $\lambda : E_0 \to E$ and an element $x$ in $H_{\mathbf{f}}^\times$ so that $\lambda x$ induces an isomorphism of enhanced Tate modules. Using the strong approximation theorem we find an element $\beta$ in $xUH_p^\times \cap H^\times$ so that $\nu(\beta)^{-1}$ is the degree of $\lambda$. If $n$ is sufficiently large and $p^n \equiv 1 \mod N$, then $p^n \lambda \beta$ is an isogeny $\mathbf{E}_0 \to \mathbf{E}$ whose degree is an even power of $p$. We then proceed as in [14], page 452 to conclude that $\delta_p$ is surjective. (Note that $\mathbf{E}_{2j}$ gets replaced by $\langle p \rangle^{-j} \mathbf{E}_{2j}$.)

Next we consider the action of $\mathbf{T} = \mathbf{T}_1(N, pq)$ on $L_q$. Write $Y$ for the kernel of $\delta_p$. We distinguish $T_p$ in $\mathbf{T}$ from the $p^{\text{th}}$ Hecke operator in $\mathbf{T}_1(N, q)$, denoting the latter by $\tau_p$. As in [14], (3.19), we find that $T_p$ preserves $Y$, that $T_p = -w_p$ on $Y$ and that $T_p$ on $X_q$ is represented by the matrix

$$\begin{pmatrix} \tau_p & -1 \\ \langle p \rangle p & 0 \end{pmatrix}$$

of endomorphisms of the character group obtained from $J_1(N, q)$. Define in the obvious way various "old" and "new" subspaces of $S = S_2(\Gamma_1(N, pq))$ and quotients of $\mathbf{T}$. With the evident changes in notation, we have [14], (3.20), (3.21).

We now turn our attention to the Shimura curve analogous to the one considered in [14], Section 4. We fix a quaternion algebra $B$ over $\mathbf{Q}$ of discriminant $pq$ and a maximal order $L$ in $B$. Also fix a cyclic left $L$-submodule $Q \subset L/NL$ of exponent $N$ and order $N^2$. Let $\Gamma$ denote the set of $g \in L$ of reduced norm 1 satisfying $xg = x$ for all $x$ in $Q$. Choosing an isomorphism $B \otimes \mathbf{R} \cong M_2(R)$ we obtain an action of $\Gamma$ on the upper half-plane $\mathcal{Z}$. The Shimura curve $\Gamma \backslash \mathcal{Z}$ has a canonical model over $\mathbf{Q}$ parametrizing triples $(A, i, \alpha)$ where $A$ is an abelian surface, $i$ is an action $L \hookrightarrow \text{End}(A)$ and and $\alpha$ is an $L$-linear embedding $Q \hookrightarrow A$. We have also the usual Hecke correspondences $T_n$ for integers $n \geq 1$ and $\langle m \rangle$ for integers $m$ relatively prime to $N$.

The theory of Cerednik and Drinfeld [7], Section 4 provides a model $\mathcal{C}$ over $\mathbf{Z}_p$ for this Shimura curve. The formal scheme associated to this model is as in [14], pages 457-8, except that in the description of the $p$-adic space $X$, an "enhancement" should now be an $L$-linear embedding of $Q$. Let $A$ be an abelian surface over $\overline{\mathbf{F}}_p$ with an action of $L$ defined by $i$, an enhancement $\alpha : Q \hookrightarrow A$, and an $L \otimes \mathbf{Z}_p$-isomorphism $\iota$ between the formal group of $A$ and the formal module $\Phi$ defined in [7]. We let $\mathcal{M} = \text{End}_L(A)$ and $\mathcal{H} = \mathcal{M} \otimes \mathbf{Q}$. Thus $\mathcal{H}$ is a definite quaternion algebra over $\mathbf{Q}$ with discriminant $q$, $\mathcal{M}$ is a maximal order in $\mathcal{H}$ and $\iota$ induces an isomorphism $\mathcal{M}_p \cong M_2(\mathbf{Z}_p)$. Let $K$ denote the subgroup of prime-to-$p$ ideles in $\mathcal{M}_{\mathbf{f}}^\times$ which induce automorphisms of the enhanced Tate module of $(A, \alpha)$. The construction described in [14] then gives a bijection between $X$ and the double coset space

$$K \backslash \mathcal{H}_{\mathbf{f}}^\times / \mathcal{H}^\times$$

compatible with the action of $GL_2(\mathbf{Q}_p) \cong \mathcal{H}_p^\times$.

Let $r$ be the order of $p$ in $(\mathbf{Z}/N\mathbf{Z})^\times / \{\pm 1\}$, and let $G$ denote the subgroup of $GL_2(\mathbf{Q}_p)$ consisting of those $g$ for which $v_p(\det g)$ is divisible by $2r$. By the method of [11], Section 4 we find that the dual graph $\mathcal{G}$ associated to the special fiber of $\mathcal{C}$ is isomorphic to $G\backslash(\Delta \times X)$, where $\Delta$ is the tree associated with $SL_2(\mathbf{Q}_p)$. Now let $V = K\mathcal{M}_p^\times$ and let $W = KW_p$ where $W_p$ is the subgroup of $\mathcal{M}_p^\times$ corresponding to

$$\left\{ \begin{pmatrix} a & b \\ c & d \end{pmatrix} \in GL_2(\mathbf{Z}_p) \middle| c \equiv 0 \bmod p \right\}$$

under the isomorphism induced by $\iota$. Setting $\mathcal{V} = V\backslash \mathcal{H}_\mathbf{f}^\times/\mathcal{H}^\times$, $\mathcal{E} = W\backslash \mathcal{H}_\mathbf{f}^\times/\mathcal{H}^\times$ and defining two degeneracy maps $\alpha$ and $\beta$ from $\mathcal{E}$ to $\mathcal{V}$ as in [14], we obtain the description of $\mathcal{G}$ in [14], (4.4). (Use that $G\mathbf{Q}_p^\times = GL_2(\mathbf{Q}_p)^+$.) This gives the description in [14], (4.5) of the character group of the connected component of the special fiber of the Neron model over $\mathbf{Z}_p$ of the Jacobian $J$ of the Shimura curve. We will denote this group, analogous to $Z$ in [14], by $Y_p$. Instead of [14], (4.7) we have

**Proposition 3.1** *There is a supersingular elliptic curve $E$ over $\overline{\mathbf{F}}_q$, with a point $P$ of order $N$, a subgroup $C$ of order $p$ and an isomorphism $\mathcal{M} \cong \mathrm{End}\,(E)$ such that $W \subset \mathcal{M}_\mathbf{f}^\times$ corresponds to the stabilizer of $(P, C)$ in $(\mathrm{End}\,(E) \otimes \hat{\mathbf{Z}})^\times$.*

The existence of the curve $E$ and the isomorphism $\mathcal{M} \cong \mathrm{End}\,(E)$ is proved in [14], (3.6). To choose $P$ note that if $r$ is a prime dividing $N$, then there is an isomorphism $\mathcal{M}_r \cong M_2(\mathbf{Z}_r)$ such that $W_r$ corresponds to

$$\left\{ \begin{pmatrix} a & b \\ c & d \end{pmatrix} \in GL_2(\mathbf{Z}_r) \middle| a - 1 \equiv c \equiv 0 \bmod N\mathbf{Z}_r \right\}.$$

We construct $C$ similarly.

Using $(E, P, C)$ as $\mathbf{E}_0$ in the double coset description of $\Sigma_1(N, p)$, the isomorphism $\mathcal{M} \cong R$ induces a bijection $\iota : \Sigma_1(N, p) \to \mathcal{E}$. Similarly using $(E, P, 0)$ we construct $\lambda : \Sigma_1(N, 1) \to \mathcal{V}$. We have the same compatibilities as in [14], (4.8). This establishes an isomorphism $Y \cong Y_p$. (Recall that $Y$ is the kernel of the map $\delta_p : L_q \to X_q$, whereas $Y_p$ is defined as a character group.) That this isomorphism commutes with the Hecke operator $T_p$ follows from the argument of [14], pages 463-4; one need only take care to distinguish between the various automorphisms and their inverses. (In particular $T_p$ coincides with Frob$_p$ on $\mathcal{C}(\overline{\mathbf{F}}_p)$. This induces the inverse of the automorphism of $\mathcal{G}$ denoted $F(\mathcal{C}/\mathbf{Z}_p)$ in [11]. This in turn coincides with the automorphism of $G\backslash(\Delta \times X)$ induced by an element $m$ of $GL_2(\mathbf{Q}_p)$ with $\nu_p(\det m) = 1$.) It is straighforward to verify the compatibility with $T_r$ for primes $r \neq p$, and with $\langle m \rangle$ for integers $m$ prime to $N$. Letting $\tilde{\mathbf{T}}$ denote the polynomial ring over $\mathbf{Z}$ generated by the variables $T_n$ for $n \geq 1$, we conclude that $\tilde{\mathbf{T}}$ acts on $Y_p$ via its quotient $\mathbf{T}$ and that

**Theorem 3.1.** *There is a $\mathbf{T}$-isomorphism $Y_p \to Y$.*

**4 Lowering the level.** We first follow Ribet's argument in Sections 6-8 of [15] to use the above results on character groups to prove

**Theorem 4.1.** *Suppose that $\ell$ is odd and that $\rho : \mathrm{Gal}\,(\overline{\mathbf{Q}}/\mathbf{Q}) \to GL_2(\mathbf{F})$ is an irreducible representation arising from $S_2(\Gamma_1(N,p))$ for some prime $p$ not dividing $N\ell$. If $\rho$ is unramified at $p$, then $\rho$ arises from $S_2(\Gamma_1(N))$.*

With $\Gamma_1(N,p)$ replaced by $\Gamma_0(Np)$ and $\Gamma_1(N)$ replaced by $\Gamma_0(N)$, this is [15], (1.5). Ribet's proof builds on Mazur's Principle and the work in [14] where the theorem is proved under the additional hypothesis that $p \not\equiv 1 \bmod \ell$ or that $\ell$ does not divide $N$. Note that [15], (1.5) already gives Theorem 4.1 in the case $N \leq 3$, so we assume that $N > 3$ until we have completed the proof of Theorem 4.1.

Before explaining the proof of Theorem 4.1, we continue the discussion of the Jacobians and character groups from the preceding section. Recall that we denote the $p^{\mathrm{th}}$ Hecke operator on $J_1(N,pq)$ by $T_p$ and that on $J_1(N,q)$ by $\tau_p$. We use $\delta_p$ to denote the homomorphism of Jacobians

$$J_1(N,q)^2 \to J_1(N,pq),$$

as well as that on the character groups $L_q \to X_q$, induced by the two degeneracy maps $X_1(N,pq) \to X_1(N,q)$. Thus the maps denoted $\delta_p$ become **T**-equivariant if we define the action of $T_p$ on $J_1(N,q)^2$ by $\begin{pmatrix} \tau_p & \langle p \rangle p \\ -1 & 0 \end{pmatrix}$ and on $X_q$ by $\begin{pmatrix} \tau_p & -1 \\ \langle p \rangle p & 0 \end{pmatrix}$. We let $\eta_p = T_p^2 - \langle p \rangle$. The proof of [15], (6.1) gives now a **T**-equivariant homomorphism

$$\sigma_p : J_1(N,pq) \to J_1(N,q)^2$$

such that $\sigma_p \delta_p = \eta_p$, and we let $\kappa_p$ denote the dual of $\sigma_p$. Theorem 3.1 gives **T**-equivariant exact sequences

$$0 \to Y_q \xrightarrow{\iota_q} L_p \xrightarrow{\delta_q} X_p \to 0, \quad 0 \to Y_p \xrightarrow{\iota_p} L_q \xrightarrow{\delta_p} X_q \to 0$$

analogous to [15], (6.2), (6.3).

Next we recall that there are natural monodromy pairings on the above character groups. In the case of $L_p$ and $L_q$ these are the restrictions of the standard Euclidean pairings on $\mathbf{Z}^{\Sigma_1(N,q)}$ and $\mathbf{Z}^{\Sigma_1(N,p)}$ (the former is a group of divisors on supersingular $\overline{\mathbf{F}}_p$ points, the latter on supersingular $\overline{\mathbf{F}}_q$ points). In the case of $Y_p$ and $Y_q$ the pairings are induced by the standard ones on $\mathbf{Z}^{\mathcal{E}}$ where $\mathcal{E}$ is the corresponding set of edges. (This follows from the fact that $H^\times \cap xUx^{-1}$ is trivial for $x \in H_{\mathbf{f}}^\times$ under the assumption $N > 3$.) Hence $\iota_p$ and $\iota_q$ are compatible with the monodromy pairings.

For an abelian variety $A$ over $\mathbf{Q}$ with an action of $\mathbf{T}$, we denote the dual abelian variety with its functorial **T**-action by $A'$. If $Z_p$ is the character group associated to the mod $p$ reduction of $A$, then we write $Z'_p$ the corresponding character group obtained from $A'$, and we use the same convention for component groups. (Thus for example there is a canonical isomorphism of the underlying groups $L_p \to L'_p$ which is not necessarily **T**-equivariant.) If $A$ has semistable reduction mod $p$, then the associated component group is canonically isomorphic as a **T**-module to the cokernel of the map $Z'_p \to \mathrm{Hom}(Z_p, \mathbf{Z})$

induced by the monodromy pairing. We write $\Phi_p$ and $\Theta_p$ for the component groups obtained from $J_1(N,p) \times J_1(N,p)$ and $J_1(N,pq)$ mod $p$, and $\Psi_q$ for the one obtained from $J$ mod $q$ where $J$ is the Jacobian of the Shimura curve considered above. In the statements of [15], (6.4),(6.5) we should subscript $\iota$ and $\kappa$ with $q$. As in [15], (6.6) we have a **T**-equivariant exact sequence

$$0 \to K' \to X_p/\eta_p X_p \to \Psi_q' \to C' \to 0$$

where $K'$ and $C'$ the kernel and cokernel of a **T**-equivariant map $\Phi_p' \to \Theta_p'$. The argument of [15], pages 672-3 shows that $\Phi_p$, $\Theta_p$, $\Phi_p'$ and $\Theta_p'$ are Eisenstein in the sense that $T_r = r+1$ and $\langle r \rangle = 1$ for primes $r$ not dividing $Npq$. No changes are needed in the statements of [15], (6.7),(6.8).

Recall that to a maximal ideal **m** of **T** such that **T**/**m** has characteristic $\ell$, we associate a semisimple representation

$$\rho_{\mathbf{m}} : \operatorname{Gal}(\overline{\mathbf{Q}}/\mathbf{Q}) \to GL_2(\mathbf{T}/\mathbf{m}).$$

If $r$ is a prime not dividing $Npq\ell$, then $\rho_{\mathbf{m}}$ is unramified at $r$ and $\rho_{\mathbf{m}}(\operatorname{Frob}_r)$ has characteristic polynomial $X^2 - T_r X + r\langle r \rangle$ mod **m**. This characterizes $\rho_{\mathbf{m}}$ up to isomorphism. The representations $\rho : \operatorname{Gal}(\overline{\mathbf{Q}}/\mathbf{Q}) \to GL_2(\mathbf{F})$ arising from $S_2(\Gamma_1(N,pq))$ are those isomorphic to $\rho_{\mathbf{m}} \otimes \mathbf{F}$ for some maximal ideal **m** and some embedding $\mathbf{T}/\mathbf{m} \to \mathbf{F}$. If **m** is Eisenstein in the sense that it contains $T_r - (r+1)$ and $\langle r \rangle - 1$ for all but finitely primes $r$, then $\rho_{\mathbf{m}}$ is reducible. (As noted in Section 2, a weaker notion of Eisenstein implies reducibility of $\rho_{\mathbf{m}}$.) Thus we have [15], (6.9) as stated except that we are writing $\eta_q$ rather than $\eta$.

Now let us assume that $\rho$ satisfies the hypotheses of Theorem 4.1, namely that $\rho : \operatorname{Gal}(\overline{\mathbf{Q}}/\mathbf{Q}) \to GL_2(\mathbf{F})$ is irreducible, arises from $S_2(\Gamma_1(N,p))$ for some prime $p$ not dividing $N\ell$, and is unramified at $p$. We assume also that $\ell$ is odd. As in [15], page 654 consider the three-dimensional representation $\sigma = \rho \times \chi$ where $\chi$ is the mod $\ell$ cyclotomic character, and choose a prime $q$ not dividing $Np\ell$ so that the conjugacy class of $\sigma(\operatorname{Frob}_q)$ coincides with that of $\sigma(c)$ where $c$ is complex conjugation. We then find a maximal ideal **m** of $\mathbf{T} = \mathbf{T}_1(N,pq)$ and an embedding $\mathbf{T}/\mathbf{m} \to \mathbf{F}$ so that $\rho$ is obtained from $\rho_{\mathbf{m}}$ by extension of scalars. Moreover $q \equiv -1 \bmod \ell$ and

$$T_q^2 \equiv \langle q \rangle \equiv 1 \bmod \mathbf{m},$$

so that in particular $\eta_q \in \mathbf{m}$.

Next consider the kernel of **m** in $J_1(N,pq)$, denoted $J_1(N,pq)[\mathbf{m}]$. For our contravariant Hecke operators and arithmetic Frobenius $\operatorname{Frob}_r$, the Eichler-Shimura relation gives

$$\operatorname{Frob}_r^2 - \langle r \rangle^{-1} \operatorname{Frob}_r + \langle r \rangle^{-1} r = 0$$

on $J_1(N,pq)[\ell]$ for primes $r$ not dividing $Npq\ell$. It therefore follows from [2] that $J_1(N,pq)[\mathbf{m}]$ is isomorphic as a $\mathbf{T}/\mathbf{m}[\operatorname{Gal}(\overline{\mathbf{Q}}/\mathbf{Q})]$-module to a direct sum of copies of $W = \operatorname{Hom}(V, \mu_l)$ where $V$ is a model for $\rho_{\mathbf{m}}$. We define $\lambda$ as the number of copies, so that $\lambda > 0$ and $\dim_{\mathbf{T}/\mathbf{m}}(J_1(N,pq)[\mathbf{m}]) = 2\lambda$. We have similarly that $J[\mathbf{m}]$ is a direct sum of copies of $\mu$ copies of $W$ where $\dim_{\mathbf{T}/\mathbf{m}}(J[\mathbf{m}]) = 2\mu$.

Under the assumption that $\rho$ does not arise from $S_2(\Gamma_1(N,q))$, we have equalities and inequalities identical to [**15**], (7.4)-(7.8). (For the analogues of [**15**], (7.7),(7.8), use that Frob$_q$ acts as $T_q$ on $Y_q$ and $L_q$. Since $T_q^2 = \langle q \rangle$ on these groups it follows that Frob$_q$ acts as $q\langle q \rangle^{-1} T_q$ on $\text{Hom}(Y_q, \mu_\ell)$ and $\text{Hom}(L_q, \mu_\ell)$, and as $\langle q \rangle^{-1} T_q$ on $Y'_q$.) The resulting contradiction shows that $\rho$ arises from $S_2(\Gamma_1(N,q))$.

To complete the proof of Theorem 4.1, we appeal to Mazur's Principle [**15**], (8.1) to conclude that $\rho$ arises from $S_2(\Gamma_1(N))$. We remark on a potentially confusing point in the proof of Mazur's Principle in [**15**]. Ribet considers a maximal ideal $\mathbf{m}$ in $\mathbf{T}_1(N,q)$ such that $\rho_\mathbf{m}$ provides a model $V$ for $\rho$ and asserts that $J_1(N,q)[\mathbf{m}]$ is isomorphic to a direct sum of copies of $V$. Implicit in this assertion is a choice of conventions different from the one used above. For example, one could use the above definitions for the curve $X_1(N,q)$ and the Hecke correspondences $T_n$, but define the action of $T_n$ on $J_1(N,q)$ using Albanese functoriality. In that case, Frob$_q$ would act on the character group $X$ as $\langle q \rangle^{-1} T_q$ rather than $T_q$ as stated on [**15**], page 674. Here $X$ denotes the character group of the torus $T$ in the reduction mod $q$ of $J_1(N,q)$, so Frob$_q$ would then act as $qT_q$ on $T(\overline{\mathbf{F}}_q)$. With our conventions, however, $J_1(N,q)[\mathbf{m}]$ is isomorphic to a direct sum of copies of $W = \text{Hom}(V, \mu_\ell)$, Frob$_q$ coincides with $T_q$ on $X$, and with $q\langle q \rangle^{-1} T_q$ on $T(\overline{\mathbf{F}}_q)$. The square of this last scalar is $q^2\langle q \rangle^{-1}$ while the determinant of Frob$_q$ on $W$ is $q\langle q \rangle^{-1}$, yielding Mazur's Principle just the same. Yet another set of conventions is used in [**9**] where the modular curve parametrizes elliptic curves together with embeddings of $\mu_N$ and the Hecke operators on the Jacobian are defined using Albanese functoriality. (Note in particular [**9**], (3.12).)

**5  Proof of Theorem 1.1.** As explained in [**15**], Theorem 1.1 follows from Theorem 4.1 if $\ell > 3$. Very few changes are needed to Sections 2, 3 and 4 of [**15**] to include the case $\ell = 3$. We record a slightly stronger version of [**15**], (2.2):

**Theorem 5.1.** *Let $f$ be an eigenform in $S_k(\Gamma_1(N))$. Suppose that $\ell$ does not divide $N$, that $2 \leq k \leq \ell + 1$ and that $\rho = \rho_f \otimes_{R_f} \mathbf{F}$ is irreducible. Then there is an eigenform $g$ in $S_2(\Gamma_1(N\ell))$ such that $\alpha(a_n(f)) = \alpha(a_n(g))$ for all $n \geq 1$. In particular $\rho \cong \rho_g \otimes_{R_g} \mathbf{F}$ arises from $S_2(\Gamma_1(N\ell))$.*

For $\ell > 3$, this follows from [**1**], (3.5), and for $N > 4$ from [**9**], (9.3). Note that the eigenform $g$ must be cuspidal since $\rho$ is irreducible. Note also that $S_k(\Gamma_1(N)) = 0$ if $2 \leq k \leq 4$ and $N \leq 4$.

Appealing to the work of Gross [**9**], Edixhoven [**8**] and Coleman and Voloch [**4**] on the weight in Serre's conjecture, we see that [**15**], (3.3) is valid for odd $\ell$ so that Theorem 1.1 will be a consequence of the following.

**Theorem 5.2.** *Let $\ell$ be an odd prime and $N$ a positive integer. If $\ell = 3$, then suppose also that 9 divides $N$. Let $\rho$ be an irreducible representation arising from $S_2(\Gamma_1(N))$. Suppose that $p \neq \ell$ is a prime dividing $N/N(\rho)$. Then $\rho$ arises from $S_2(\Gamma_1(N/p))$.*

We proceed as in [15], Section 4, combining methods and results of Carayol [3] with Theorem 4.1. We first regard $\rho$ as arising from an automorphic representation $\pi$ of $GL_2(\mathbf{A})$ associated to a weight two newform of conductor dividing $N$. We may assume that the power of $p$ dividing $N$ coincides with the conductor of $\pi_p$. We distinguish three cases according to whether $\pi_p$ is principal series, special or cuspidal.

Suppose that $\pi_p$ is principal series. Let $d$ be the highest power of $\ell$ dividing $N$. Now apply Lemma 2.1 to conclude that $\rho$ arises from $S_k(\Gamma_1(N))^\chi$ where $\chi$ is a mod $N/d$ Dirichlet character of order prime to $\ell$. The conclusion of Theorem 5.2 follows from Carayol's argument [15], (4.4).

Suppose that $\pi_p$ is special. The discussion in [15] explains why the desired conclusion follows from Theorem 4.1. Indeed Theorem 4.1 provides a solution to [15], (4.6). The discussion applies also to the case $\ell = 3$.

Suppose that $\pi_p$ is cuspidal, in which case $p^2$ divides $N$. We first remark that if $k = 2$, then the hypothesis that $\ell$ does not divide $N$ is not needed in [15], (5.1). Indeed [13], (4.2) provides the exact sequence [15], (5.3) of free $\mathbf{Z}_\ell$-modules. So we may choose a prime $q$ not dividing $N\ell$ such that $\rho$ arises from a newform of weight two and conductor dividing $Nq$ and for which the associated automorphic representation $\pi'$ is special at $q$. In view of the cases considered above, we need only prove that if $\pi'_p$ is cuspidal, then $\rho$ arises from $S_2(\Gamma_1(Nq/p))$. For $\ell > 3$ this case is treated by Carayol [3], Section 5. If $\ell = 3$, then the hypothesis that 9 divides $N$ ensures that the stabilizers in [3], Lemme 2 are trivial so that Carayol's argument applies here as well. This completes the proof of Theorem 1.1.

As a final remark we mention an alternative to the above argument treating the cuspidal case. Recall that Carayol uses the Jacquet-Langlands correspondence to switch to an automorphic representation on $(B \otimes \mathbf{A})^\times$, where $B$ is a quaternion algebra over $\mathbf{Q}$ of discriminant $pq$, and he then applies an analogue of Lemma 2.1. One can instead switch to a definite quaternion algebra of discriminant $p$ without introducing an auxiliary prime $q$. As an analogue of Carayol's lemma in the setting of a definite quaternion algebra, we have [6], Theorem 7. (The hypothesis that $\ell$ does not divide $N$ appears in the statement, but in fact it is never used in the proof.) To obtain the desired conclusion, one can apply [6], Theorem 7 with $T = S(D) = \{p\}$ and $\psi'_p$ as the trivial character.

## 6 Wiles' Variant.

We now explain how to deduce a variant of Theorem 1.1 proposed by Wiles [19]. We wish to prove that if

$$\rho : \text{Gal}\,(\overline{\mathbf{Q}}/\mathbf{Q}) \to GL_2(\mathbf{F})$$

is irreducible and modular, then $\rho$ arises from an eigenform of weight two for which the associated $\ell$-adic representation is of the same "deformation type" as $\rho$. This means that there is a weight two eigenform $f$ such that $\rho \cong \rho_f \otimes_{R_f} \mathbf{F}$, subject to some additional constraints on the local representations $\rho_f|_{D_p}$. Here $D_p$ denotes a decomposition group at $p$ and the constraints are determined by $\rho|_{D_p}$. For primes $p$ not dividing $\ell N(\rho)$ we will require simply that $\rho_f$ be unramified at $p$. To describe the restrictions imposed at the ramified primes, we will need more notation and definitions.

Let $\sigma : G \to GL_2(R)$ be a continuous homomorphism, where $G$ is a profinite group and $R$ is a complete local Noetherian ring with finite residue field of characteristic $\ell$. Let $\mathfrak{m}$ denote the maximal ideal of $R$ and write $\sigma_n$ for $\sigma \otimes_R R/\mathfrak{m}^{n+1}$. Then $\sigma$ is called a lifting of $\sigma_0$. Two such liftings are equivalent if they are conjugate by an element of the kernel of $GL_2(R) \to GL_2(R/\mathfrak{m})$, and an equivalence class of liftings is called a deformation of $\sigma_0$ (see [12]).

Consider now the case of $G = D_p$. Let $I_p$ denote the inertia subgroup of $D_p$ and $H_p$ the wild inertia subgroup. We will write $\chi_\ell$ for the composition of the $\ell^{\text{th}}$ cyclotomic character with the natural homomorphism $\mathbf{Z}_\ell^\times \to R^\times$. We recall the types of local representations considered by Wiles. For $p \neq \ell$ we make the following definitions:

- $\sigma$ is *type A* if $\sigma_0$ is ramified and $\sigma \cong \begin{pmatrix} \psi\chi_\ell & * \\ 0 & \psi \end{pmatrix}$ for some unramified character $\psi : D_p \to R^\times$.

- $\sigma$ is *type B* if $\sigma|_{I_p} \cong \begin{pmatrix} \psi & 0 \\ 0 & 1 \end{pmatrix}$ for some nontrivial character $\psi : I_p \to R^\times$ of finite order prime to $\ell$.

For $p = \ell$ we say

- $\sigma$ is *finite* if $\det \sigma|_{I_p} = \chi_\ell|_{I_p}$ and for each $n \geq 0$, $\sigma_n$ arises from a finite flat group scheme over $\mathbf{Z}_p$. (This means that there is a finite flat group scheme $G_n$ over $\mathbf{Z}_p$ with an action of $R$ such that $G_n(\overline{\mathbf{Q}}_p)$ is free of rank two over $R/\mathfrak{m}^{n+1}$ and $\sigma_n$ is isomorphic to $D_p \to \text{Aut}\,_R(G_n(\overline{\mathbf{Q}}_p))$.)

- $\sigma$ is *Selmer* if $\sigma \cong \begin{pmatrix} \phi & * \\ 0 & \psi \end{pmatrix}$ for characters $\phi, \psi : D_p \to R^\times$ with $\psi$ unramified and $\phi|_{H_p} = \chi_\ell|_{H_p}$.

In this last case we say also that $\sigma$ is $\psi_0$-Selmer (where for a homomorphism $\psi : D_p \to R^\times$, we write $\psi_0$ for $\psi \mod \mathfrak{m}$). Note that if $\sigma$ is Selmer, then $\psi_0$ is determined by $\sigma_0$ unless $\sigma_0$ is the direct sum of two distinct unramified characters.

We remark on some slight differences between the above terminology and that of Wiles. Firstly, we use the adjectives "type A", "type B", etc., to describe local Galois representations rather than deformations of a fixed global residual representation. For $p = \ell$ there are further distinctions. Note that a representation may be both finite and Selmer, and that Wiles calls a deformation "flat" if the local representation is (according to the definitions above) finite, but not Selmer. Note also that our notion of Selmer is more general than that of [19] in that we do not require $\phi_0 \neq \psi_0$. Wiles considers also "ordinary" deformations for which there is no restriction on $\phi|_{H_p}$. We make no mention of these as the results analogous to those below are immediate from the ones in the Selmer case. Finally, Wiles calls a deformation "strict" if the local representation $\sigma$ is Selmer with $\phi = \psi\chi_\ell$ and $\sigma_0$ is not finite. We make no further mention of these either since the following proposition shows that if $\sigma_0$ is strict, then "strict" and "Selmer" are equivalent.

**Proposition 0.1.** *Suppose that $p = \ell$ and $\sigma \cong \begin{pmatrix} \phi\chi_\ell & * \\ 0 & \psi \end{pmatrix}$ for unramified characters $\phi$ and $\psi$. If $\phi \neq \psi$, then $\sigma_0$ is finite.*

*Proof.* We may assume that $R$ is Artinian and $\psi = 1$. Now $\sigma \cong \begin{pmatrix} \phi \chi_\ell & t \\ 0 & 1 \end{pmatrix}$ for a cocycle $t : D_p \to M(1)$ where $M$ is a free $R$-module of rank one on which $D_p$ acts via $\phi$. Let $u$ denote the cohomology class in $H^1(D_p, M(1))$ defined by $t$, and let $u_0$ denote its image in $H^1(D_p, M_0(1))$ where $M_0 = M/\mathfrak{m}M$. Let $G = \ker \phi$ and let $F$ be the fixed field of $G$ (so $F$ is a finite unramified extension of $\mathbf{Q}_p$). Choose $n$ so that $\ell^n R = 0$. Since $H^2(G, \mu_{\ell^r}) \to H^2(G, \mu_{\ell^s})$ is injective for $r \leq s$, we see that the natural map of $R[D_p/G]$-modules $H^1(G, \mu_{\ell^n}) \otimes_{\mathbf{Z}_\ell} M \to H^1(G, M(1))$ is an isomorphism. So by Kummer theory we have $H^1(G, M(1)) \cong F^\times/(F^\times)^{\ell^n} \otimes_{\mathbf{Z}_\ell} M$. Now consider the commutative diagram

$$\begin{array}{ccccc} H^1(G, M(1))^{D_p} & \xrightarrow{\sim} & ((F^\times/(F^\times)^{\ell^n}) \otimes_{\mathbf{Z}_\ell} M)^{D_p} & \to & M^{D_p} \\ \downarrow & & \downarrow & & \downarrow \\ H^1(G, M_0(1)) & \xrightarrow{\sim} & (F^\times/(F^\times)^\ell) \otimes_{\mathbf{F}_\ell} M_0 & \to & M_0, \end{array}$$

where the rightmost horizontal maps are induced by $v_p : F^\times \to \mathbf{Z}$. If $\phi \neq 1$, then $M^{D_p} \subset \mathfrak{m}M$, so that the element res $u_0$ of $H^1(G, M_0(1))$ is in the image of $(\mathcal{O}_F^\times/(\mathcal{O}_F^\times)^\ell) \otimes_{\mathbf{F}_\ell} M_0$. But this means that $\sigma_0$ is "peu ramifié" in the sense of [16] and therefore $\sigma_0$ is finite (see [8], (8.2)). □

We remark that the same argument for $p \neq \ell$ proves the following:

**Proposition 6.2.** *Suppose $p \neq \ell$, $\sigma|_{I_p} \cong \begin{pmatrix} 1 & * \\ 0 & 1 \end{pmatrix}$ and $\sigma_0$ is ramified. Then $\sigma$ is type A.*

We now return to the setting of $\ell$-adic representations arising from modular forms. We let $f$ be a newform of weight $k$ and conductor $N$. Suppose that $\rho = \rho_f \otimes_{R_f} \mathbf{F}$ is irreducible. Write $N(\rho) = \prod_p p^{\alpha_p}$ and $N = \prod_p p^{\beta_p}$. We consider first $\sigma = \rho_f|_{D_p}$ for primes $p \neq \ell$. Recall that $p^{\beta_p}$ is the conductor of $\sigma \otimes_{R_f} \overline{K}_{f,\lambda}$, that $p^{\alpha_p}$ is the conductor of $\sigma_0 \otimes_{R_f} \mathbf{F}$ and that $\alpha_p \leq \beta_p$. Also note that if $\sigma$ is type A then $\alpha_p = 1$ and that if $\sigma$ is type B then $p^{\alpha_p}$ is the conductor of $\det \sigma_0$.

**Proposition 6.3.** *Suppose that $\alpha_p = \beta_p$, $\det \sigma|_{I_p}$ has order prime to $\ell$ and $\sigma_0$ is type A (resp. type B). Then $\sigma$ is type A (resp. type B).*

*Proof.* For $\ell$ odd, it is immediate from the classification in [6], Section 1 that $\tau_p = \sigma \otimes_{R_f} \overline{K}_{f,\lambda}$ has the desired form. It follows easily that so does $\sigma$.

For $\ell = 2$, the classification of [6], pp. 255-6 is valid except for the conclusion in case 3 that $F$ is unramified if $\tau_p$ is induced from a character whose reduction is equal to its conjugate. Note in this case however that the semisimplification of $\sigma \otimes_{R_f} \mathbf{F}$ is scalar so that $\sigma_0$ cannot be type B. The argument is then the same as for $\ell$ odd. □

We remark that the above proposition is implicit in [19] where results of Langlands are used to obtain information about the structure at $p$ of the automorphic representation corresponding to $f$, and consequently information about $\sigma$.

Now we consider $\sigma = \rho_f|_{D_p}$ for $p = \ell$.

It is well-known that $\sigma$ is finite if $k = 2$ and $\beta_\ell = 0$. Indeed $\det \sigma$ is the product of $\chi_\ell$ with an unramified character, and the $\sigma_n$ occur in the $\ell^n$-torsion of an abelian variety with good reduction at $p = \ell$.

Recall ([18], Theorem 2) that if $k = 2$, $\beta_\ell \leq 1$ and $a_\ell(f)$ is a unit in $R_f$, then $\rho_f$ is $\psi_0$-Selmer where $\psi_0$ is the unramified character $D_\ell \to (\mathcal{O}_f/\lambda)^\times$ defined by $\mathrm{Frob}\,_\ell \mapsto a_\ell(f) \mod \lambda$. Recall also that if $2 \leq k \leq \ell + 1$, $\beta_\ell = 0$ and $\sigma_0$ is Selmer, then the eigenvalue $a_\ell(f)$ is a unit in $R_f$ and $\sigma_0$ is $\psi_0$-Selmer where $\psi_0$ is again defined by $\mathrm{Frob}\,_\ell \mapsto a_\ell(f) \mod \lambda$. This is a consequence of results of Deligne and Fontaine stated in Edixhoven's article [8], (2.5), (2.6) using Katz's definition of modular forms over $\mathbf{F}$. If $\ell$ does not divide $N$ then the reduction of an eigenform $f$ gives (according to the definitions in [8]) an eigenform $f'$ over $\mathbf{F}$ of type $(N, k)$, and $f'$ is characterized by $a_n(f') = \alpha(a_n(f))$. Conversely, if $k \geq 2$, $N\ell > 4$ and $\ell$ does not divide $N$, then an eigenform over $\mathbf{F}$ of type $(N, k)$ is the reduction of a characteristic zero eigenform of weight $k$ and level $N$. Recall also that if $\rho_{f,0}$ is irreducible then $f$ must be a cusp form.

We now prove

**Theorem 6.4.** *Let $\mathbf{F}_0$ be a finite subfield of $\mathbf{F}$ and let $\rho_0 : \mathrm{Gal}\,(\overline{\mathbf{Q}}/\mathbf{Q}) \to GL_2(\mathbf{F}_0)$ and $\psi_0 : D_\ell \to \mathbf{F}_0^\times$ be continuous homomorphisms. Suppose that $\rho = \rho_0 \otimes_{\mathbf{F}_0} \mathbf{F}$ is irreducible and modular. Assume also that $\ell > 3$ or that $\ell = 3$ and $\rho$ is not induced from a character of $\mathrm{Gal}\,(\overline{\mathbf{Q}}/\mathbf{Q}(\sqrt{-3}))$. Then there is a newform $f$ of weight two and conductor $N$ dividing $\ell^2 N(\rho)$ so that $\rho \cong \rho_f \otimes_{R_f} \mathbf{F}$ and $\det \rho_f = \chi \chi_\ell$ for a character $\chi$ of order prime to $\ell$. Moreover*

- *if $\rho_0|_{D_\ell}$ is finite, then $f$ can be chosen so that $\ell$ does not divide $N$;*
- *if $\rho_0|_{D_\ell}$ is $\psi_0$-Selmer, then $f$ can be chosen so that $\alpha(a_\ell(f)) = \psi_0(\mathrm{Frob}\,_\ell)$ and $\ell^2$ does not divide $N$.*

*Proof.* The first assertion is immediate on combining Theorem 1.1, Theorem 5.1 and Lemma 2.1.

For the second assertion, let $g$ be an eigenform in $S_{k(\rho)}(\Gamma_1(N(\rho)))$ such that $\rho \cong \rho_g \otimes_{R_g} \mathbf{F}$. Recall that if $\rho_0|_{D_\ell}$ is finite, then $k(\rho) = 2$, and that if $\rho_0|_{D_\ell}$ is Selmer, then $2 \leq k(\rho) \leq \ell + 1$. We claim that if $\rho_0|_{D_\ell}$ is $\psi_0$-Selmer, then $g$ can be chosen so that $\alpha(a_\ell(g)) = \psi_0(\mathrm{Frob}\,_\ell)$. This is clear unless $\rho_0|_{D_\ell}$ is the direct sum of distinct unramified characters $\phi_0$ and $\psi_0$ and $\alpha(a_\ell(g)) = \phi_0(\mathrm{Frob}\,_\ell)$. But in that case $k(\rho) = \ell$ and [8], (2.9) gives a "companion form" $g'$ over $\mathbf{F}$ of type $(N(\rho), 1)$. We also see from the proof of [8], (2.8) that $X^2 - a_\ell(g')X + \det(\rho_0(\mathrm{Frob}\,_\ell))$ has distinct roots $\psi_0(\mathrm{Frob}\,_\ell)$ and $\phi_0(\mathrm{Frob}\,_\ell)$ and that there is an eigenform $g''$ of type $(N(\rho), \ell)$ satisfying

- $a_p(g'') = \alpha(a_p(g))$ for $p \neq \ell$;
- $a_\ell(g'') = \psi_0(\mathrm{Frob}\,_\ell)$.

Since $S_\ell(\Gamma_1(N(\rho))) \neq 0$, we see that $N(\rho)\ell > 4$, so $g''$ is the reduction of an eigenform with which we replace $g$.

If $\rho_0|_{D_\ell}$ is finite, then let $f = g$. If $\rho_0|_{D_\ell}$ is Selmer but not finite, then by Theorem 5.1, there is an eigenform $f$ in $S_2(\Gamma_1(\ell N(\rho))$ satisfying $\alpha(a_n(f)) = \alpha(a_n(g))$ for all $n \geq 1$. Therefore $\rho \cong \rho_f \otimes_{R_f} \mathbf{F}$ and $\alpha(a_\ell(f)) = \psi_0(\mathrm{Frob}\,_\ell)$. Finally by the variant (Lemma 2.2) of Carayol's result, we see that $f$ can be chosen with character of order prime to $\ell$. $\square$

**Corollary 6.5.** *Under the hypotheses of the theorem, there is a weight two newform $f$ such that*
- *$\rho_f$ is unramified outside $\ell N(\rho)$;*
- *$\rho \cong \rho_f \otimes_{R_f} \mathbf{F}$;*
- *for each prime $p$ such that $\rho_0|_{D_p}$ is type A (resp. type B, resp. finite, resp. $\psi_0$-Selmer), $\rho_f|_{D_p}$ is also type A (resp. type B, resp. finite, resp. $\psi_0$-Selmer).*

*References*

[1] A. Ash and G. Stevens, *Modular forms in characteristic $\ell$ and special values of their L-functions*, Duke Math. J. **53** (1986), 849-868.

[2] N. Boston, H.W. Lenstra and K. Ribet, *Quotients of group rings arising from two-dimensional representations*, C. R. Acad. Sci. Paris, Série I **312** (1991), 323-328.

[3] H. Carayol, *Sur les représentations galoisiennes modulo l attachées aux formes modulaires*, Duke Math. J. **59** (1989), 785-801.

[4] R.F. Coleman and J.F. Voloch, *Companion forms and Kodaira-Spencer theory*, Invent. Math. **110** (1992), 263-281.

[5] P. Deligne and M. Rapoport, *Les schémas modulaires de courbes elliptiques*, Lecture Notes in Math. **349** (1973), 143-316.

[6] F. Diamond and R. Taylor, *Lifting modular mod l representations*, Duke Math. J. **74** (1994), 253-269.

[7] V.G. Drinfeld, *Coverings of p-adic symmetric regions* (in Russian), Funkts. Anal. Prilozn. **10** (1976), 29-40. Translation in Funct. Anal. Appl. **10** (1976), 107-115.

[8] B. Edixhoven, *The weight in Serre's conjectures on modular forms*, Invent. Math. **109** (1992), 563-594.

[9] B.H. Gross, *A tameness criterion for Galois representations associated to modular forms mod p*, Duke math. J. **61** (1990), 445-517.

[10] H. Hida, *Galois representations into $GL_2(\mathbf{Z}_p[[X]])$ attached to ordinary cusp forms*, Invent. Math. **85** (1986), 545-613.

[11] B. Jordan and R. Livné, *Local diophantine properties of Shimura curves*, Math. Ann. **270** (1985), 235-248.

[12] B. Mazur, *Deforming Galois representations*, in Galois Groups over $\mathbf{Q}$, MSRI Publ. **16** (1989), 385-437.

[13] K. Ribet, *Congruence relations between modular forms*, Proc. I.C.M. (1983), 503-514.

[14] K. Ribet, *On modular representations of $\mathrm{Gal}\,(\overline{\mathbf{Q}}/\mathbf{Q})$ arising from modular forms*, Invent. Math. **100** (1990), 431-476.

[15] K. Ribet, *Report on mod l representations of $\mathrm{Gal}\,(\overline{\mathbf{Q}}/\mathbf{Q})$*, in Motives, Proc. Symp. Pure Math. **55**:2 (1994), 639-676.

[16] J.-P. Serre, *Sur les représentations modulaires de degré 2 de $\mathrm{Gal}\,(\overline{\mathbf{Q}}/\mathbf{Q})$*, Duke Math. J. **54** (1987), 179-230.

[17] G. Shimura, *Introduction to the arithmetic theory of automorphic functions*, Iwanami Shoten and Princeton University Press, 1971.

[18] A. Wiles, *On ordinary λ-adic representations associated to modular forms*, Invent. Math. **94** (1988), 529-573.

[19] A. Wiles, Course at Princeton University, February–April, 1994.

Conference on Elliptic Curves and Modular Forms
Hong Kong, December 18-21, 1993
Copyright ©1995 International Press

# Wiles minus epsilon implies Fermat

Noam D. Elkies

For an elliptic curve $E/Q$ and for each prime $l$ define the group $G_l(E) \subseteq GL_2(Z/l)$ as the image of the representation of $\mathrm{Gal}(\bar{Q}/Q)$ on $E[l]$, that is, $G_l(E)$ is the Galois group of the extension of $Q$ obtained by adjoining the coordinates of all the $l$-torsion points of $E$. By "Wiles minus epsilon" I mean the following recently re-announced result [4, 3]:

**Theorem.** *Let $E/Q$ be a semistable elliptic curve. Assume that $G_3(E) = GL_2(Z/3)$. Then $E$ is modular.*

The "epsilon" is then the concluding part of Wiles' monumental paper, wherein he removes the hypothesis on $G_3(E)$ by cleverly switching between $G_3$ and $G_5$ on $E$ and an auxiliary semistable curve. Since Ribet had already shown [1], building on observations of Serre [2], that the semistable Frey curve associated to a counterexample to Fermat's Last Theorem (FLT) could not be modular, that "epsilon" also completed the proof of FLT.

In this note we show that the above theorem together with the ideas of Frey-Serre-Ribet already implies FLT. The argument is unabashed ad-hockery; in our defense we note that the same is true of Wiles' $G_3/G_5$ trick (which among other things depends crucially on the fact that the modular curve $X(5)$ has genus zero!), and that though our aims are more modest (we cannot conclude that every semistable curve admitting a rational 3-isogeny is modular), our approach is almost entirely elementary.

Recall that the Frey curve $E = E(A, B, C)$ associated to a solution of

(1) $$A + B + C = 0$$

in relatively prime nonzero integers $A, B, C$ is the elliptic curve

(2) $$Y^2 = X(X + A)(X - B)$$

or its $\sqrt{-1}$-twist $Y^2 = X(X - A)(X + B)$.

Now fix a prime $p > 3$. Then it is a corollary of the work of Serre and Ribet [1, 2] that if $(A, B, C)$ is a solution to (1) in integers, each of the form

(3) $$2^r 3^s m^p \quad \text{with} \quad \text{either} \quad r = 0 \text{ or } r \geq 4,$$

then the Frey curve $E = E(A, B, C)$ is semistable but <u>not</u> modular. This is because if $E$ were modular then the $p$-torsion points of $E$ would yield a weight-2 cusp form mod $p$ on $X_0(N)$ for $N = 1, 2, 3,$ or $6$, and no such form exists. [The only possibly new wrinkle here is that I must allow $r = 4$. Fortunately in that case $E$ actually has good reduction at 2, with odd $j$-invariant. The examples $1 + 1 = 2$ (or $1 + 2 = 3$), $1 + 3 = 4$, and $1 + 8 = 9$ show that we cannot allow $r = 1, 2$, or 3. There are infinitely many examples such as $3 + 125 = 128$ that preclude taking $p = 3$. In [2] the condition $p > 3$ assures that $G_p[E]$ acts irreducibly on $E[p]$, i.e. that $E$ does not admit a rational $p$-isogeny, and indeed the Frey curve associated to $3 + 125 = 128$ admits a rational 3-isogeny; this is curve #30F in the Antwerp tables. Our argument also requires $p > 3$ as will be seen later.]

Thus by the above Theorem $G_3(E)$ must be strictly smaller than $\mathrm{GL}_2(Z/3)$. We shall show that this is impossible for such a Frey curve, or rather (which is good enough), that given a putative counterexample we shall construct a possibly different counterexample whose associated Frey curve has maximal $G_3$. This will prove that (1) has no solutions with each of $A, B, C$ of the form (3). In particular this will prove FLT while circumventing the $G_3/G_5$ trick.

Note that $G_3(E)$ cannot be contained in $\mathrm{SL}_2(Z/3)$ because $Q$ does not contain the cube roots of unity. Thus if $G_3(E)$ is properly contained in $\mathrm{GL}_2(Z/3)$ then either it is a 2-group, and so contained in a 2-Sylow subgroup of $\mathrm{GL}_2(Z/3)$, or it acts reducibly on $E[3]$, whence $E$ admits a rational 3-isogeny.

We dispense with the first possibility by noting that, for any elliptic curve $E$, if $G_3(E)$ is a 2-group then the $j$-invariant of $E$ is a cube (because $j^{1/3}$ parametrizes the modular curve corresponding to the 2-Sylow subgroup of $\mathrm{GL}_2(Z/3)$). But the $j$-invariant of the Frey curve $E(A, B, C)$ is $2^8(A^2 + AB + B^2)^3/(ABC)^2$, which is a cube if and only if $ABC = 2D^3$ for some integer $D$. This, however, is incompatible with $A + B + C = 0$ unless $(A, B, C)$ is proportional to a permutation of $(1, 1, -2)$: the elliptic curve $ABC = 2D^3, A + B + C = 0$ has Weierstrass model $y^2 = x^3 + 1$, conductor 36 and rank zero; its six rational points yield the permutations of $(1, 0, -1)$ and $(1, 1, -2)$. But $(1, 1, -2)$ does not satisfy the condition on the 2-valuations of $A, B, C$, and anyway yields a Frey curve with complex multiplication ($j = 12^3$) which is thus modular.

We may thus assume that $E$ admits a 3-isogeny. In that case some twist of $E$ has a subgroup of rational points isomorphic to $(Z/2)^2 \times (Z/3)$. Writing that twist of $E$ as $Y^2 = (X - x_1)(X - x_2)(X - x_3)$, and translating $X$ to put the 3-torsion points at $X = 0$, we calculate that $(x_1 : x_2 : x_3)$ must lie on the rational plane quartic nicely parametrized as

$$(4) \qquad (x_1 : x_2 : x_3) :: \left((ab)^2 : (bc)^2 : (ca)^2\right) \quad with \quad a + b + c = 0.$$

[In effect[1] $(a : b : c)$ is a uniformizer for $X_0(12)$; the modular automorphisms of that curve coming from $\Gamma(1)/\Gamma(2)$ are visible in the permutations of $a, b, c$, and the involution $w_3$ comes from $(a : b : c) \leftrightarrow (b - c : c - a : a - b)$.] Thus $A, B, C$ are proportional to $a^2(c^2 - b^2)$, $b^2(a^2 - c^2)$, and $c^2(b^2 - a^2)$, or equivalently to $a^3(b-c), b^3(c-a), c^3(a-b)$, for some rational $(a, b, c)$ with $a+b+c = 0$. We may take $a, b, c$ to be coprime integers, and assume without loss of generality that $a$

---

[1] The arithmetic subgroup $\Gamma(2) \cap \Gamma_0(3)$ of $\mathrm{PSL}_2(Z)$ is conjugate in $\mathrm{PGL}_2^+(Q)$ with $\Gamma_0(12)$ (because the groups $\Gamma(2)$ and $\Gamma_0(4)$ are conjugate); thus the corresponding modular curves are isomorphic.

is even and $b, c$ are odd. Then $a, b, c, b-c, c-a, a-b$ are pairwise coprime except for $\gcd(a, b-c) = 2$, and possibly a common factor of 3 among $b-c, c-a, a-b$. Thus each of $a, b, c, b-c, c-a, a-b$ is a $p$-th power times some powers of 2 and 3. (This is where we use $p > 3$; our earlier example of $3 + 125 = 128$ with $p = 3$ yields $(a, b, c) = (4, 1, -5)$ with $c$ not of the form (3).)

So $A$ is even, say with 2-valuation $v_2(A) = r \geq 4$, and $B, C$ are odd. Furthermore, either $a$ is singly even and $v_2(b - c) = r - 3$, or $b - c$ is singly even and $v_2(a) = (r - 1)/3$ — but not both, since $a$ and $b - c$ cannot both be singly even under the hypotheses of the previous paragraph. This immediately excludes the case $r = 4$. (That case could also be ruled out by noting that the (good) reduction mod 2 of $E$ would have both a 2- and a 3-torsion point, but an elliptic curve mod 2 cannot have 6 rational points!). When $r$ is larger, we'll use the fact that $a, b, c, c - a, a - b, b - c$ are each a $p$-th power times some powers of 2 and 3 to produce a new $A, B, C$ as above, except with smaller $r$, and in a finite number of steps reach $r = 4$ and obtain our contradiction.

This can probably be done in several ways. Here's one: the identity

$$(5) \qquad a(b - c) + b(c - a) + c(a - b) = 0$$

gives new $A, B, C$ with $r$ replaced by either $r - 2$ or $(r + 2)/3$, both $< r$. The only problem is that if $r$ was 5 or 7 then the new $r$ will be 3, which is forbidden. In the first case, I use instead the identity

$$(6) \qquad (b - c)^2 - (c - a)^2 = (b - a)(b + a - 2c) = 3c(a - b),$$

and in the second case the identity

$$(7) \qquad a^2 - b^2 = (a + b)(a - b) = c(b - a)$$

(the $w_3$ image of (6)), to get new $A, B, C$ with $r = 4$. This completes the proof of the impossibility of nonzero coprime integers $A, B, C$ satisfying (1) and (3), and thus in particular of FLT, given of course that "Wiles minus epsilon" is confirmed.

**Acknowledgements.** Thanks to Ken Ribet for explaining and discussing parts of [4], and to Barry Mazur for several corrections and improvements of exposition.

I am grateful to the National Science Foundation and the Packard Foundation for partial support during the preparation of this note.

*References*

[1] Ribet, K.A., *On modular representations of $Gal(\bar{Q}/Q)$ arising from modular forms.* Invent. Math. **100** (1990), 431–476.

[2] Serre, J.-P., *Sur les représentations modulaires de degré 2 de $Gal(\bar{Q}/Q)$*, Duke Math. J. **54** (1987), 179–230.

[3] Taylor, R., Wiles, A., *Ring Theoretic Properties of Certain Hecke Algebras*, Preprint, 1994.

[4] Wiles, A., *Modular elliptic curves and Fermat's Last Theorem*, Preprint, 1994.

# Geometric Galois Representations

JEAN-MARC FONTAINE
UNIVERSITÉ DE PARIS-SUD
MATHÉMATIQUE-BÂTIMENT 425, 91405-ORSAY CEDEX FRANCE

BARRY MAZUR
HARVARD UNIVERSITY
DEPARTMENT OF MATHEMATICS
1 OXFORD STREET, CAMBRIDGE, MA 02138

The main point of this paper [1] is to present in detail some of our conjectures concerning $p$-adic representations of the Galois groups of number fields (cf. [10] for a brief synopsis). The motivating idea behind our conjectures is simply the hope that, given an irreducible $p$-adic representation $\rho$ of $G_K$, the Galois group of a number field $K$ (assumed unramified except at a finite number of places), one can find necessary and sufficient local conditions (on the restriction of the representation to the decomposition groups $G_{K_V}$ at the finite set of places $v$ of $K$ of residual characteristic $p$) for the representation $\rho$ to "come from algebraic geometry". The local condition we have in mind is that $\rho$ restricted to $G_{K_V}$ be *potentially semi-stable*. This will be made precise in Conjecture 1 of § 1 below.

There is a well known conjecture which, vaguely put, asserts that a pure motive of rank 2 and of Hodge type $(0,r),(r,0)$ "comes from" a modular newform of weight $k = r+1$ (for related conjectures, see [28] and the large literature concerning the connection between elliptic curves and modular curves; for spectacular recent work in the way of proving such conjectures, see [31] and [29]). Combining our Conjecture 1 with this "well known conjecture" leads to a conjectural necessary and sufficient condition, stated only in terms of a local condition at $p$, for an irreducible representation $\rho : G_{\mathbb{Q}} \to GL_2(\overline{\mathbb{Q}}_p)$ to be the representation associated to a cuspidal newform.

**Conjecture:** *Let*

$$\rho : G_{\mathbb{Q}} \to GL_2(\overline{\mathbb{Q}}_p)$$

*be an irreducible representation which is unramified except at a finite number of primes and which is not the Tate twist of an even representation which factors through a finite quotient group of $G_{\mathbb{Q}}$.*

*Then $\rho$ is associated* [2] *to a cuspidal newform $f$ if and only if $\rho$ is potentially semi-stable at $p$.*

---

[1] Research supported in part by the NSF, Contract No. DMS-89-05205; by the Institute Universitaire de France and the CNRS (URA D 0752).

[2] recall that "associated" is a technical term which means that one can find an isomorphism between $\overline{\mathbb{Q}}_p$ and $\mathbb{C}$ such that for all but a finite number of prime numbers $l$, the above isomorphism brings the trace of $\rho(\text{Frob}_l)$ to the eigenvalue of the Hecke operator $T_l$ on the newform $f$. Here, $\text{Frob}_l$ denotes the Frobenius element at $l$, and of course, one need only consider prime numbers $l$ which do not divide the level of $f$, and at which the representation $\rho$ is unramified.

For a refinement of the above conjecture, see Conjecture 3c below.

We also include certain "finiteness conjectures" for irreducible representations which "come from algebraic geometry", which have a fixed level of ramification and a fixed Hodge-Tate type. These "finiteness conjectures" are akin to the assertion (in view of the Conjecture displayed above) that there are only a finite number of newforms of given level and weight. This discussion is collected in §3. In §7 and §8 we examine Conjecture 1 in the case where the representation is potentially abelian, and potentially everywhere unramified.

Part II of our paper is devoted to a deformation - theoretic study of the condition of *potential semi-stability* for degree two, $p$-adic representations of $G_{\mathbf{Q}_p}$ which are residually absolutely irreducible. The main results here are due to Ramakrishna [23], but we revisit Ramakrishna's theory to give a slightly more detailed picture, particularly in the case when the representations are assumed to be potentially Barsotti-Tate.

Here is our motivation for doing such a study. Fixing an absolutely irreducible representation

$$\bar{\rho}_p : G_{\mathbf{Q}_p} \to GL_2(\mathbf{F}_p),$$

Ramakrishna has shown that the universal deformation ring $R(\bar{\rho}_p)$ is isomorphic to a power series ring in five variables $\mathbf{Z}_p[[T_1, T_2, T_3, T_4, T_5]]$ (and therefore the $p$-adic variety $X(\bar{\rho}_p) := \text{Hom}(R(\bar{\rho}_p), \mathbf{Z}_p)$ which classifies lifts of $\bar{\rho}_p$ to $GL_2(\mathbf{Z}_p)$ is smooth on five parameters). Fix a finite set of primes $S$ including $p$, let $G_{\mathbf{Q},S}$ denote the Galois group of the maximal algebraic extension of $\mathbf{Q}$, unramified outside $S$, and fix an imbedding of $\bar{\mathbf{Q}}$ in $\bar{\mathbf{Q}}_p$ and hence a homomorphism $i : G_{\mathbf{Q}_p} \to G_{\mathbf{Q},S}$. Let $\bar{\rho}_{\mathbf{Q},S} : G_{\mathbf{Q},S} \to GL_2(\mathbf{F}_p)$ be any representation (if such exists) which extends $\bar{\rho}_p$. That is, $\bar{\rho}_p = \bar{\rho}_{\mathbf{Q},S} \circ i$. One knows that the universal deformation ring $R(\bar{\rho}_{\mathbf{Q},S})$ is of Krull dimension $\geq 4$, and if smooth, is a power series ring over $\mathbf{Z}_p$ on three parameters. It is conjectured always to have Krull dimension 4.

Now, since $5 - 3 = 2$, for our "finiteness conjectures" to be feasible it would be nice to prove that all *potential semi-stable* liftings

$$\rho_p : G_{\mathbf{Q}_p} \to GL_2(\mathbf{Z}_p),$$

of $\bar{\rho}$ with fixed Hodge-Tate weights $(r, s)$ lie in a finite union of (at most!) two-parameter subspaces of the universal deformation space $X(\bar{\rho}_p)$. Indeed, the best would be if there were a finite quotient ring, call it $R(\bar{\rho}_p)_{r,s}$, of $R(\bar{\rho}_p)$ through which all homomorphisms $R(\bar{\rho}_p) \to \mathbf{Z}_p$ classifying all potentially semi-stable liftings of $\bar{\rho}_p$ (with Hodge-Tate weights $r, s$) factors, and such that $R(\bar{\rho}_p)_{r,s}$ is of Krull dimension $\leq 3$, and whose generic fiber over $\mathbf{Q}_p$ is formally smooth of dimension two.

This would indeed be nice, for then we could entertain the hope that our "finiteness conjecture" is a phenomenon related to a possible transversality question; namely, do the generic fibers of $\text{Spec} R(\bar{\rho}_p)_{r,s}$ and of $R(\bar{\rho}_{\mathbf{Q},S})$ have images in the generic fiber of $\text{Spec} R(\bar{\rho}_p)$ which intersect transversally?

We do not know whether rings $R(\bar{\rho}_p)_{r,s}$ as described above exist for all Hodge-Tate weights $r, s$. But Ramakrishna established that the locus of crystalline liftings possessing Hodge-Tate weights contained in the interval $[0, p-1]$ indeed forms a subscheme of $\text{Spec}(R(\bar{\rho}_p))$, smooth on two formal parameters over $\mathrm{Spec}(\mathbb{Z}_p)$. A Hodge-Tate lifting with Hodge-Tate weights $r, s$ contained in $[0, p-1]$ will be said to be of **moderate Hodge-Tate type**. We complement Ramakrishna's theory by classifying (at least for $p \geq 5$) all weakly admissible two-dimensional irreducible potentially semi-stable filtered $(\varphi, N, G_{\mathbb{Q}_p})$-modules (Thm. A of § 11) and then by constructing explicitly the universal such module which classifies crystalline liftings of $\overline{\rho}_p$ to $GL_2(\mathbb{Z}_p)$ of moderate Hodge-Tate type (Thm. B2 of § 12).

We also discuss the interesting case of **potential Barsotti-Tate liftings** of $\overline{\rho}_p$ to $GL_2(\mathbb{Z}_p)$. The aim here is, firstly, to preview a theory (classifying Barsotti-Tate groups "strictly of slope 1/2") which will be expounded more fully in further publications by one of us, and, secondly, to formulate a consequence of this theory which gives a classification of potentially Barsotti-Tate liftings of $\overline{\rho}_p$ to $GL_2(\mathbb{Z}_p)$. This classification is put forward in Thm. C3 of § 13 below; we give a series of hints towards its proof in Appendix C. Specifically, the locus of all liftings of $\overline{\rho}_p$ to $GL_2(\mathbb{Z}_p)$ which are potentially Barsotti-Tate (PBT, for short) is either empty, or else it is a union of two disjoint smooth two-parameter spaces in $X(\overline{\rho}_p)$. More exactly, there is a quotient-ring $R_{\mathrm{PBT}}$ of $R(\overline{\rho}_p)$ though which any homomorphism $R(\overline{\rho}_p) \to \mathbb{Z}_p$ which classifies a PBT lifting factors, and

$$R_{\mathrm{PBT}} \cong \mathbb{Z}_p[[Y_1, Y_2]] \times_{\mathbf{F}_p} \mathbb{Z}_p[[Y'_1, Y'_2]],$$

i.e., the fiber product of two power series rings over $\mathbb{Z}_p$ each in two variables, the fiber product being taken over their common residue field $\mathbf{F}_p$.

The PBT liftings whose classifying homomorphisms factor through $\mathbb{Z}_p[[Y_1, Y_2]]$ are quite different from those which factor through $\mathbb{Z}_p[[Y'_1, Y'_2]]$. For certain residual representations $\overline{\rho}_p$ (specifically, for those $\overline{\rho}_p$ whose invariants [3] $j_1, j_2$ have the property that $j_2 - j_1 = 1$) this difference can readily be seen by considering their associated admissible pst modules:

Those PBT liftings whose classifying homomorphisms factor through $\mathbb{Z}_p[[Y_1, Y_2]]$ have associated pst modules occuring in our "type I" series (cf. Thm. A of §11) and the two $p$-adic parameters $Y_1$ and $Y_2$ correspond to a two-parameter variation of (the two) coefficients of the characteristic polynomial $X^2 - aX + d$ of the "Frobenius" operator $\varphi$ acting on the associated pst modules of these liftings. The datum of the filtration on these pst modules contributes, however, no further variation.

Those PBT liftings whose classifying homomorphisms factor through $\mathbb{Z}_p[[Y'_1, Y'_2]]$ have associated pst modules occurring in our "type IV" series (cf. Thm. A of §11) and here the two $p$-adic parameters $Y'_1$ and $Y'_2$ correspond to somewhat different features of the associated pst modules. The characteristic polynomial of the "Frobenius" operator $\varphi$ acting on the associated pst modules of these liftings have the form $X^2 + d$, and the $p$-adic parameter $Y'_1$ corresponds to the possibility of varying the coefficient $d$. The datum of the filtration on these pst modules is important, and contributes another one-parameter of $p$-adic variation corresponding to the second variable $Y'_2$.

Full proofs of Thm. C1-3 will be given in later publications. Beyond this PBT case, however there still remains a good deal of work to do to get a completely satisfactory, and completely general, local picture of pst liftings, even in the

---

[3] cf. Thm. C3 of §13

special context of two-dimensional representations of $G_{\mathbb{Q}_p}$.

**Notation:**
$K :=$ number field; $\overline{K} :=$ an algebraic closure of $K$, $G_K = \text{Gal}(\overline{K}/K)$. $S :$ a finite set of finite places of $K$; $G_{K,S} :=$ the Galois group of the maximal algebraic extension of $K$ in $\overline{K}$ unramified outside $S$; $p :=$ a fixed prime number; $E :=$ a field containing $\mathbb{Q}_p$; $N :=$ a fixed positive integer.

For each finite place $v$ of $K$, we denote by $K_v$ the completion of $K$ at $v$. We choose an algebraic closure $\overline{K}_v$ of $K_v$ and an embedding of $\overline{K}$ into $\overline{K}_v$. This gives us an identification of $G_v = \text{Gal}(\overline{K}_v/K_v)$ with a decomposition subgroup of $G_K$. We denote by $I_v$ the inertia subgroup of $G_v$.

If $G$ is a profinite group, a *p*-**adic representation of** $G$ is a finite dimensional $\mathbb{Q}_p$-vector space equipped with a continuous and linear action of $G$. Similarly, an **E-representation** of $G$ is a $\mathbb{Q}_p$-vector space $V$ of finite dimension (called the **degree of the representation**) equipped with a linear action of $G$ which has the property that, if we choose a basis of $V$, the image of the corresponding map
$$\rho : G \to GL_N(E)$$
is contained in $GL_N(E_0)$ where $E_0$ is a suitable finite $\mathbb{Q}_p$-algebra contained in $E$ and the map $G \to GL_N(E_0)$ is continuous (this doesn't depend on the choice of the basis).

## Part I. The conjectures

### §1. Geometric representations and representations coming from algebraic geometry.

A *p*-adic representation of $G_K$ is called **geometric** if
(a) it is unramified outside a finite set of places of $K$,
(b) its restriction to every decomposition group $G_v$ (for $v$ ranging through all non-archimedean places of $K$) is *potentially semi-stable* in the sense of [13] (see also [12] for $v$'s dividing $p$).

Equivalently, we may ask that its restriction to the decomposition groups for $v|p$ are potentially semi-stable, since, by a theorem of Grothendieck, the restriction to the other decomposition groups are automatically potentially semi-stable.

**Remark:** We do not know whether the condition (a) is satisfied for all semi-simple representations.

A continuous *irreducible* $\mathbb{Q}_p$-representation of $G_K$ is said to **come from algebraic geometry** if it is isomorphic to a subquotient of an é-tale cohomology group with coefficients in $\mathbb{Q}_p(r)$ for some Tate twist $r \in \mathbb{Z}$, of an algebraic variety over $K$ (equivalently [4] : of a projective smooth algebraic variety over $K$).

**Conjecture 1:** *An irreducible p-adic representation is geometric if and only if it comes from algebraic geometry.*

---
[4] using resolutions of singularities.

**Evidence:** The part of conjecture 1 saying that "irreducible representations coming from algebraic geometry are "geometric" is in no way original to this article, and goes in a direction where much more precise results are expected to be true about varieties defined over local fields. See, for example, Conjecture $\underline{C}_{\text{pst}}$ of [**12**] (6.2.1) and the further conjectures in (6.2.4) there. This part of Conjecture 1 has been known for a long time to hold for abelian varieties, and, more recently, [**6**] for varieties which have good reduction at all places dividing $p$ (in which case the representations of $G_v$ are crystalline) and also in slightly more general instances (see [**15**] and [**19**] for a survey, see also the forthcoming work of Tsuji).

The part of Conjecture 1 saying that irreducible "geometric" representations "come from algebraic geometry" is presently known for irreducible potentially abelian representations (see §6 below). The recent work of Wiles [**31**] and Taylor-Wiles [**29**] establishes this also for a very significant class of irreducible $p$-adic representations $V$ of $G_\mathbb{Q}$, of dimension two over $\text{End}_{\mathbb{Q}_p[G_\mathbb{Q}]}(V)$.

## §2. Hodge-Tate representations.

If the extension $E/\mathbb{Q}_p$ is finite, an $E$-representation of degree $N$ of $G_K$ is called **geometric** if it is geometric as a $p$-adic representation of degree $N \cdot \dim_{\mathbb{Q}_p} E$. Generally, an $E$-representation $V$ of degree $N$ of $G_K$ is called **geometric** if there exist a finite $\mathbb{Q}_p$-algebra $E_0$ contained in $E$, a geometric $E_0$-representation $V_0$ of $G_K$ and an isomorphism of $E$-representations $E \otimes_{E_0} V_0 \simeq V$.

Let $V$ be an $E$-representation of $G_K$ of degree $N$ which is geometric. If $v$ is a place of $K$ dividing $p$ and if the representation is potentially semi-stable at $v$ then $V$, as a representation of $G_v$, is Hodge-Tate. Let us recall what this means. Put $\mathbb{C}_v :=$ the completion of $\overline{\mathbb{Q}}_v$ (on which $G_v$ acts by continuity), and $B_{HT,v} :=$ the ring $\oplus_{r \in \mathbb{Z}} \mathbb{C}_v(r)$, with $\mathbb{C}_v(r)$ the usual Tate twist. The ring $B_{\text{HT},v}$ is the ring of Laurent polynomials in the indeterminate $t$, a generator of $\mathbb{Z}_p(1)$, with coefficients in $\mathbb{C}_v$ and $(B_{\text{HT},v})^{G_v} = K_v$. The $G_v$-representation $V$ is **Hodge-Tate** in the sense that, if $\underline{D}^*_{\text{HT},v}(V) = \text{Hom}_{\mathbb{Q}_p[G_v]}(V, B_{\text{HT},v})$, the natural map

$$B_{\text{HT},v} \otimes_{K_v} \underline{D}^*_{\text{HT},v}(V) \to \text{Hom}_{\mathbb{Q}_p}(V, B_{\text{HT},v})$$

is an isomorphism. This implies that $\underline{D}^*_{\text{HT},v}(V)$ is a free $E \otimes_{\mathbb{Q}_p} K_v$-module of rank $N$, equipped with a $\mathbb{Z}$-gradation by sub-$E \otimes_{\mathbb{Q}_p} K_v$-modules

$$\text{gr}^r \underline{D}^*_{\text{HT},v}(V) = \text{Hom}_{\mathbb{Q}_p[G_v]}(V, \mathbb{C}_v(r)).$$

By the $(E,v)$-**Hodge-Tate type** $h(v)$ of $V$, we mean the isomorphism class of the graded $E \otimes_{\mathbb{Q}_p} K_v$-module $\underline{D}^*_{\text{HT},v}(V)$, and we call $N$ the **degree** of $h(v)$. For instance, if $K_v = \mathbb{Q}_p$, $\underline{D}^*_{\text{HT},v}(V)$ is just a graded $E$-vector space and to know $h(v)$ amounts to the same as to know the **Hodge-Tate numbers**, that is, the non-negative integers $h_r(v) = \dim_E (\mathbb{C}_v(r) \otimes_{\mathbb{Q}_p} V)^{G_v} = \dim_E \text{gr}^{-r} \underline{D}^*_{\text{HT},v}(V)$ (these are almost all 0 and $\sum h_r(v) = N$).

**Remark:** If $v$ divides $p$, the **trivial $(E,v)$-Hodge-Tate type of degree** $N$ is the isomorphism class of $D = (E \otimes_{\mathbb{Q}_p} K_v)^N$ with $\text{gr}^0 D = D$. An $E$-representation $V$ of $G_{K_v}$ is Hodge-Tate of trivial type if and only if the image of $I_v$ in $\text{Aut}_E(V)$ is finite ([**24**], th. 1 and [**25**], th. 11, cor.). In particular, if

$V$ is a geometric representation of $G_{K,S}$ and if $v$ is not in $S$ but divides $p$, the $(E,v)$-Hodge-Tate type of $V$ is trivial.

By an **$E$-Hodge-Tate type of $K$**, we mean a function $h$ assigning to each place $v$ dividing $p$ an $(E,v)$-Hodge-Tate type $h(v)$, all those $h(v)$ having the same degree $N$. Let **Geom**$(K,S,h;E)$ denote the set of isomorphism classes of geometric irreducible $E$-representations of $G_{K,S}$ with the indicated Hodge-Tate type $h(v)$ for each $v$ dividing $p$ (such a representation is necessarily of dimension $N$).

## §3. The finiteness conjectures.

If $V$ is a geometric $E$-representation of $G_K$, for any finite place $v$ of $K$, there is an unique invariant open subgroup $\mathfrak{L}_v(V)$ of $I_v$ which is such that, if $L$ is a finite extension of $K_v$ contained in $\overline{K}_v$, then $V$, when viewed as a representation of $\text{Gal}(\overline{K}_v/L)$, is semi-stable if and only if $\text{Gal}(\overline{K}_v/L) \cap I_v \subset \mathfrak{L}_v(V)$. If $v \notin S$, we have $\mathfrak{L}_v(V) = I_v$.

By an **inertial level for** $S$, we mean a rule $\mathfrak{L}$ which assigns to each $v \in S$ an open invariant subgroup $\mathfrak{L}_v$ of $I_v$.

Let **Geom**$(K,S,\mathfrak{L},h;E)$ denote the set of isomorphism classes of geometric irreducible $E$-representations $V$ of $G_{K,S}$ in the set **Geom**$(K,S,h;E)$ such that, for each $v \in S$, $\mathfrak{L}_v \subset \mathfrak{L}_v(V)$.

**Conjecture 2a:** *For any finite set of places $S$ of $K$, inertial level $\mathfrak{L}$ for $S$, $\overline{\mathbb{Q}}_p$-Hodge-Tate type $h$ of $K$, the set **Geom**$(K,S,\mathfrak{L},h;\overline{\mathbb{Q}}_p)$ is finite.*

**Example [14], prop. 1** (see [1] for other examples of this kind): If $K = \mathbb{Q}, S = \{7\}, \mathfrak{L}_7 = I_7$ and $h$, given by the Hodge numbers $(h_r)_{r \in \mathbb{Z}}$, is such that $h_r h_s \neq 0 \Rightarrow s - r \leq 3$, then **Geom**$(K,S,Z,h;\overline{\mathbb{Q}}_7)$ is empty, unless there is $i \in \mathbb{Z}$ such that $h_{-i} = 1$ and $h_r = 0$ if $r \neq i$, in which case **Geom**$(K,S,Z,h;\overline{\mathbb{Q}}_7)$ has one element which is the class of $\overline{\mathbb{Q}}_7(i)$.

**Conjecture 2b:** *For any finite set of places $S$, inertial level $\mathfrak{L}$ for $S$, finite extension field $E$ of $\mathbb{Q}_p$, $E$-Hodge-Tate type $h$ of $K$, the set **Geom**$(K,S,\mathfrak{L},h;\overline{\mathbb{Q}}_p)$ is finite.*

**Conjecture 2c:** *For any finite set of places $S$, finite extension field $E$ of $\mathbb{Q}_p$, $E$-Hodge-Tate type $h$ of $K$, the set **Geom**$(K,S,h;E)$ is finite.*

## §4. Remarks:

### (a) Concerning the finiteness conjectures.

Conjecture 2c is in the spirit of the finiteness conjectures of Shafarevich (proved by Faltings [5]).

Obviously, *conjecture 2a implies conjecture 2b. Conjecture 2b and 2c are equivalent*: the implication $2c \Rightarrow 2b$ is obvious. Conversely, attached to a potentially semi-stable representation of degree $N$ (at $v$) one has (for this theory, see [12]) a representation $D_v$ of the Weil-Deligne group of $K_v$ of degree $N$ with coefficients in a field $E'$ which is $E$ if $v$ does not divide $p$ and a finite unramified extension of $E$ if $v$ divides $p$. In particular, if we choose a basis of $D_v$, we have an homomorphism $\rho_v : I_v \to GL_N(E')$ whose kernel is the invariant open subgroup $\mathfrak{L}_v(V)$.

If $v$ does not divide $p$, $E' = E$, and, because the order of the finite subgroups of $GL_N(E)$ is bounded, there is an integer $M_v(E, N)$ such that $[I_v : \mathfrak{L}_v(V)]$ divides $M_v(E, N)$. If $v$ divides $p$, one can check (cf. [**12**]) that, for any $g \in I_v$, the characteristic polynomial of $g$ acting on $D_v$ has coefficients in $E$. This implies [**7**] that there exists a linear representation of degree $N$ of $I_v/\mathfrak{L}_v(V)$ with coefficients in the field $E_1 = E(^{p-1}\sqrt{1})$ if $p \neq 2$ (resp. $E(^4\sqrt{1})$ if $p = 2$) which has the same character as $\rho_v$. Therefore, because the order of the finite subgroups of $GL_N(E_1)$ is bounded, in this case as well, there is an integer $M_v(E, N)$ such that $[I_v : \mathfrak{L}_v(V)]$ divides $M_v$.

For any $v \in S$, it is easy to see that, given the integer $M_v(E, N)$, there is an open invariant subgroup $\mathfrak{L}_v(E, N)$ of $I_v$ which is contained in all open invariant subgroups whose order divides $M_v(E, N)$. Therefore $\mathbf{Geom}(K, S, h; E) \subset \mathbf{Geom}(K, S, \mathfrak{L}, h; E)$ where $\mathfrak{L}$ is given by $v \mapsto \mathfrak{L}_v(E, N)$.

**(b) Changing $K$.** Let $L$ be a finite extension of $K$ contained in $\overline{K}$. It is easy to check that an $E$-representation $V$ of $G_K$ is geometric if and only if its restriction $\mathrm{Res}_K^L V$ to $G_L$ is geometric; similarly an $E$-representation $W$ of $G_L$ is geometric if and only if the induced representation $\mathrm{Ind}_L^K W$ of $G_K$ is. It is also easy to compute the Hodge-Tate types and the inertial levels of $\mathrm{Res}_K^L V$ (resp. $\mathrm{Ind}_L^K W$) from those of $V$ (resp. $W$). From that, using adjunction, we see easily that Conjecture 1 (resp. 2a, 2b) is true for $K$ if and only if it is true for $L$.

**(c) Semi-stable representations.** A geometric $E$-representation $V$ of $G_K$ is said to be *semi-stable* if, for each finite place $v$ of $K$, $V$ is semi-stable as a representation of $G_v$, i.e. if $\mathfrak{L}_v(V) = I_v$. Let $\mathbf{Geom}_{st}(K, S, h; \overline{\mathbb{Q}}_p)$ denote the set of isomorphism classes of geometric irreducible $\overline{\mathbb{Q}}_p$-representations $V$ of $G_{K,S}$ in the set $\mathbf{Geom}(K, S, h; \overline{\mathbb{Q}}_p)$ which are semi-stable. The following is a special case of conjecture 2a (corresponding to choosing $\mathfrak{L}$ such that $\mathfrak{L}_v = I_v$ for all $v \in S$):

**Conjecture 2a':** *For any finite set of places $S$ of $K$, $\overline{\mathbb{Q}}_p$-Hodge-Tate type $h$ of $K$, the set $\mathbf{Geom}_{st}(K, S, h; \overline{\mathbb{Q}}_p)$ is finite.*

Conversely, it is easy to check that conjecture 2a' for all number fields implies conjecture 2a.

**(d) Concerning geometric $p$-adic representations, and "compatible families" of representations.**

The curious implication of Conjecture 1, taken together with standard conjectures concerning the étale cohomology of algebraic varieties of number fields, is that any irreducible $p$-adic geometric representation $V$ of $G_{K,S}$ has these ("motivic") properties:
(1) For all nonarchimedean places $v \notin S$, let $\Phi_v \in G_v \subset G_K$ be a lifting of the geometric Frobenius $f_v \in G_v/I_v$ and let $P_v(V; T) \in \overline{\mathbb{Q}}_p[T]$ denote the characteristic polynomial of $\Phi_v$ acting on $V$. Then, there exists $r \in \mathbb{Z}$ and a finite extension $E$ of $\mathbb{Q}$ contained in $\mathbb{Q}_p$ such that, for all $v \notin S$, $P_v(V, T) \in E[T]$, the complex roots of $P_v(V, T)$ (for a chosen embedding of $E$ into $\mathbb{C}$) have their complex absolute values equal to $q_v^{r/2}$ where $q_v$ is the cardinality of the residue field of $v$.

(2) There exists a finite extension $E'$ of the above $E$ and, for any finite place $\lambda$ of $E'$, a geometric $E'_\lambda$-representation $V_\lambda$ of $G_{K,S_\lambda}$ where $S_\lambda = S \cup \{v|v$ has the same residual characteristic as $\lambda\}$ such that, if $v \notin S_\lambda, P_v(V_\lambda, T)(:=$ the characteristic polynomial of $\Phi_v$ acting on the $E'_\lambda$-vector space $V_\lambda) = P_v(V, T)$.

**(e) $L$-functions and weights.**

Assume $K = \mathbb{Q}$ and let $V$ be an irreducible geometric $\overline{\mathbb{Q}}_p$-representation of $G_\mathbb{Q}$ of degree $N$. For each prime $l$, one can associate to $V$ an $N$-dimensional $\overline{\mathbb{Q}}_p$-linear representation of the Weil-Deligne group $W'_l$ of $\mathbb{Q}_l$ (cf. [**13**] or [**17**]). Hence if we choose an imbedding of $\overline{\mathbb{Q}}_p$ into $\mathbb{C}$, one can define in the usual way a local factor $L_l(V, s)$ for each prime $l$ and one can define the global $L$-function $L(V, s)$ as being the formal product of all those local factors. One conjectures that this Dirichlet series converges for $\Re(s) >> 0$ and admits a meromorphic continuation in the whole complex plane; moreover, one can give a conjectural interpretation of the order of the zero or pole at $s = 0$ in terms of Galois cohomology (see [**17**], n°3.4).

One can define the **weight** $w(V)$ of $V$ : if $\dim_{\overline{\mathbb{Q}}_p} V = 1$, there is an unique integer $i$ such that the action of $G_\mathbb{Q}$ on $V(i)$ is finite and $w(V) = 2i$; if $\dim_{\overline{\mathbb{Q}}_p} V = N, w(V) = w(\wedge^N V)/N$; hence this is a rational number. For $r, s \in \mathbb{Z}$, define the **Hodge numbers** $h_{r,s}(V)$ of $V$ as being 0 unless $r + s = w(V)$ in which case $h_{r,s}(V)$ is the Hodge-Tate number $h_r(V)$.

If we choose an imbedding of $\overline{\mathbb{Q}}$ into $\mathbb{C}$, this defines the complex conjugation $c \in G_\mathbb{Q}$, hence we can define

$$h^+(V) = \dim_{\overline{\mathbb{Q}}_p}\{v \in V|cv = v\} \text{ and } h^-(V) = \dim_{\overline{\mathbb{Q}}_p}\{v \in V|cv = -v\}.$$

Define also $h^+_{r,s}(V) = h^-_{r,s}(V) = h_{r,s}(V)/2$ if $r \neq s$ and, if $j = w(V)/2 \in \mathbb{Z}$,

$$h^+_{j,j}(V) = h^+(V) - \sum_{r \neq s} h^+_{r,s}(V), \quad h^-_{j,j}(V) = h_{j,j}(V) - h^+_{j,j}(V).$$

**Conjecture 3a:** *Let $V$ be an irreducible geometric $\overline{\mathbb{Q}}_p$-representation of $G_\mathbb{Q}$. Then the weight $w(V)$ of $V$ is an integer. Moreover, for $r, s \in \mathbb{Z}$ such that $r + s = w(V)$, the numbers $h^+_{r,s}(V)$ and $h^-_{r,s}(V)$ are non negative integers and $h_{r,s}(V) = h_{s,r}(V)$.*

Assuming this conjecture, these numbers define in the usual way an isomorphism class of a linear representation of the Weil group $W_\mathbb{R}$, hence a $\Gamma$-factor $L_\infty(V, s)$ and we can consider the complete $L$-function $\Lambda(V, s) = L_\infty(V, s) \cdot L(V, s)$.

Because we have a representation of $W'_l$ for each prime number $l$ and a representation of $W_\mathbb{R}$, we can define in the usual way the **conductor** $N_v$ and the **$\epsilon$-factor** $\epsilon(V, s)$ (cf. e.g. [**3**]). We thus have the conjectural functional equation:

**Conjecture 3b:** *Let $V$ be as in 3a. Then $\Lambda(V, s)$ converges for $\Re(s) >> 0$ and admits a meromorphic continuation in the whole complex plane, satisfying*

$$\Lambda(V, s) = \epsilon(V, s) \cdot L(V^*, 1 - s).$$

**(f) Modular forms.**

Combining conjecture 1 with classical conjectures (e.g. [**26**]), we obtain

**Conjecture 3c:** *Let $V$ be an irreducible geometric $\overline{\mathbb{Q}}_p$-representation of $G_{\mathbb{Q}}$ of degree two which is not a Tate twist of a finite representation. Then there is an integer $i \in \mathbb{Z}$ such that $V(-i)$ is isomorphic to the representation associated to a "new" modular form.*

Using the previous discussion, one can be more precise: the integer $i$ must be the smallest integer such that $h_j(V) \neq 0$. Twisting by $\mathbb{Z}_p(-i)$ if necessary, we can assume $i = 0$. Then, if $w$ is the biggest integer such that $h_w(V) \neq 0$, the weight of $V$ is $w$ and we see that the weight of the corresponding modular form $f$ must be $w + 1$. The conductor of $f$ must be the conductor $N_V$ of $V$. The nebentypus can be also computed in the usual way using the representations of the Weil-Deligne groups. To prove this conjecture one "need only" to prove that the $L$-function of $V$ is the Mellin transform of a modular form.

Observe also
i) that, because of the finiteness of the dimension of the space of modular forms of fixed weight and level, conjecture 2a and conjecture 2b in the two dimensional case follows from conjecture 3c;
ii) that, because of the fact that, if $X$ is an elliptic curve over $\mathbb{Q}$, if $T_p(X)$ is its Tate module and if $V = \mathbb{Q}_p \otimes_{\mathbb{Z}_p} T_p(X)$, $V$ is geometric [**12**], conjecture 3c implies the Shimura-Taniyama-Weil conjecture.

**(g) Tannakian categories.** (for a discussion of tannakian categories, see, for instance, [**12**]). Let $K$ be a number field, and let $\underline{\text{Rep}}_{\mathbb{Q}_p}(G_K)$ denote the tannakian category of $p$-adic representations of $G_K$. A tannakian subcategory is a full subcategory containing an object of positive dimension, and which is stable under passage to sub-object, quotient, direct sum, tensor product and dual.

Let $\underline{\text{Rep}}_{\mathbb{Q}_p,g}(G_K)$ (resp. $\underline{\text{Rep}}^{ss}_{\mathbb{Q}_p,g}(G_K)$) denote the tannakian subcategory of $\underline{\text{Rep}}_{\mathbb{Q}_p}(G_K)$ whose objects are geometric (resp. semi-simple geometric) representations of $G_K$.

**Conjecture 4a.** *The category $\underline{\text{Rep}}^{ss}_{\mathbb{Q}_p,g}(G_K)$ is the smallest tannakian subcategory of $\underline{\text{Rep}}_{\mathbb{Q}_p}(G_K)$ containing all representations of the form $H^m_{\text{ét}}(X_{/\overline{K}}, \mathbb{Q}_p)$ where $X$ ranges through all proper smooth varieties over $K$.*

Note that Conjecture 4a is equivalent to the union of Conjecture 1 and the classical conjecture, due to Tate, that if $X$ is smooth and proper over $K$, the $G_K$-representation $H^m_{\text{ét}}(X_{/\overline{K}}, \mathbb{Q}_p)$ is semi-simple.

Although we don't have a precise definition of "mixed motives" in mind, it is tempting to us to ask the somewhat vague question:

**Question 4b:** *Is the category $\underline{\text{Rep}}_{\mathbb{Q}_p,g}(G_K)$ the smallest tannakian subcategory of $\underline{\text{Rep}}_{\mathbb{Q}_p}(G_K)$ containing the $p$-adic realization of all mixed motives over $K$.*

One concrete interpretation of the above question is the following:

**Question 4c:** *Is the category* $\underline{\operatorname{Rep}_{\mathbb{Q}_p,g}}(G_K)$ *the smallest tannakian subcategory of* $\underline{\operatorname{Rep}_{\mathbb{Q}_p}}(G_K)$ *containing all representations of the form* $H^m_{ét}(X_{/\overline{K}}, \mathbb{Q}_p)$ *where $X$ ranges through all simplicial schemes of finite type over $K$ ?*

## §5. "Universal" geometric deformation rings.

**A preliminary remark on deformations:** Let $G$ be a profinite group and $\wedge$ a local complete noetherian ring, of residue field $k$. Let $GSW$ be the category of local complete noetherian $\wedge$-algebras, with residue field $k$. For short, by $\wedge$-*algebra* we mean an object of $\mathfrak{S}_\wedge$.

If $A$ is any $\wedge$-algebra, an $A$-**representation** of $G$ is an $A$-module of finite type equipped with a linear and continuous action of $G$. We say that such a representation is *flat* if the underlying $A$-module is flat.

We give ourselves a finite dimensional $k$-representation $\overline{V}$ of $G$, which is such that the natural map $k \to \operatorname{End}_{k[G]}(\overline{V})$ is an isomorphism[5].

For each $\wedge$-algebra $A$, an $A$-**deformation of** $\overline{V}$ is a flat $A$-representation $V$ of $G$ such that $k \otimes_A V$ is isomorphic to $\overline{V}$. Denote by $\Psi_{\overline{V}}(A)$ the set of isomorphism classes of $A$-deformations of $\overline{V}$. We get in this way a functor

$$\Psi_{\overline{V}} : \mathfrak{S}_\wedge \to \underline{\operatorname{Sets}}.$$

Then [22][6], if $\dim_k H^1(G, \operatorname{End}_k(\overline{V})) < +\infty$, this functor is representable. Therefore, one can define the universal $\wedge$-deformation ring $R_\wedge(\overline{V})_G$ together with a $R_\wedge(\overline{V})_G$-deformation of $\overline{V}$, well defined up to isomorphism, with the obvious universal property.

**Remark:** If we choose a basis of $\overline{V}$ over $k$, the action of $G$ on $\overline{V}$ gives us an homomorphism $\overline{\rho} : G \to GL_d(k)$. Usually, one considers continuous homomorphisms from $G \to GL_d(A)$ lifting $\overline{\rho}$ rather than deformations of $\overline{V}$. Recall that if $\rho, \rho' : G \to GL_d(A)$ are two such homomorphisms, one says that $\rho$ and $\rho'$ are isomorphic (resp. strictly isomorphic) if there exists $\alpha \in GL_d(A)$ (resp. $\alpha \in GL_d(A)$ lifting the identity in $GL_d(k)$) such that $\rho'(g) = \alpha \rho(g) \alpha^{-1}$, for all $g \in G$. From the surjectivness of $A^*$ onto $k^*$ and the fact that $\operatorname{End}_{k[G]}(\overline{V}) = k$, we see that the two equivalence relations are actually the same. If we denote by $\psi_{\overline{\rho}}(A)$ the set of equivalence classes, we get also in this way a functor $\psi_{\overline{\rho}} : \mathfrak{S}_\wedge \to \underline{\operatorname{Sets}}$. It is an easy exercise to build a natural isomorphism between the functors $\psi_\rho$ and $\psi_{\overline{V}}$ and we will use this natural isomorphism to identify $\Psi_{\overline{\rho}}$ and $\psi_{\overline{V}}$.

Now let $\mathfrak{D}$ be a strictly full subcategory of the category $\underline{\operatorname{Rep}^f_\wedge}(G)$ of $\wedge$-modules of finite length equipped with a linear and continuous action of $G$ which is stable under subobjects, quotients and direct sums. Assume that $\overline{V}$ is an object of $\mathfrak{D}$. Then we can define the subfunctor $\psi_{\overline{V},\mathfrak{D}}$ of $\psi_{\overline{V}}$ by the condition that, for any $\wedge$-algebra $A$, $\psi_{\overline{V},\mathfrak{D}}(A)$ consists of elements which are represented by representations $V$ such that, for any artinian quotient $A'$ of $A$, $A' \otimes_A V$ is an object of $\mathfrak{D}$. One sees [23] that, if $\dim_k H^1(G, \operatorname{End}_k(\overline{V})) < +\infty$, then

---

[5]actually, we will be mostly interested in the absolutely irreducible case.
[6]at least in the absolutely irreducible case, but the same proof works in general.

$\psi_{\overline{V}, \mathfrak{D}}$ is representable by a quotient $R_\wedge(\overline{V})_\mathfrak{D}$ of $R_\wedge(\overline{V})_G$. We call $R_\wedge(\overline{V})_\mathfrak{D}$ the **universal ring of deformations of $\overline{V}$ lying in $\mathfrak{D}$.**

We can apply that to $G = G_{K,S}$ and $\wedge = \mathcal{O}_E$ the ring of the integers of a finite extension of $\mathbb{Q}_p$. Hence, given $\overline{V}$, we can consider the noetherian and complete $\mathcal{O}_E$-algebra $R_{\mathcal{O}_E}(\overline{V})_S := R_{\mathcal{O}_E}(\overline{V})_{G_{K,S}}$. Now, if moreover, we fix an inertia level $\mathfrak{L}$ for $S$ and two integers $a, b \in \mathbb{Z}$ with $a \leq b$, we can consider the full subcategory $\mathfrak{D} = \underline{\mathrm{Rep}}^f_{\mathcal{O}_E}(G_{K,S})_{\mathfrak{L},[a,b],\mathrm{st}}$ (resp. $\underline{\mathrm{Rep}}^f_{\mathcal{O}_E}(G_{K,S})_{\mathfrak{L},[a,b],\mathrm{cris}}$) of $\underline{\mathrm{Rep}}^f_{\mathcal{O}_E}(G_{K,S})$ whose objects are the $T$'s such that, for each $v \in S$,
i) if $v$ doesn't divides $p$, $\mathfrak{L}_v$ acts trivially on $T$,
ii) if $v$ divides $p$, one can find a semi-stable (resp. crystalline) $p$-adic representation $V$ of $\mathfrak{L}_v$ satisfying $(\mathbb{C}_v(r) \otimes_{\mathbb{Q}_p} V)^{\mathfrak{L}_v} = 0$ such that $T$ is isomorphic to a sub-quotient of $V$ as a $\mathbb{Z}_p[\mathfrak{L}_v]$-module.

We denote by $R_{\mathcal{O}_E}(\overline{V})_{S,\mathfrak{L},[ab],\mathrm{st}}$ (resp. $R_{\mathcal{O}_E}(\overline{V})_{S,\mathfrak{L},[ab],\mathrm{cris}}$) the corresponding quotient of $R_{\mathcal{O}_E}(\overline{V})_S$. Observe that the second is a quotient of the first.

**Conjecture 5:** *For any finite set of places $S$ of $K$, finite extension $E$ of $\mathbb{Q}_p$ of residue field $k$, inertial level $\mathfrak{L}$, integers $a \leq b$, finite dimensional $k$-representation $\overline{V}$ of $G_{K,S}$ belonging to $R_{\mathcal{O}_E}(\overline{V})_{S,\mathfrak{L},[ab],\mathrm{st}}$, which is such that the natural map $k \to \mathrm{End}_{k[G]}(\overline{V})$ is an isomorphism, the ring $R_{\mathcal{O}_E}(\overline{V})_{S,\mathfrak{L},[ab],\mathrm{st}}$ is a finite $\mathcal{O}_E$-algebra.*

This conjecture implies conjecture 2b: if $S, \mathfrak{L}$ and $h$ are given as in conjecture 2a, we can choose $a \leq b$ such that, for all $v$'s dividing $p$, if $D$ is a graded $(\overline{\mathbb{Q}}_p \otimes_{\overline{\mathbb{Q}}_p} K_v)$-module representing the class $h(v)$, we have $\mathrm{gr}^{-r} D = 0$ if $r \notin [a, b]$. Then, if $V$ is an $E$-representation of $G_{K,S}$, say of degree $N$, representing an element of $\mathbf{Geom}(K, S, \mathfrak{L}, h; E)$, the fact that $V$ is irreducible implies that one can find an $\mathcal{O}_E$-lattice $T$ of $V$ stable under $G_{K,S}$ and such that, if $\overline{V} = k \otimes_{\mathcal{O}_E} T$, the map $k \to \mathrm{End}_{k[G]}(\overline{V})$ is an isomorphism. Obviously, each finite quotient of $T$ lies in $\underline{\mathrm{Rep}}^f_{\mathcal{O}_E}(G_{K,S})_{\mathfrak{L},[a,b],\mathrm{st}}$, hence the natural homomorphism from $R_{\mathcal{O}_E}(\overline{V})_S$ to $\mathcal{O}_E$ factors through $R_{\mathcal{O}_E}(\overline{V})_{S,\mathfrak{L},[a,b],\mathrm{st}}$. Therefore the conjecture implies that, given $\overline{V}$, there are only finitely many elements of $\mathbf{Geom}(K, S, \mathfrak{L}, h; E)$ within a representative of which we can find a $G_{K,S}$-stable $\mathcal{O}_E$-lattice $T$ such that $k \otimes_{\mathcal{O}_E} T \cong \overline{V}$. The result follows from the fact, easy to check, that there are only finitely many isomorphism classes of $k$-representations $\overline{V}$ of degree $N$ of $G_{K,S}$ such that the map $k \to \mathrm{End}_{k[G]}(\overline{V})$ is an isomorphism.

**Remarks: a)** Let $V$ be an $E$-representation of $G_{K,S}$ which admits a lattice $T$ stable under $G_{K,S}$ such that each finite quotient lies in $\underline{\mathrm{Rep}}^f_{\mathcal{O}_E}(G_{K,S})_{\mathfrak{L},[a,b],\mathrm{st}}$ (resp. $\underline{\mathrm{Rep}}^f_{\mathcal{O}_E}(G_{K,S})_{\mathfrak{L},[a,b],\mathrm{cris}}$). It should not be very hard to prove that $V$ is geometric. Up to now, we have checked this property only in special cases (for instance, this is OK for $\underline{\mathrm{Rep}}^f_{\mathcal{O}_E}(G_{K,S})_{\mathfrak{L},[a,b],\mathrm{cris}}$ if $b - a \leq p - 1$ and $K/\mathbb{Q}$ is unramified at $p$ or if $b - a \leq 1$ and, for all $v$ dividing $p$, the ramification index of $K_v/\mathbb{Q}_p$ is $\leq p - 1$).

**(b) Universal geometric deformation rings and the Langlands program:** Let $R$ be either $R_{\mathcal{O}_E}(\overline{V})_{S,\mathfrak{L},[a,b],\mathrm{st}}$ or $R_{\mathcal{O}_E}(\overline{V})_{S,\mathfrak{L},[a,b],\mathrm{cris}}$ as above and $T_R$ the corresponding free $R$-module with action of $G_{K,S}$ (well defined up to isomorphism). For any place $v$ of $K$ not contained in $S$, there is an element which might be called a **"Hecke element"** $\tau_v$ in $R$ given by taking the trace of Frob$_v$ acting on $T_R$, where Frob$_v$ is any choice of Frobenius element at $v$ in $G_{K,S}$.

Can one "reconstuct" the rings $R_{\mathcal{O}_E}(\overline{V})_{S,\mathfrak{L},[a,b],\mathrm{st}}$ and $R_{\mathcal{O}_E}(\overline{V})_{S,\mathfrak{L},[a,b],\mathrm{cris}}$ and their systems of Hecke elements $v \mapsto \tau_v$ as completions (at maximal ideals of residual characteristic $p$) of rings which are generated by Hecke operators and which act (faithfully) on automorphic representations spaces for specific reductive groups (notably $GL_{N/K}$) ?

## §6. The special case where $\rho$ is potentially abelian.

A $p$-adic representation $V$ of $G_K$ is **potentially abelian** if there is an open subgroup of $G_K$ which operates on $V$ through an abelian quotient group.

For any abelian variety $A$ over $K$, let $A[p^n](\overline{K})$ denote the group of $\overline{K}$-rational points of $A$ annihilated by multiplication by $p^n$, let $T_p(A)=\lim.\mathrm{proj}.$ $A[p^n](\overline{K})$
and put $V_p(A) = \mathbb{Q}_p \otimes_{\mathbb{Z}_p} T_p(A)$. Recalling the terminology of Remark (g) of § 4, let $\underline{\mathrm{Rep}}_{\mathbb{Q}_p,CM}(G_K)$ denote the smallest Tannakian subcategory of $\underline{\mathrm{Rep}}_{\mathbb{Q}_p}(G_K)$ containing all representations of $G_K$ which factor through finite groups, and also containing the $V_p(A)$ for $A$ ranging through all abelian varieties over $K$ which are potentially of CM type.

**Proposition:** *Let $V$ be a potentially abelian p-adic representation of $G_K$ and, for each place $v$ of $K$, let $V_v$ be the underlying representation of $G_v$. The following are equivalent:*
*1) For every place $v$ of $K$ dividing $p, V_v$ is of Hodge-Tate type and the action of $I_v$ is semi-simple;*
*2) for every place $v$ of $K$ dividing $p, V_v$ is of Hodge-Tate type;*
*3) for every place $v$ of $K$ dividing $p, V_v$ is of de Rham type;*
*4) for every place $v$ of $K$ dividing $p, V_v$ is potentially semi-stable;*
*5) for every place $v$ of $K$ dividing $p, V_v$ is potentially crystalline;*
*6) the representation $V$ is geometric;*
*7) the representation $V$ is an object in $\underline{\mathrm{Rep}}_{\mathbb{Q}_p,CM}(G_K)$.*

*Proof.* Since potentially abelian representations are unramified outside a finite set of places ([**27**], Cor. of p. III.11) we have (4) ⇔ (6). One knows that every abelian variety $A$ which is potentially of CM type has potentially good reduction. It follows that if $V = V_p(A)$, then $V$ satisfies (5). Moreover, any finite representation satisfies (5), and (5) is stable under tannakian operations. Therefore (7) ⇒ (5).

The implications (5) ⇒ (4) ⇒ (3) ⇒ (2) and (1) ⇒ (2) are all trivial.

From ([**27**], Thm. 3 p. III-52) one has that (1) implies that $V$ is "locally algebraic" which, in turn, implies (7) by ([**4**], Prop. D1).

It remains to establish the implication (2) ⇒ (1) which is an exercise, using the Theorem of Sen [**25**] that a representation, which is Hodge-Tate and whose only Hodge-Tate weight is 0 (with any multiplicity) must factor through a finite quotient group. □

## § 7. The special case where $p$ is potentially unramified.

After a finite base change we can make $p$ unramified, i.e., we can view our representation as being a representation of $G_{K,S}$ where $S$ is a finite set of primes, none of which have residual characteristic $p$.

**Conjecture 5a:** *If $p$ is distinct from all of the residual characteristics of $S$, then any $p$-adic representation of $G_{K,S}$ factors through a finite quotient group of $G_{K,S}$.*

**Remark:** Conjecture 5a for all $p$-adic representations follows from the same conjecture, but stated only for semi-simple geometric $p$-adic representations. This latter conjecture follows from Conjecture 1 in conjunction with the Tate Conjecture about the subspace of étale cohomology generated by algebraic cycles.

An equivalent way of stating Conjecture 5a is that if $p$ is distinct from all of the residual characteristics of $S$, then any quotient group of $G_{K,S}$ which is a $p$-adic analytic group, is finite. Conjecture 5a for $S$ empty bears on the structure of the Galois group of a Golod-Shafarevich $p$-tower: Let $\Gamma(K,p)$ denote the Galois group of the maximal everywhere unramified pro-$p$-extension of the number field $K$. Conjecture 5a implies.

**Conjecture 5b:** *Any quotient group of $\Gamma(K,p)$ which is a $p$-adic analytic group, is finite.*

For some partial corroboration of this conjecture, see the recent work of N.Boston ([2], Theorem 1), and F.Hajir [18].

## Part II. Representations of dimension 2 of $\mathrm{Gal}(\overline{\mathbb{Q}}_p/\mathbb{Q}_p)$.

In this part, $G_p = G_{\mathbb{Q}_p} = \mathrm{Gal}(\overline{\mathbb{Q}}_p/\mathbb{Q}_p)$ and $I_p$ is the inertia subgroup. For any $\mathbb{Z}_p$-module $V$ and any $r \in \mathbb{Z}$, we denote by $V(r) = V \otimes_{\mathbb{Z}_p} \mathbb{Z}_p(r)$ the usual Tate twist.

For $\epsilon \in \overline{\mathbb{F}}_p$, the residue field of $\overline{\mathbb{Q}}_p$, we denote $[\epsilon]$ its Teichmüller representative in $\overline{\mathbb{Q}}_p$.

For simplicity, we will assume $p \neq 2$ (and sometimes $p \geq 5$).

## §8. The representations of $G_p$ of dimension 1.

Let $A$ be a complete noetherian local ring with finite residue field. For any profinite group $G$, a *character with values in $A$* is a continuous homomorphism $\eta : G \to A^*$. If $N$ is any closed invariant subgroup of $G$ contained in the kernel of $\eta$, we still denote $\eta : G/N \to A^*$ the character we get by factoring, and conversely.

If $V$ is an $A$-module equipped with an action of $G$, $V(\eta)$ denotes the same $A$-module with the action of $G$ twisted by $\eta$.

If $a \in A^*$, we denote by $\eta_a : G_p/I_p \to A^*$ the unique character such that $\eta_a(\text{arith.frob.}) = a$.

Let $\xi_1 : G_p \to \mathbb{F}_p^*$ be the character giving the action of $G_p$ on the $p$-th roots of 1. For $i \in \mathbb{Z}/(p-1)\mathbb{Z}$ and $\epsilon \in \mathbb{F}_p^*$, denote by $\overline{U}_{i,\epsilon}$ the one dimensional $\mathbb{F}_p$-representation of $G_p$ which is $\mathbb{F}_p$ on which $G_p$ acts through $\eta_\epsilon \xi_1^i$. Then any one dimensional $\mathbb{F}_p$-representation of $G_p$ is isomorphic to one and only one of the $\overline{U}_{i,\epsilon}$.

Define $\hat{\xi}_1(g) = [\xi_1(g)]$ and $\chi : G_p \to \mathbb{Z}_p^*$ the cyclotomic character which we can write as $\chi = \hat{\xi}_1 \cdot \chi_0$ with $\chi_0$ taking values in $1 + p\mathbb{Z}_p$.

If we denote $R(\overline{U}_{i,\epsilon}) = R_{\mathbb{Z}_p}(\overline{U}_{i,\epsilon})_{G_p}$ the universal $\mathbb{Z}_p$-deformation ring, one sees that one can write $R(\overline{U}_{i,\epsilon}) = \mathbb{Z}_p[\![T_1, T_2]\!]$, in such a way that the corresponding character $\eta : G_p \to \mathbb{Z}_p[\![T_1, T_2]\!]^*$ giving the universal deformation has the property that if, for $r \in \mathbb{Z}_p$

$$e_r : \mathbb{Z}_p[[T_1, T_2]] \longrightarrow \mathbb{Z}_p[[T_1]]$$

is the ring-homomorphism to $\ker_\rho T_1$ to $T_1$ and $T_2$ to $(1+p)^r - 1$, then

$$e_r \cdot \eta = \eta_{[\epsilon](1+T_1)} \cdot \hat{\xi}_1^i \cdot \chi_0^{T_2}$$

for all $r \in \mathbb{Z}_p$.

Now, let $U$ be a one dimensional $\mathbb{Q}_p$ representation of $G_p$. To give $U$ up to isomorphism is the same as to give the unique $(i, \epsilon, t_1, t_2) \in ((\mathbb{Z}/(p-1)\mathbb{Z}) \times \mathbb{F}_p^* \times p\mathbb{Z}_p \times p\mathbb{Z}_p$ such that $G_p$ acts on $U$ via $\eta_{[\epsilon](1+t_1)} \cdot \hat{\xi}_1^i \cdot \chi_0^r$ where $(1-p)^r - 1 = t_2 r$. If $T$ is a $\mathbb{Z}_p$-lattice and if $\overline{U} = T/pT$, then $\overline{U} \simeq \overline{U}_{i,\epsilon}$; moreover $U$ is potentially semi-stable if it is Hodge-Tate, which amounts to requiring that $r$ be in $\mathbb{Z}$. In this case $r$ is the (unique) Hodge-Tate weight of $U$.

Now fix $\overline{U}_{i,\epsilon}$ and $r \in \mathbb{Z}$. We observe that, if $U$ is potentially semi-stable of Hodge-Tate weight $r$ and such that $\overline{U} \simeq \overline{U}_{i,\epsilon}$, then there is a unique $t_1 \in p\mathbb{Z}_p$ such that $G_p$ acts on $U$ via $\eta_{[\epsilon](1+t_1)} \cdot \hat{\xi}_1^i \cdot \chi_0^r = \hat{\xi}_1^{i-\overline{r}} \cdot \eta_{[\epsilon](1+t_1)} \cdot \chi^r$ (where $\overline{r}$ denotes the image of $r$ in $\mathbb{Z}/(p-1)\mathbb{Z}$), hence $U(\hat{\xi}_1^{\overline{r}-i})$ is a crystalline representation with $r$ as unique Hodge-Tate weight. Now, if we consider the full subcategory $\mathfrak{D}$ of the category of finite $\mathbb{Z}_p$-representations $T$ of $G_p$ which are such that $T(\hat{\xi}_1^{\overline{r}-i})$ is isomorphic to a subquotient of a crystalline representation of unique Hodge-Tate weight $r$, $\mathfrak{D}$ is stable under subobjects, quotients and direct sums, so we can speak of the universal deformation of $\overline{U}_{i,\epsilon}$ "lying in $\mathfrak{D}$". The corresponding ring $R_{\mathbb{Z}_p}(\overline{U}_{i,\epsilon})_\mathfrak{D}$ is isomorphic to $\mathbb{Z}_p[\![T_1]\!]$ with the corresponding character $\eta_{[\epsilon](1+T_1)} \cdot \hat{\xi}_1^i \cdot \chi_0^r$.

## §9. The absolutely irreducible $\mathbb{F}_p$-representations of $G_p$ of dimension 2 and their deformations.

By convention, for any positive integer $n, \mathbb{Q}_{p^n} \subset \overline{\mathbb{Q}}_p$ will denote the unramified extention of $\mathbb{Q}_p$ in $\overline{\mathbb{Q}}_p$ of degree $n$ and $\mathbb{F}_{p^n}$ the corresponding residue field. We choose $\pi_1, \pi_2 \in \overline{\mathbb{Q}}_p$ such that $\pi_1^{p-1} = -p$ and $\pi_2^{p+1} = \pi_1$ and, for $j = 1, 2$, let $F_j =$ the Galois closure over $\mathbb{Q}_p$ of $\mathbb{Q}_p(\pi_j)$; let $L = \mathbb{Q}_{p^{2(p-1)}}(\pi_2)$ and $L_0 = \mathbb{Q}_{p^{2(p-1)}}$ so that we have the diagram

$$\begin{array}{ccccccc}
 & & \mathbb{Q}_p(\pi_2) & - & F_2 = \mathbb{Q}_{p^2}(\pi_2) & - & L = \mathbb{Q}_{p^{2(p-1)}}(\pi_2) \\
 & & | & & | & & | \\
F_1 & = & \mathbb{Q}_p(\sqrt[p-1]{1}) = \mathbb{Q}_p(\pi_1) & - & \mathbb{Q}_{p^2}(\pi_1) & - & \mathbb{Q}_{p^{2(p-1)}}(\pi_1) \\
 & & | & & | & & | \\
 & & \mathbb{Q}_p & - & \mathbb{Q}_{p^2} & - & L_0 = \mathbb{Q}_{p^{2(p-1)}}
\end{array}$$

The $\mathrm{Gal}(L/\mathbb{Q}_p(\pi_2))(\simeq \mathrm{Gal}(L_0/\mathbb{Q}_p))$ is cyclic of order $2(p-1)$ generated by the "Frobenius element" $\tau$ satisfying $\tau x = x^p$ if $x \in \mathbb{F}_{p^{2(p-1)}}$, the residue field of

$L$. The group $\mathrm{Gal}(L/L_0)$ is canonically isomorphic to the multiplicative group of $\mathbb{F}_{p^2}$, via the "**fundamental character**" $\xi_2 : \mathrm{Gal}(L/L_0) \to \mathbb{F}_{p^2}^*$ associated to $g \in \mathrm{Gal}(L/L_0)$, the image of $g\pi_2/\pi_2$ in $\mathbb{F}_{p^2}$. The group $\mathrm{Gal}(L/\mathbb{Q}_p)$ is a semi-direct product of $\mathrm{Gal}(L/\mathbb{Q}_p(\pi_2))$ and the normal subgroup $\mathrm{Gal}(L/L_0)$, with $\tau g \tau^{-1} = g^p$ for any $g \in \mathrm{Gal}(L/L_0)$.

Consider $(\iota, \varepsilon)$ where $\iota \in \mathbb{Z}/(p^2-1)\mathbb{Z}$ and $\varepsilon \in \mathbb{F}_p^*$. Given such data, *we choose* $\zeta \in \mathbb{F}_{p^2}$ such that $\zeta^{p+1} = -\varepsilon$. Attached to the above data, *and this choice*, we have a unique homomorphism

$$\overline{\rho}_{\iota,\varepsilon} : \mathrm{Gal}(L/\mathbb{Q}_p) \to \mathrm{Aut}_{\mathbb{F}_p}(\mathbb{F}_{p^2})$$

such that the restriction to $\mathrm{Gal}(L/L_0)$ consists in $\mathbb{F}_{p^2}$-linear automorphisms given by the one-dimensional $\mathbb{F}_{p^2}$-character $\xi_2^\iota$, and $\tau$ acts on $\mathbb{F}_{p^2}$ via the $\mathbb{F}_{p^2}$-semilinear automorphism $x \mapsto \zeta \cdot x^p$. We denote also $\overline{\rho}_{\iota,\varepsilon} : G_p \to \mathrm{Aut}_{\mathbb{F}_p}(\mathbb{F}_{p^2})$ the natural lifting and $\overline{V}_{\iota,\varepsilon}$ the 2-dimensional $\mathbb{F}_p$-representation of $G_p$ which is $\mathbb{F}_{p^2}$ on which $G_p$ acts through $\overline{\rho}_{\iota,\varepsilon}$.

**Proposition .** *i) The isomorphism class of the $\mathbb{F}_p[G_p]$-module $\overline{V}_{\iota,\varepsilon}$ is independent of the choice of $\zeta$ such that $\zeta^{p+1} = -\varepsilon$. The "determinant" $\wedge^2 \overline{V}_{\iota,\varepsilon}$ is isomorphic to $\overline{U}_{2\overline{\iota},\varepsilon}$ (where $\overline{\iota}$ is the image of $\iota$ in $\mathbb{Z}/(p-1)\mathbb{Z}$). One has $\overline{V}_{\iota',\varepsilon'} \simeq \overline{V}_{\iota,\varepsilon}$ if and only if $\varepsilon' = \varepsilon$ and $\iota' \in \{\iota, p\iota\}$. The representation $\overline{V}_{\iota,\varepsilon}$ is absolutely irreducible if and only if $p\iota \neq \iota$ (in $\mathbb{Z}/(p^2-1)\mathbb{Z}$).*

*ii) Let $\overline{V}$ be a 2-dimensional absolutely irreducible $\mathbb{F}_p$-representation of $G_p$. Then $\overline{V} \simeq \overline{V}_{\iota,\varepsilon}$ for a suitable $(\iota, \varepsilon) \in (\mathbb{Z}/(p^2-1)\mathbb{Z}) \times \mathbb{F}_p^*$.*

*Proof.* exercise. □

Let $\iota \in \mathbb{Z}/(p^2-1)\mathbb{Z}$ such that $p\iota \neq \iota$. The residual representation $\overline{V}_{\iota,\varepsilon}$ being absolutely irreducible possesses a universal deformation ring $R_{\iota,\varepsilon} = R(\overline{V}_{\iota,\varepsilon})$ as in [**22**], and we have the theorem due to R. Ramakrishna ([**23**] Th.4.1):

**Theorem .** *For any absolutely irreducible $\overline{V} = \overline{V}_{\iota,\varepsilon}$, the "deformation problem is smooth" and the ring $R_{\iota,\varepsilon}$ is isomorphic to a power series ring in five variables over $\mathbb{Z}_p$, i.e. to $\mathbb{Z}_p[\![T_1, T_2, \cdots, T_5]\!]$.*

## §10. Potentially semi-stable $p$-adic representations and pst-modules.

It is known[**12**] that, using the ring $B_{\mathrm{st}}$ [**11**], one gets an equivalence of categories between potentially semi-stable $p$-adic representations of $G_p = \mathrm{Gal}(\overline{\mathbb{Q}}_p/\mathbb{Q}_p)$ and the category of "admissible filtered $(\varphi, N, G_p)$-modules". We want to recall a bit of this theory.

(a) **Filtered modules** : Let $F$ be a finite Galois extension of $\mathbb{Q}_p$ contained in $\overline{\mathbb{Q}}_p$ and $F_0$ the maximal unramified extension of $\mathbb{Q}_p$ contained in $F$. A **filtered** $(\varphi, N, \mathrm{Gal}(F/\mathbb{Q}_p))$-**module** is a finite dimensional $F_0$-vector space $D$ endowed with

- a "Frobenius" endomorphism $\varphi : D \to D$ bijectiv and semi-linear via the action of the absolute Frobenius on $F_0$,

- a "monodromy" operator $N : D \to D$ linear and satisfying $N\varphi = p\varphi N$,
- an action of $\mathrm{Gal}(F/\mathbb{Q}_p)$ semi-linear with respect to the natural action of this group (via its quotient $\mathrm{Gal}(F_0/\mathbb{Q}_p)$) on $F_0$ and commuting with $\varphi$ and $N$,
- a decreasing filtration $(\mathrm{Fil}^r D_F)_{r \in \mathbb{Z}}$ of the $F$-vector space $D_F = F \otimes_{F_0} D$ stable under the natural action of $\mathrm{Gal}(F/\mathbb{Q}_p)$ (acting via $(g, \lambda \otimes d) \mapsto g\lambda \otimes gd$) verifying $\mathrm{Fil}^r D_F = 0$ for $f \gg 0$ and $= D_F$ for $r \ll 0$. Observe that, if we put $D_{dR} = (D_F)^{\mathrm{Gal}(F/\mathbb{Q}_p)}$, the obvious map $F \oplus_{\mathbb{Q}_p} D_{dR} \to D_F$ is an isomorphism and the condition that the $\mathrm{Fil}^r D_F'$s are stable under $\mathrm{Gal}(F/\mathbb{Q}_p)$ means exactly that the filtration comes from a filtration of the $\mathbb{Q}_p$-vector space $D_{dR}$.

Whenever $F = \mathbb{Q}_p$, $D$ is just a $\mathbb{Q}_p$-vector space together with two linear endomorphisms $\varphi$ and $N$ (with $\varphi$ bijective and $N\varphi = p\varphi N$) and a filtration and we call $D$ a **filtered $(\varphi, N)$-module**.

The filtered $(\varphi, N, \mathrm{Gal}(F/\mathbb{Q}_p))$-modules are the objects of an additive $\mathbb{Q}_p$-linear category (which is not abelian) where "morphisms" are defined in the evident way. One can define in a natural way the dual $D^*$ of an object of this category and the tensor product $D_1 \otimes D_2$ of two objects.

Recall ([9], [12]) that for such a $D$, one can define its Hodge polygon (associated to the Hodge-Tate type, i.e. to the number $h^r = \dim gr^r D_F$) and its Newton polygon (associated to the slope of Frobenius). One says that $D$ is **weakly admissible** (or w.a. for short) if

i) for any sub-$(\varphi, N, \mathrm{Gal}(F/\mathbb{Q}_p))$-module $D'$ of $D$, with $D'_F \subset D_F$ equipped with the induced filtration, the Hodge polygon of $D'$ lies above the Newton polygon of $D'$;

ii) the Hodge polygon and the Newton polygon of $D$ ends up at the same point.

The weakly admissible modules form an abelian category [12]. If $D$ is an object of this category, sub-objects of $D$ are sub-$(\varphi, N, \mathrm{Gal}(F/\mathbb{Q}_p))$-modules such that the Hodge polygon and Newton polygon of $D$ end up at the same point.

**Changing $F$:** Call $\mathfrak{F}$ the set of finite Galois extensions of $\mathbb{Q}_p$ contained in $\overline{\mathbb{Q}}_p$. Given $F, F' \in \mathfrak{F}$ with $F \subset F'$, we say that a $(\varphi, N, \mathrm{Gal}(F'/\mathbb{Q}_p))$-module $D'$ is $F$-**semi-stable** if the natural map $F'_0 \otimes_{F_0} (D')^{\mathrm{Gal}(F'/F)} \to D'$ is an isomorphism. To any filtered $(\varphi, N, \mathrm{Gal}(F/\mathbb{Q}_p))$-module $D$, one can associate a filtered $(\varphi, N, \mathrm{Gal}(F'/\mathbb{Q}_p))$-module $D_{/F'}$ in an obvious way (the underlying $F'_0$-vector space is $F'_0 \otimes_{F_0} D$). We get in this way an equivalence of categories between filtered $(\varphi, N, \mathrm{Gal}(F/\mathbb{Q}_p))$-modules and the full subcategory of filtered $(\varphi, N, \mathrm{Gal}(F'/\mathbb{Q}_p))$-modules consisting of $F$-semi-stable ones. Thus we can form the inductive limit over $\mathfrak{F}$ of the category of filtered $(\varphi, N, \mathrm{Gal}(F/\mathbb{Q}_p))$-modules; we call an object of this $\mathbb{Q}_p$-linear category a **pst-module**.

Each time we have a pst-module $\Delta$, we may choose an $F \in \mathfrak{F}$ such that this object is $F$-semi-stable and we can speak of its $F$-**realisation** which is a filtered $(\varphi, N, \mathrm{Gal}(F/\mathbb{Q}_p))$-module $D$. On this $D$ we have an $F_0$-linear action of the inertia subgroup of $\mathrm{Gal}(F/\mathbb{Q}_p)$ that we can view as a continuous action of $I_p$ whose kernel contains $I_p \cap \mathrm{Gal}(\overline{\mathbb{Q}}_p/F)$. If $F' \in \mathfrak{F}$, $\Delta$ is $F'$-semi-stable if and only if $I_p \cap \mathrm{Gal}(\overline{\mathbb{Q}}_p/F')$ acts trivially on $D$. To know $\Delta$ is the same as to know $D$; the choice of $F$ doesn't matter and we will sometimes speak of the pst-module $D$.

If $D$ is a filtered $(\varphi, N, \mathrm{Gal}(F/\mathbb{Q}_p))$-module and if $F \subset F'$, $D_{F'}$ is weakly admissible if and only if $D$ is. Then it makes sense to speak of a **weakly admissible pst-module**. These modules form an *abelian $\mathbb{Q}_p$-linear category*.

(b) **Representations**: If $F \in \mathfrak{F}$, we have a natural functor $\underline{D}_{\mathrm{st},F}$ from the category of $p$-adic representations of $G_p$ to the category of filtered $(\varphi, N, \mathrm{Gal}(F/\mathbb{Q}_p))$-modules:

$$V \mapsto D = \underline{D}_{\mathrm{st},F}(V) = (B_{\mathrm{st}} \otimes_{\mathbb{Q}_p} V)^{\mathrm{Gal}(\overline{\mathbb{Q}}_p/F)}.$$

We have $\dim_{F_0} \leq \dim_{\mathbb{Q}_p} V$ and $V$ is said to be $F$-**semi-stable** if equality holds. The property to be $F$-semi-stable is stable under taking subobject, quotient, direct sum, tensor product, dual. Say a filtered $(\varphi, N, \mathrm{Gal}(F/\mathbb{Q}_p))$-module is **admissible** if it is isomorphic to $\underline{D}_{\mathrm{st},F}(V)$ for an $F$-semi-stable V. *It is known that admissible implies weakly admissible and conjectured that the converse is true*. The restriction of $\underline{D}_{\mathrm{st},F}$ to $F$-semi-stable representations is fully faithful, hence induces an *equivalence of categories between p-adic F-semi-stable representations of $G_p$ and* **admissible filtered** $(\varphi, N, \mathrm{Gal}(F/\mathbb{Q}_p))$-**modules**. There is a natural quasi-inverse $\underline{V}_{\mathrm{st}}$ of $\underline{D}_{\mathrm{st},F}$ ($\underline{V}_{\mathrm{st}}(D)$ is defined as a suitable sub-$\mathbb{Q}_p$-vector space of $B_{\mathrm{st}} \otimes_{F_0} D$). This is an equivalence of tannakian categories, i.e. $\underline{D}_{\mathrm{st},F}$ is compatible in a natural way with duality and tensor product.

If $D$ is admissible and if $D'$ is a sub-$(\varphi, N, \mathrm{Gal}(F/\mathbb{Q}_p))$-module of $D$, then $D'$ is admissible if and only if it is weakly admissible.

It is sometimes convenient to use a contravariant version of this equivalence. If $V$ is $F$-semi-stable, one can define $\underline{D}_{\mathrm{st},F}^*(V)$ in three different ways (there are canonical isomorphisms between them [**12**])

$$\underline{D}_{\mathrm{st},F}^*(V) = \mathrm{Hom}_{\mathbb{Q}_p[\mathrm{Gal}(\overline{\mathbb{Q}}_p/F)]}(V, B_{\mathrm{st}}) = \underline{D}_{\mathrm{st},F}(V^*) = (\underline{D}_{\mathrm{st},F}(V))^*$$

Let $V$ be $F$-semi-stable and $D = \underline{D}_{\mathrm{st},F}(V)$. Then $V$ is *de Rham* and the filtered $\mathbb{Q}_p$-vector space $\underline{D}_{dR}(V) = (B_{dR} \otimes_{\mathbb{Q}_p} V)^{G_p}$ can be identified with $D_{dR} = (D_F)^{\mathrm{Gal}(F/\mathbb{Q}_p)}$. Hence $V$ is also Hodge-Tate and the Hodge-Tate multiplicities $h^r(V) = \dim_{\mathbb{Q}_p} \mathrm{Hom}_{\mathbb{Q}_p[G_p]}(V, \mathbb{C}_p(r))$ satisfy

$$h^r(V) = \dim_F \mathrm{Fil}^{-r} D_F / \mathrm{Fil}^{-r+i} D_F = \dim_F \mathrm{Fil}^r D_F^* / \mathrm{Fil}^{r+1} D_F^*.$$

**Tate twists**: For any filtered $(\varphi, N, \mathrm{Gal}(F/\mathbb{Q}_p))$-module $D$ and for $i \in \mathbb{Z}$, we denote by $D\{i\}$ the filtered $(\varphi, N, \mathrm{Gal}(F/\mathbb{Q}_p))$-module which is $D$ as a $F_0$-vector space with the same action of $N$ and of $\mathrm{Gal}(F/\mathbb{Q}_p)$, and

$$\varphi_{\mathrm{new}} = p^{-i} \varphi_{\mathrm{old}} \quad \text{and} \quad \mathrm{Fil}^r D_{F,\mathrm{new}} = \mathrm{Fil}^{r+1} D_{F,\mathrm{old}}.$$

If $V$ is any $p$-adic representation of $G_p$ and if $i \in \mathbb{Z}$, the usual Tate twist $V(i)$ is $F$-semi-stable if and only if $V$ is. In this case, $\underline{D}_{\mathrm{st},F}(V(i))$ can be identified with $\underline{D}_{\mathrm{st},F}(V)\{i\}$ and $\underline{D}_{\mathrm{st},F}^*(V(i))$ with $\underline{D}_{\mathrm{st},F}^*(V)\{-i\}$.

**Changing $F$**: A $p$-adic representation $V$ of $G_p$ is said to be **potentially semi-stable** if it is $F$-semi-stable for some $F \in \mathcal{F}$. If $V$ is $F$-semi-stable, with $D = \underline{D}_{\mathrm{st},F}(V)$, and if $F' \in \mathcal{F}$ contains $F$, $V$ is also $F'$-semi-stable and

$\underline{D}_{\mathrm{st},F'}(V)$ can be identified with $D_{/F'}$. Hence we have an evident notion of **admissible** pst-**modules** and we get an equivalence between potentially semi-stable representations and the full subcategory of weakly admissible pst-modules whose objects are admissible ones (and conjecturally they are all).

**Remark**: A $p$-adic representation $V$ of $G_p$ is **semi-stable** if it is $\mathbb{Q}_p$-semi-stable, **$F$-crystalline** if it is $F$-semi-stable and $N = 0$ on $\underline{D}_{\mathrm{st},F}(V)$, **potentially crystalline** if it is $F$-crystalline for some $F$, **crystalline** if it is $\mathbb{Q}_p$-crystalline.
**The one dimensional case** : Consider a one dimensional pst-module $\Delta$, choose $F \in \mathcal{F}$ such that $\Delta$ is $F$-semi-stable and let $D$ be its $F$-realisation. This is a one dimensional $F_0$-vector space on which $I_p$ acts linarly via a character of finite order with values in $\mathcal{O}_{F_0}$. The fact that this action commutes with $\varphi$ implies that it takes values in $\mathbb{Z}_p$, hence its order divides $p - 1$. This means that $\Delta$ is $F_1$-semi-stable (recall $F_1 = \mathbb{Q}_p(\sqrt[p]{1})$), hence we can choose $F = F_1$ and $D$ is a one dimensional $\mathbb{Q}_p$-vector space; it is an easy exercise to check that
i) there is a unique $(r, a, i) \in \mathbb{Z} \times \mathbb{Q}_p^* \times (\mathbb{Z}/p - 1)\mathbb{Z}$ such that $D \simeq D(r; a; i)$, where $D(r; a; i)$ is defined as follows. Its underlying $\mathbb{Q}_p$-vector space we take to be $\mathbb{Q}_p$, and hence $D_{F_1} = F_1$. Put

$$\varphi 1 = a, N1 = 0, g1 = \hat{\xi}_1(g)^i \text{ if } g \in \mathrm{Gal}(F_1/\mathbb{Q}_p), \mathrm{Fil}^r F_1 = F_1, \mathrm{Fil}^{r+1} F_1 = 0.$$

ii) the module $D(r; a; i)$ is weakly admissible if and only if $v_p(a) = r$; in this case it is also admissible and $G$ acts on $\underline{V}_{\mathrm{st}}(D(r; a; i))$ via the character $\chi^{-r} \cdot \eta_{(a/p^r)^{-1}} \cdot \hat{\xi}_1^i$.

All the one dimensional potentially semi-stable $p$-adic representations of $G_p$ are $F_1$-crystalline.

## §11. The irreducible 2-dimensional weakly admissible pst-modules.

If $F \in \mathcal{F}$ and $D$ is a 2-dimensional filtered $(\varphi, N, \mathrm{Gal}(F/\mathbb{Q}_p))$-module, $D_F$ is a 2-dimensional filtered $F$-vector space and there are well defined integers $r, s$ satisfying $r \leq s$ such that, if $i \in \mathbb{Z}$, $\mathrm{Fil}^i D_F = D_F \Leftrightarrow i \leq r$ and $\mathrm{Fil}^i D_F = 0 \Leftrightarrow i > s$. We call $(r, s)$ the Hodge-Tate type of $D$.

**We fix** $(r, s)$ and we want to describe some 2-dimensional filtered $(\varphi, N, \mathrm{Gal}(F/\mathbb{Q}_p))$-modules, $D$ of Hodge-Tate type $(r, s)$. If $r = s$, there is no choice for the filtration. If $r < s$, the filtration of $D_F$ is determined by the knowledge of the $F$-line $\mathrm{Fil}^{r+1} D_F = \mathrm{Fil}^s D_F$ and we will denote it by $\Delta$. The underlying $F_0$-vector space of each $D$ will be taken to be $(F_0)^2$ and we denote $\{e_1, e_2\}$ the canonical basis.
(a) Let's first choose $F = F_1$, hence $F_0 = \mathbb{Q}_p, \Gamma_1 = \mathrm{Gal}(F/\mathbb{Q}_p)$ and the characters of $\Gamma_1$ with values in $\mathbb{Z}_p$ are the $\hat{\xi}_1^i$ for $0 \leq i < p - 1$.
*Type I* : Choose $(a, d, i) \in \mathbb{Q}_p \times \mathbb{Q}_p^* \times \mathbb{Z}$ with $0 \leq i < p - 1$; define $D = D_1(r, s; a, d, i)$ by $\varphi e_1 = e_2, \varphi e_2 = -de_1 + ae_2, Ne_1 = e_2 = 0, gx = \hat{\xi}_1^i(g).x$ ($x \in D, g \in \Gamma_1$), and if $r < s$, $\Delta = F_1.1 \otimes e_1$.
*Type II* : Choose $(b, c, i) \in \mathbb{Q}_p^* \times \mathbb{Q}_p \times \mathbb{Z}$ with $0 \leq i < p - 1$; define $D = D_{II}(r, s; b, c; i)$ by $\varphi e_1 = pbe_1, \varphi e_2 = be_2, Ne_1 = e_1, Ne_2 = 0, gx = \hat{\xi}_1^i(g).x$ ($x \in D, g \in \Gamma_1$), and if $r < s, \Delta = F_1.(1 \otimes e_1 + c \otimes e_2)$.
*Type III* ; Choose $(a_1, a_2, i_1, i_2) \in (\mathbb{Q}_p^*)^2 \times \mathbb{Z}^2$, with $0 \leq i_1 < i_2 \leq p - 1$; define $D = D_{III}(r, s; a_1, a_2; i_1, i_2)$ by $\varphi e_1 = ae_1, \varphi e_2 = a_2 e_2, Ne_1 = Ne_2 = 0, ge_1 = \hat{\xi}_1^{i_1}.e_1, \quad ge_2 = \hat{\xi}_1^{i_2}.e_2$ ($g \in \Gamma_1$), and, if $r < s, \Delta = F_1(\pi_1^{i_2} \otimes e_1 + \pi_1^{i_1} \otimes e_1)$.

(b) We now choose $F = F_2$ with, as in §9, $F_2 = \mathbb{Q}_{p^2}(\pi_2)$ where we have chosen $\pi_2 \in \overline{\mathbb{Q}}_p$ such that $\pi_2^{p^2-1} = -p$. Then $F_0 = \mathbb{Q}_{p^2}$. Let $\Gamma_2 = \text{Gal}(F_2/\mathbb{Q}_p)$ and $I\Gamma_2$ the inertia subgroup. We denote by $\hat{\xi}_2 : I\Gamma_2 \to \mu_{p^2-1}(\mathbb{Q}_{p^2})$ the isomorphism defined by $\hat{\xi}_2(g) = g\pi_2/\pi_2$. The group $\Gamma_2$ is the semi-direct product of the invariant subgroup $I\Gamma_2$ by the subgroup of order 2 generated by the unique nontrivial element $\overline{\tau}$ of $\Gamma_2$ such that $\overline{\tau}\pi_2 = \pi_2$; moreover, if $g \in I\Gamma_2, \overline{\tau}g\overline{\tau} = g^p$.
*Type IV*; Choose $d \in \mathbb{Q}_p^*$, integers $i_1, i_2$ satisfying $0 \leq i_1 < i_2 \leq p-1$. If $s > r$, choose also $\alpha \in \mathbb{P}^1(\mathbb{Q}_p)$. If $s = r$ (resp. $s > r$) define $D = D_{IV}(r,r;d;i_1,i_2)$(resp. $D_{IV}(r,s;d;i_1,i_2;\alpha)$) by $\varphi e_1 = e_2, \varphi e_2 = -de_1, Ne_1 = Ne_2 = 0, \overline{\tau}e_1 = e_1, \overline{\tau}e_2 = e_2, ge_1 = \hat{\xi}_2^{i_1+pi_2}(g)e_1$ and $ge_2 = \hat{\xi}_2^{i_2+pi_1}(g)e_2$ for $g \in I\Gamma_2$ and, if $r < s, \Delta = F_2.(\pi_2^{(p-1)i_1} \otimes e_1 + \alpha.\pi_2^{(p-1)i_2} \otimes e_2)$ if $\alpha \neq \infty$ and $\Delta = F_2.1 \otimes e_2$ if $\alpha = \infty$.

**Theorem A** : i) *The following pst-modules are weakly admissible and irreducible:*
1) $D_I(r,r;a,d,i)$ *if* $v_p(a) \geq r, v_p(d) = 2r$ *and* $X^2 - aX + d$ *irreducible over* $\mathbb{Q}_p$,
1') $D_I(r,s;a,d,i)$ *for* $r < s$, *if* $v_p(a) > r$ *and* $v_p(d) = r + s$,
2) $D_{II}(r,s;b,c;i)$ *for* $s - r \geq 3$ *and odd, if* $v_p(b) = (s - r - 1)/2$,
3) $D_{III}(r,s;a_1,a_2;i_1,i_2)$ *for* $s - r \geq 2$, *if* $v_p(a_1) > r, v_p(a_2) > r, v_p(a_1 a_2) = r + s$,
4) $D_{IV}(r,r;d;i_1,i_2)$(*resp.* $D_{IV}(r,s;d;i_1,i_2;\alpha)$) *if* $v_p(d) = 2r$(*resp.* $r + s$).
ii) *The above objects are all absolutely irreducible exept* $D_I(r,r;a,d,i)$ *for which the ring of endomorphisms is isomorphic to* $\mathbb{Q}_p[X]/(X^2 - aX + d)$.
iii) *Any irreducible 2-dimensional w.a. pst-module, which is* $F_2$-*semi-stable, is isomorphic to one and only one object of the lists 1)-4).*
iv) *If* $p \geq 5$, *any irreducible 2-dimensional w.a. pst-module is* $F_2$-*semi-stable.*

*Proof.* See App., §A. □

This theorem reduces the problem of finding the complete list of isomorphism classes of irreducible two dimensional $p$-adic potentially semi-stable representations of $G_p$ to finding out which are the pst-modules in the above list which are admissible (and conjecturally they all are). Each time that we discover that one of those $D$'s is admissible we will denote the corresponding representation $\underline{V}_{st}(D)$ (resp. the dual representation $\underline{V}^*_{st}(D) = \underline{V}_{st}(D)^*$) using the same notation but replacing $D$ with $V$ (resp. $V^*$) (e.g. $V_{IV}(r,s;d;i_1,i_2;\alpha) = \underline{V}_{st}(D_{IV}(r,s;d;i_1,i_2;\alpha))$ and $V^*_{IV}(r,s;d;i_1,i_2;\alpha)$ is the dual of this representation).

Given a two dimensional $p$-adic representation $V$ of $G_p$ and two integers $r \leq s$, we say that $V$ **is of Hodge-Tate type** $(r, s)$ if $V$ is Hodge-Tate and $\text{Hom}_{\mathbb{Q}_p[G_p]}(V, \mathbb{C}_p(i)) \neq 0$ if and only if $i \in \{r, s\}$. We see that if $V = \underline{V}^*_{st}(D)$ for some admissible pst-module $D$, than the Hodge-Tate type of $V$ is the same as the Hodge-Tate type of $D$. Using this fact and checking the list of $D$'s occurring in the previous theorem, we get the following result:

**Proposition 1.** *Let $V$ be a two dimensional irreducible $p$-adic representation of $G_p$ which is potentially semi-stable of Hodge-Tate type* $(r, s)$. *Then, we*

are in one and only one of the following cases:
1) There is a character of finite order $\nu$ such that $V(\nu)$ is crystalline, in which case there is a unique integer $i$ satisfying $0 \leq i < p-1$ such that one can choose $\nu = \hat{\xi}_1^i$ and a unique $(a,d) \in \mathbb{Q}_p \times \mathbb{Q}_p^*$ such that $D_I(r,s;a,d;i)$ is admissible and $V \simeq V_I^*(r,s;a,d;i)$;

2) The representation is not potentially crystalline, in which case there is a unique integer $i$ satisfying $0 \leq i < p-1$ such that $V(\hat{\xi}_1^i)$ is semi-stable and a unique $(b,c) \in \mathbb{Q}_p^* \times \mathbb{Q}_p$ such that $D_{II}(r,s;b,c;i)$ is admissible and $V \simeq V_{II}^*(r,s;b,c;i)$;

3) The representation is $F_1$-crystalline, but there is no character $\nu$ such that $V(\nu)$ is crystalline, in which case there is a unique couple $(i_1, i_2)$ of integers satisfying $0 \leq i_1 < i_2 \leq p-1$ and a unique $(a_1, a_2) \in (\mathbb{Q}_p^*)^2$ such that $D_{III}(r,s;a_1,a_2;i_1,i_2)$ is admissible and $V \simeq V_{III}^*(r,s;a_1,a_2;i_1,i_2)$;

4) The representation is not $F_1$-semi-stable, in which case it is $F_2$-crystalline, there is a unique couple $(i_1, i_2)$ of integers satisfying $0 \leq i_1 < i_2 \leq p-1$, a unique $d \in \mathbb{Q}_p^*$ and, if $r < s$ a unique $\alpha \in \mathbb{P}^1(\mathbb{Q}_p)$, such that, if $r = s$ (resp. $r < s$), $D_{IV}(r,r;d;i_1,i_2)$ (resp. $D_{IV}(r,s;d;i_1,i_2;\alpha)$) is admissible and $V \simeq V_{IV}^*(r,r;d;i_1,i_2)$ (resp. $V \simeq V_{IV}^*(r,s;d;i_1,i_2;\alpha)$).

From the fact that we have an equivalence of tannakian categories, one sees that, if $D$ is admissible, then $\wedge^2 D$ is also and that $\underline{V}_{st}(\wedge^2 D) \simeq \wedge^2 \underline{V}_{st}(D)$. An easy computation gives us the following result:

**Proposition 2.** *Let $D$ be one of the w.a. pst-modules listed in the previous theorem. Assume it is admissible and let $V = \underline{V}_{st}^*(D)$. Then $G_p$ acts on $\det V$, via $\eta_u \chi^{r+s} \hat{\xi}_1^{-j}$,*
1) *with $u = d/p^{r+s}$ and $j = 2i$ if $V = V_I^*(r,s;a,d,i)$,*
2) *with $u = b^2/p^{r+s-1}$ and $j = 2i$ if $V = V_{II}^*(r,s;b,c;i)$,*
3) *with $u = a_1 a_2/p^{r+s}$ and $j = i_1 + i_2$, if $V = V_{III}^*(r,s;a_1,a_2;i_1,i_2)$,*
4) *with $u = d/p^{r+s}$ and $j = i_1 + i_2$, if $V = V_{IV}^*(r,s;d;i_1,i_2;\alpha)$ or $V_{IV}^*(r,s;d;i_1,i_2)$ with $s = r$.*

## §12. Ramakrishna's theorem : crystalline representations of dimension 2 and their deformations.

We say that a $p$-adic representation $V$ of $G_p$ **is absolutely irreducible mod p** if there exists a $\mathbb{Z}_p$-lattice $T$ of $V$ stable under $G_p$ such that the $G_p$-module $T/pT$ is absolutely irreducible; we put $\overline{V} = T/pT$ and call it the **reduction mod p** of $V$. This is well defined, because the other lattices stable under $G_p$ are the $p^n T$, for $n \in \mathbb{Z}$. Of course, any $p$-adic representation which is absolutely irreducible mod $p$ is absolutely irreducible, but the converse is not true.

**Theorem B1.** *The $D$'s of type IV which $s - r \leq p - 1$ listed in theorem A are admissible. Moreover*
i) *If $d$ is a $p$-adic unit, if $X^2 - aX + d$ is irreducible over $\mathbb{Q}_p$ and if $\lambda$ is a root of this polynomial, $V_I^*(r,r;a,d;i)$ is isomorphic to a one dimensional $\mathbb{Q}_p(\lambda)$-vector space on which $G_p$ acts through the character $\eta_\lambda . \chi^r . \hat{\xi}_1^{-i}$;*

ii) If $1 \leq s-r \leq p-1$, $V_I^*(r,s;a,d;i)$ is absolutely irreducible mod $p$ and its reduction mod $p$ is isomorphic to $\overline{V}_{\iota,\varepsilon}$ where $\iota$ is the image of $rp + s - (p+1)i$ in $\mathbb{Z}/(p^2-1)\mathbb{Z}$ and $\varepsilon$ the image of $d/p^{r+s}$ in $\mathbb{F}_p^*$.

*Proof.* See App., §B. □

Let $\underline{\mathrm{Rep}}_{\mathbb{Q}_p}(G_p)$ be the category of (finite dimensional) $p$-adic representations of $G_p$ and $\underline{\mathrm{Rep}}_{\mathbb{Z}_p}^f(G_p)$ the category of $\mathbb{Z}_p$-modules of finite length equipped with a linear and continuous action of $G_p$. If $a \leq b$ are integers, we denote by $\underline{\mathrm{Rep}}_{\mathbb{Q}_p}(G_p)_{cr,[a,b]}$ the full subcategory of $\underline{\mathrm{Rep}}_{\mathbb{Q}_p}(G_p)$ consisting of those $V$ which are crystalline and such that $h^r(V)(= \dim_{\mathbb{Q}_p} gr^r \underline{D}_{HT}^*(V)) = 0$ if $r \notin [a,b]$. Denote also by $\underline{\mathrm{Rep}}_{\mathbb{Z}_p}^f(G_p)_{cr,[a,b]}$ the full subcategory of $\underline{\mathrm{Rep}}_{\mathbb{Z}_p}^f(G_p)$ whose objects are $T$'s for which one can find an object $V$ of $\underline{\mathrm{Rep}}_{\mathbb{Q}_p}(G_p)_{cr,[a,b]}$ such that $T$ is isomorphic to a subquotient of $V$. This is stable under subobjects, quotients, direct sums. Hence, for any $(\iota,\varepsilon)$ as in §9 such that $\overline{V}_{\iota,\varepsilon}$ is absolutely irreducible one can speak of the ring $R_{\iota,\varepsilon}(cr,[a,b])$ of the universal $\mathbb{Z}_p$-deformation of $\overline{V}_{\iota,\varepsilon}$ lying in $\underline{\mathrm{Rep}}_{\mathbb{Z}_p}^f(G_p)_{cr,[a,b]}$ as soon as $\overline{V}_{\iota,\varepsilon}$ itself is an object of this category.

**Theorem B2.** *Let $\overline{V} = \overline{V}_{\iota,\varepsilon}$ be absolutely irreducible.*
i) *The representation $\overline{V}$ lies in $\underline{\mathrm{Rep}}_{\mathbb{Z}_p}^f(G_p)_{cr,[0,p-1]}$;*
ii) *if $r,s$ are the unique integers such that $0 \leq r < s \leq p-1$ and the image of $r + ps$ mod $p^2 - 1$ is equal to $\iota$ or $p\iota$, there is a unique isomorphism $R_{\iota,\varepsilon}(cr,[0,p-1]) \simeq \mathbb{Z}_p[\![Y_1,Y_2]\!]$ such that, for any morphism $Y_1 \mapsto y_1, Y_2 \mapsto y_2$ of $R_{\iota,\varepsilon}(cr,[0,p-1])$ to $\mathbb{Z}_p$, if $T$ is representative of the corresponding isomorphism class of $\mathbb{Z}_p$-deformation, then*

$$\mathbb{Q}_p \otimes_{\mathbb{Z}_p} T \simeq V_I^*(r,s;p^r y_2, p^{r+s}[\varepsilon](1+y_1),0).$$

*Proof.* See App., §B. This theorem is a slight refinement of a result of Ramakrishna [23] who proved the fact that $R_{\iota,\varepsilon}(cr,[0,p-1]) \simeq \mathbb{Z}_p[\![Y_1,Y_2]\!]$. □

**Remark** If $0 < s - r \leq p - 1$, $u \in \mathbb{Z}$ and if $V_I^*(r,s;a,d;i)$ is as in theorem B1, then $V_I^*(r+u,s+u;a,d;i) \simeq V_I^*(r,s;p^u a, p^{2u} d; 0)(\chi^u \hat{\xi}_1^{-i})$. If $\overline{u}$ denotes the remainder of euclidean division of $u$ by $p-1$, the character $\chi^u \hat{\xi}_1^{-\overline{u}} = \chi_0^u$ takes values in $1 + p\mathbb{Z}_p$. For each $u \in \mathbb{Z}$, twisting the representation on $\mathbb{Z}_p[\![Y_1,Y_2]\!]$ which is the "universal deformation of $\overline{V}_{\iota,\varepsilon}$ for which all finite quotients lie in $\underline{\mathrm{Rep}}_{\mathbb{Z}_p}^f(G_p)_{cr,[0,p-1]}$" by $\chi_0^u$, we get the "universal deformation of $\overline{V}_{\iota,\varepsilon}$ for which all finite quotients $T$ are such that $T(\tilde{\xi}_1^{\overline{u}})$ lie in $\underline{\mathrm{Rep}}_{\mathbb{Z}_p}^f(G_p)_{cr,[u,u+p-1]}$". Any $V = \underline{V}_{st}^*(D)$ for $D$ absolutely irreducible of type I and Hodge-Tate type $(r',s')$ with $s'-r' \leq p-1$ such that $\overline{V} \simeq \overline{V}_{\iota,\varepsilon}$ can be obtained by a suitable specialization of one of those universal deformations.

## §13. Potentially Barsotti-Tate representations of dimension 2 and their deformations.

*We will only give a sketch of the proof of the main results of this section (see Appendix, §C). Details will be published elsewhere.*

Consider $(\iota, \varepsilon)$ where $\iota \in \mathbb{Z}/(p^2-1)\mathbb{Z}$ and $\varepsilon \in \mathbb{F}_p^*$. We denote by $V_{\iota,\varepsilon}$ the $p$-adic representation of $G_p$ which is the "canonical lifting" of $\overline{V}_{\iota,\varepsilon}$: this is $\mathbb{Q}_{p^2}$ on which $G_p$ acts through the homomorphism

$$\rho_{\iota,\varepsilon} : \operatorname{Gal}(L/\mathbb{Q}_p) \to \operatorname{Aut}_{\mathbb{Q}_p}(\mathbb{Q}_{p^2})$$

such that the restriction to $\operatorname{Gal}(L/L_0)$ consists in $\mathbb{Q}_{p^2}$-linear automorphisms given by the one-dimensional $\mathbb{Q}_{p^2}$-character $\hat{\xi}_2^\iota$, and $\tau$ acts on $\mathbb{Q}_{p^2}$ via the $\mathbb{Q}_{p^2}$-semilinear automorphism $x \mapsto [\zeta] \cdot \sigma x$ (where, as in §9, $\zeta^{p+1} = -\varepsilon$).

Let $\mathbb{Q}_{p^2}^{ab}$ be the maximal abelian extension of $\mathbb{Q}_{p^2}$ contained in $\overline{\mathbb{Q}}_p$ and $\theta : \mathbb{Q}_{p^2}^* \to \operatorname{Gal}(\mathbb{Q}_{p^2}^{ab}/\mathbb{Q}_{p^2})$ the inverse of the local reciprocity map[7]. The restriction $\theta_0$ of $\theta$ to the group of units is an isomorphism onto the inertia subgroup $\operatorname{In}(\mathbb{Q}_{p^2}^{ab}/\mathbb{Q}_{p^2})$ of $\operatorname{Gal}(\mathbb{Q}_{p^2}^{ab}/\mathbb{Q}_{p^2})$; as the natural map $I_p \to \operatorname{In}(\mathbb{Q}_{p^2}^{ab}/\mathbb{Q}_{p^2})$ is onto, the inverse of $\theta_0$ can be viewed as a character

$$\nu : I_p \to \mathbb{Q}_{p^2}^*.$$

Given $d \in \mathbb{Q}_p$ with $v_p(d) = 1$ and $\iota \in \mathbb{Z}/(p^2-1)\mathbb{Z}$, one sees easily that, up to conjugacy, there is one and only one continuous homomorphism

$$\rho_{LT,d,\iota} : G_p \to \operatorname{Aut}_{\mathbb{Q}_p}(\mathbb{Q}_{p^2})$$

("$LT$" for "Lubin-Tate") such that the restriction of $\rho_{LT,d,\iota}$ to $I_p$ is the character $\nu \cdot \hat{\xi}_2^\iota$ and the determinant of $\rho_{LT,d,\iota}$ is the character $\eta_d/p \cdot \chi \cdot \hat{\xi}_1^\iota$. We denote by $V_{LT,d,\iota}$ a chosen representative of the corresponding isomorphism class of $p$-adic representations. One sees that, if $(d', \iota',) \neq (d, \iota)$, then $V_{LT,d',\iota'} \not\simeq V_{LT,d,\iota}$ and that $V_{LT,d,\iota}$ is absolutely irreducible mod $p$ if and only if $p+1$ doesn't divide $\iota+1$ in which case $\overline{V}_{LT,d,\iota} \simeq \overline{V}_{\iota+1,\varepsilon}$ with $\varepsilon$ the image of $d/p$ in $\mathbb{F}_p^*$. also $V_{LT,d,0} \simeq V_{IV}^*(0,1;0,d;0) \simeq \mathbb{Q}_p \otimes_{\mathbb{Z}_p} T_p(J_d)$ where $J_d$ is a $p$-divisible group over $\mathbb{Z}_p$ which viewed as a $p$-divisible over the ring $\mathbb{Z}_{p^2}$ of the integers of $\mathbb{Q}_{p^2}$ is a Lubin-Tate formal group for $\mathbb{Q}_{p^2}$.

Observe that $D_{IV}(r,r;d;i_1,i_2) \simeq D_{IV}(0,0;p^{-2r}d;i_1,i_2)\{-r\}$, hence one of those modules is admissible if and only if the other is, in which case $V_{IV}^*(r,r;d;i_1,i_2) \simeq V_{IV}^*(0,0;p^{-2r}d;i_1,i_2)(r)$. When $r < s$, we have a similar statement for $D_{IV}(r,s;d;i_1,i_2;\alpha) \simeq D_{IV}(0,s-r;p^{-2r}d;i_1,i_2;p^r\alpha)\{-r\}$ with, in case of admissiblity, $V_{IV}^*(r,s;d;i_1,i_2;\alpha) \simeq V_{IV}^*(0,s-r;p^{-2r}d;i_1,i_2;p^r\alpha)(r)$. In particular, the first representation is absolutely irreducible mod $p$ if and only if the second is.

**Theorem C1.** *The $D$'s of type 4 with $s - r \leq 1$ listed in theorem A are admissible. Moreover, if $0 \leq i_1 < i_2 \leq p-1$ and if $\bar{\iota}$ denotes the image of $i_1 + p i_2$ in $\mathbb{Z}/(p^2-1)\mathbb{Z}$, then*

  i) *if $d = [\varepsilon] u^2$ with $\varepsilon \in \mathbb{F}_p^*$ and $u \in 1 + p\mathbb{Z}_p$, $V_{IV}^*(0,0;d;i_1,i_2) \simeq V_{-\bar{\iota},\varepsilon}(\eta_u)$ (hence this representation is absolutely irreducible mod $p$ and its reduction mod $p$ is isomorphic to $\overline{V}_{-\bar{\iota},\varepsilon}$),*

---

[7]hence the image of $\theta(p)$ in $\operatorname{Gal}(\overline{\mathbb{F}}_{p^2}/\mathbb{F}_{p^2})$ is the arithmetic Frobenius

ii) if $d = p[\varepsilon]u^2$ with $\varepsilon \in \mathbb{F}_p^*, u \in 1 + p\mathbb{Z}_p$ and if $\alpha \in \mathbb{P}^1(\mathbb{Q}_p)$, then $V = V_{IV}^*(0,1;d;i_1,i_2;\alpha)$ is absolutely irreducible mod p if and only if - either $v_p(\alpha) \geq 0$, in which case $\overline{V} \simeq \overline{V}_{1-\bar{i},\varepsilon}$ - or $v_p(\alpha) \leq -2$ and $i_2 - i_1 > 1$, in which case $\overline{V} \simeq \overline{V}_{1-p\bar{i},\varepsilon}$;

iii) if $v_p(d) = 1, V_{IV}^*(0,1;d;i_1,i_2;0) \simeq V_{LT,d,1-\bar{i}}$ and $V_{IV}^*(0,1;d;i_1,i_2;\infty) \simeq V_{LT,d,1-p\bar{i}}$.

We observed that all w.a. pst modules listed in theorem A with $s - r \leq 1$ are admissible because they are either of type I or of type IV. Say that a p-adic representation $V$ of $G_p$ is **potentially Barsotti-Tate** if one can find a finite extension $E$ of $\mathbb{Q}_p$ contained in $\overline{\mathbb{Q}}_p$ and a Barsotti-Tate group $J$ defined over the ring of integers of $E$ such that $V \simeq V_p(J) = \mathbb{Q}_p \otimes_{\mathbb{Z}_p} T_p(J)$ as a $\mathbb{Q}_p[\operatorname{Gal}(\overline{\mathbb{Q}}_p/E)]$-module, where $T_p(J)$ is the Tate module of $J$. If this is the case, one knows that $V$ is potentially crystalline with Hodge-Tate weights $\in \{0,1\}$. One conjectures the converse is true and this is clearly the case whenever 0 is the only weight (this means that the image of inertia is finite, hence that $J$ is étale) or whenever 1 is the only weight (this means that the image of inertia on $V(-1)$ is finite, hence that $J$ is of multiplicative type, i.e. is the Cartier dual of an étale Barsotti-Tate group).

Assume $V$ is irreducible of dimension 2. Then the Hodge-Tate type can be $(0,0), (1,1)$ or $(0,1)$. If it is $(0,0)$ this means that $V \simeq \underline{V}_{st}^*(D)$ with $D$ equal to one of the w.a. modules $D_I(0,0;a,d,i)$ or $D_{IV}(0,0;d;i_1,i_2)$ listed in theorem A. If it is $(1,1)$ this means that $V(-1)$ is of type $(0,0)$. The problem for type $(0,1)$ is solved by the next theorem, for which we need two more definitions:

**Definitions**: Let $J_0$ be a Barsotti-Tate group over $\mathbb{F}_p$. Then $J_0$ is equipped with two endomorphisms, the Frobenius $\varphi = F$ and the Verschiebung $V$ satisfying $\varphi V = V \varphi = p$. We say that $J_0$ **is strictly of slope** $1/2$ if there is an automorphism $u$ of $J_0$ (necessarily unique) such that $\varphi = Vu$.

Let $\mathcal{O}_E$ be the ring of the integers of a finite, totally ramified extension $E$ of $\mathbb{Q}_p$. We say that a Barsotti-Tate group $J$ over $\mathcal{O}_E$ **is strictly of slope** $1/2$ if its special fiber is strictly of slope $1/2$ and if moreover, given any invariant differential form $\omega$ on $J$, one can find differential forms $\omega_1, \omega_2$ with $\omega_1$ exact and $\lambda \in \mathcal{O}_E$ with $v_p(\lambda) \geq 1/2$ such that $\omega = \omega_1 + \lambda \omega_2$.

**Theorem C2.** *Assume $p \geq 5$. Let $V$ be an irreducible two dimensional p-adic representation of $G_p$ which is Hodge-Tate of Hodge-Tate type $(0,1)$. Then the following are equivalent:*

i) *$V$ is potentially Barsotti-Tate,*

ii) *$V$ is potentially crystalline,*

iii) *$V$ is potentially semi-stable,*

iv) *There is a w.a. pst module $D$ which is either one of the modules $D_I(0,1;a,d,i)$ or one of the modules $D_{IV}(0,1;d,i_1,i_2;\alpha)$ listed in Theorem A such that $V \simeq \underline{V}_{st}^*(D)$.*

*Moreover,*

a) *if $V \simeq V_I^*(0,1;a,d,i)$, there is a Barsotti-Tate group $J$ defined over $\mathbb{Z}_p$,* strictly of slope 1/2, such that $V \simeq V_p(J)(\hat{\xi}_1^{-i})$ when viewed as a $G_p$-module;

b) Let $\pi = (\pi_2)^{p-1}$ (hence $\pi^{p+1} = -p$) and $\hat{\xi}_\pi : \text{Gal}(\mathbb{Q}_p(\pi_2)/\mathbb{Q}_p(\pi)) \to \mathbb{Z}_p^*$ the character defined by $\hat{\xi}_\pi(g) = g\pi_2/\pi_2$. If $V \simeq V_{IV}^*(0,1;d;i_1,i_2;\alpha)$ there is a Barsotti-Tate group $J$ defined over the integers of $\mathbb{Q}_p(\pi)$, strictly of slope 1/2, such that $V \simeq V_p(J)(\hat{\xi}_\pi^{-i_1-i_2})$ when viewed as a $\text{Gal}(\overline{\mathbb{Q}}_p/\mathbb{Q}_p(\pi))$-module.

From now on, $\mathcal{O}_E = \mathbb{Z}_p[\pi]$ and $E = \mathbb{Q}_p(\pi)$. We denote by $\underline{\text{Rep}}_{\mathbb{Z}_p}^f(G_p)_{BT,E,1/2}$ the full subcategory of $\underline{\text{Rep}}_{\mathbb{Z}_p}^f(G_p)$ consisting of those $T$'s for which we can find a Barsotti-Tate group $J$ over $\mathcal{O}_E$ strictly of slope 1/2 such that $T$, when viewed as a $\text{Gal}(\overline{\mathbb{Q}}_p/E)$-module, is isomorphic to a quotient of $T_p(J)$). This category is stable under passage to subobjects, quotients, direct sums. Hence for any $(\iota, \varepsilon)$ as in §9 such that $\overline{V}_{\iota,\varepsilon}$ is absolutely irreducible, one can speak of the ring $R_{\iota,\varepsilon}(BT, E, 1/2)$ of the universal $\mathbb{Z}_p$-deformation of $\overline{V}_{\iota,\varepsilon}$ lying in $\underline{\text{Rep}}_{\mathbb{Z}_p}^f(G_p)_{BT,E,1/2}$ as soon as $\overline{V}_{\iota,\varepsilon}$ itself is an object of this category.

**Theorem C3.** Let $j_1, j_2$ be integers satisfying $0 \le j_1 < j_2 \le p-1$, $\iota$ the image of $j_1 + pj_2$ in $\mathbb{Z}/(p^2-1)\mathbb{Z}$, $\varepsilon \in \mathbb{F}_p^*$ and $\overline{V} = \overline{V}_{\iota,\varepsilon}$. Then $\overline{V}$ belongs to $\underline{\text{Rep}}_{\mathbb{Z}_p}^f(G_p)_{BT,E,1/2}$. One can build a $\mathbb{Z}_p[\![Y_1, Y_2]\!] \times_{\mathbb{F}_p} \mathbb{Z}_p[\![Y_1', Y_2']\!]$-deformation $T_{\iota,\varepsilon}(BT, E, 1/2)$ of $\overline{V}$ in such a way that all finite quotients lie in $\underline{\text{Rep}}_{\mathbb{Z}_p}^f(G_p)_{BT,E,1/2}$ and that

i) if $y_1, y_2 \in p\mathbb{Z}_p$ and if $T$ is the $\mathbb{Z}_p$-deformation of $\overline{V}$ obtained from $T_{\iota,\varepsilon}(BT, E, 1/2)$ via the map $Y_1 \mapsto y_1, Y_2 \mapsto y_2, Y_1' \to 0, Y_2' \to 0$, then

$$\mathbb{Q}_p \otimes_{\mathbb{Z}_p} T \simeq V_I^*(0, 1; p[\varepsilon](1+y_1); j_2) \text{ if } j_2 - j_1 = 1 \quad {}^8$$

( resp. $V_{IV}^*(0, 1; p[\varepsilon](1+y_1); j_1+1, j_2, p^{-1}y_2)$ if $j_2 - j_1 > 1$);

ii) if $y_1', y_2' \in p\mathbb{Z}_p$ and if $T$ is the $\mathbb{Z}_p$-deformation of $\overline{V}$ obtained from $T_{\iota,\varepsilon}(BT, E, 1/2)$ via the map $Y_1 \mapsto 0, Y_2 \mapsto 0, Y_1' \mapsto y_1', Y_2' \mapsto y_2'$, then

$$\mathbb{Q}_p \otimes_{\mathbb{Z}_p} T \simeq V_{IV}^*(0, 1; p[\varepsilon](1+y_1'); 0, j_1+1, p^{-1}y_2') \text{ if } j_2 = p-1$$

( resp. $V_{IV}^*(0, 1; p[\varepsilon](1+y_1'); j_1, j_2+1, 1/y_2')$ if $j_2 < p-1$).

**Remarks** :
1) We observe that each isomorphism class of $p$-adic representation $V$ as in Theorem C2 which is absolutely irreducible mod $p$ is isomorphic to $\mathbb{Q}_p \otimes_{\mathbb{Z}_p} T$ for one and only one $T$ obtained via the construction referred to in Theorem C3.

2) Assume $\iota$ and $\varepsilon$ are as in Theorem C3. From this theorem, we get a natural homomorphism

$$R_{\iota,\varepsilon}(BT, E, 1/2) \to \mathbb{Z}_p[\![Y_1, Y_2]\!] \times_{\mathbb{F}_p} \mathbb{Z}_p[\![Y_1', Y_2']\!]$$

It is likely that this map is onto with kernel killed by $p$.

---
[8] with the convention that $V_I^*(r, s; a, d; p-1) = V_I^*(r, s; a, d; 0)$.

3) With the above conventions, we see that the determinant of the considered representation is $\eta_{[\varepsilon](1+y_1)} \cdot \chi \cdot \hat{\xi}_1^{-i}$ in case (i) and $\eta_{[\varepsilon](1+y_1')} \cdot \chi \cdot \hat{\xi}_1^{-i}$ in case (ii). Thus, if we consider $\mathbb{Z}_p$-deformations of $\overline{V}_{\iota,\varepsilon}$ with a given determinant, we have one less parameter.

4) The families of type I and of type IV are of different nature : if $y_1, y_2$ or $y_1', y_2'$ are as above, we can look at the characteristic polynomial of Frobenius acting on the Dieudonné module of the Barsotti-Tate group $J$ defined over $\mathcal{O}_E$ : if $j_2 - j_1 = 1$, for $V_I^*(0, 1; y_2, p[\varepsilon](1+y_1); j_2)$, this is $X^2 - y_2 X + p[\varepsilon](1+y_1)$, hence if we fix this polynomial, there is only one representation; if we require this polynomial to have coefficients in $\mathbb{Q}$ with roots Weil numbers, there are only finitely many representations. For all the other cases, the characteristic polynomial is $X^2 + p[\varepsilon](1+y_1)$ (or $X^2 + p[\varepsilon](1+y_1')$) and $y_2$ (or $y_2'$) may vary.

## Appendix : Proof of results on potentially semi-stable representations

### §A : Proof of theorem A.

By twisting, we see that it is enough to consider the case where $r = 0$, i.e. the case of type $(0, s)$ with $s \in \mathbb{N}$.

(a) *The case of twisted $\mathbb{Q}_p$-semi-stable modules (type I and II)*: Assume $D$ is a two dimensional filtered $(\varphi, N, \Gamma_1)$-module of type $(0, s)$ which has the property that the action of $\Gamma_1 = \text{Gal}(F_1/\mathbb{Q}_p)$ on it is diagonal. There is a unique integer $i$ satisfying $0 \le i < p-1$ such that the action is given by $\hat{\xi}_1^i$. Twisting by $\hat{\xi}_1^{-i}$, we can assume the action of $\Gamma_1$ is trivial i.e. we can suppose that $D$ is just a $(\varphi, N)$-module, i.e. there is no action of $\Gamma_1$ and the filtration is defined on $D$. This $D$ is a two dimensional $\mathbb{Q}_p$-vector space and the Frobenius $\varphi$ acts linearly on it. The fact that the Newton polygon of its characteristic polynomial $X^2 - aX + d$ must be above the Hodge polygon and ends up at the same point means $v_p(d) = s$ and $a \in \mathbb{Z}_p$.

Suppose $s = 0$. We must have $N = 0$. Such a $D$ is determined by $(a, d)$, is weakly admissible and is irreducible if and only if there is no line in $D$ stable under $\varphi$, i.e. if $X^2 - aX + d$ is irreducible, putting us in case 1 of the statement of Theorem A, part (i).

Suppose now $s \ge 1$ and $N = 0$. Observe that $\Delta = \text{Fil}^s D$ can't be stable under $\varphi$. Otherwise $\Delta$ would be stable under $\varphi$ and $N$ and one could find $c \in \mathbb{Z}_p$ with $0 \le v_p(c) \le s$ such that $\varphi x = cx$ if $x \in \Delta$. But,

-if $v_p(c) < s$ the Newton polygon (of $\varphi$ acting on $\Delta$) is not above its Hodge polygon and $D$ woould not be weakly admissible;

-if $v_p(c) = s$, $D$ would be weakly admissible but $\Delta$ would be a proper weakly admissible sub-object and $D$ would not be irreducible.

Hence, if we choose a basis $e$ of $\Delta$, $\{e, \varphi e\}$ forms a basis of $D$. Now $D$ is weakly admissible. As $\Delta$ is not stable under $\varphi$, the possible proper weakly admissible sub-objects are lines $\Delta'$ stable under $\varphi$ such that $\varphi x = ux$ for $x \in \Delta'$ and $u$ a unit. This doesn't occur if and only if $p$ divides $a$ and we are in case $1'$ of the statement of Theorem A, part (i).

The last case to consider is $s \ge 1$ and $N \ne 0$. The condition $N\varphi = p\varphi N$ implies, that $\varphi$ must have two distinct eigenvalues $b, pb \in \mathbb{Z}_p$ with $v_p(b) + v_p(pb) = s$, hence $s$ must be odd and $v_p(b) = (s-1)/2$. If $e$ is a non-zero eigenvector corresponding to the eigenvalue $pb$, then $e$ and $Ne$ form a basis of $D$, and we have $\varphi e = pbe, \varphi Ne = b \cdot Ne, N^2 e = 0$. The unique proper subobject of $D$ is the $\mathbb{Q}_p$-subspace $\Delta'$ spanned by $Ne$ and weak admissibility amounts to requiring that $\Delta \neq \Delta'$. There is a unique $c \in \mathbb{Q}_p$ such that $\Delta$ is the line spanned by $e + c \cdot Ne$. Now $D$ is weakly admissible; it is irreducible if and only if $\Delta'$ is not weakly admissible, i.e. if $v_p(b) > 0$ which amounts to saying that $s \geq 3$ and we are in case 2 of the statement of Theorem A, part(i).

Hence we have proved that *the $D$'s of type I or II listed in theorem A are irreducible weakly admissible and that they exhaust the possibilities of $D$'s which are $F_1$-semi-stable with a diagonal action of $\Gamma_1$*.

(b) *The case of $F_1$-semi-stable modules which are not twists of $\mathbb{Q}_p$-semi-stable modules (type III)* : Let $D$ be a two dimensional $(\varphi, N, \Gamma_1)$-module of type $(0, s)$. The group $\Gamma_1$ acts linearly on $D$ and because it is cyclic of degree $p - 1$, this action is diagonalisable. To ask that it is not diagonal amounts to requiring that there are interes $0 \leq i_1 < i_2 < p - 1$ and two lines $D_1$ and $D_2$ in $D$ such that $gx = (\hat{\xi}_1(g))^{i_j} \cdot x$ for all $g \in \Gamma_1$ and all $x \in D_j, j \in \{1, 2\}$.

The Frobenius $\varphi$ acts linearly on $D$. The fact that $\varphi$ commutes with the action of $\Gamma_1$ means that $D_1$ and $D_2$ are stable under $\varphi$, hence there are elements $a_1, a_2 \in \mathbb{Q}_p$ such that $\varphi x = a_j \cdot x$ for all $x \in D_j$. The fact that the Newton polygon lies above the Hodge polygon and ends at the same point implies $a_1, a_2 \in \mathbb{Z}_p$ and $v_p(a_1) + v_p(a_2) = s$. As $N$ is nilpotent, the fact that it commutes with the action of $\Gamma_1$ implies $N = 0$.

Now, if $s = 0$, one sees easily that such a $D$ is weakly admissible, but $D_1$ and $D_2$ are proper sub-objects and $D$ is not irreducible. Assume $s \geq 1$. For $j = 1, 2$, the line $F_1 \otimes D_j$ is stable under the action of $\Gamma_1$, but if $\Delta$ were this line, either $v_p(a_j) < s$ and $D$ would not be weakly admissible or $v_p(a_j) = s$ and $D$ would be weakly admissible but $D_j$ would be a proper sub-object and $D$ would not be irreducible. Therefore $\Delta$ must be a line stable under $\Gamma_1$ and $\neq D_1, D_2$. Then it is easy to see that one can choose a basis $e_1$ of $D_1$ and a basis $e_2$ of $D_2$ in such a way that $\Delta$ is generated by $\pi_1^{i_2} \otimes e_1 + \pi_1^{i_1} \otimes e_1$. It is now easy to check that such a $D$ is weakly admissible. It will be irreducible if there is no proper sub-object. A proper sub-object must be stable under $\varphi$, hence must be $D_1$ or $D_2$. But, one sees that $D_j$ is a proper subobject if and only if $v_p(a_j) = 0$.

Hence we have proved that *the $D$'s of type III listed in theorem A are irreducible weakly admissible and that they exhaust the possibilities of $D$'s which are $F_1$-semi-stable with a non-diagonal action of $\Gamma_1$*.

(c) *The case of $F_2$-, but not $F_1$-semi-stable modules (type IV)* : Let $D$ be a two dimensional $(\varphi, N, \Gamma_2)$-module of type $(0, s)$. Let $D_0$ the sub-$\mathbb{Q}_p$-vector space of $D$ consisting of those $x$ such that $\bar{\tau}x = x$. One deduces easily from the fact that $\bar{\tau}$ acts semi-linearly on the $\mathbb{Q}_{p^2}$-vector space $D$ that $D_0$ is of dimension 2 over $\mathbb{Q}_p$ and spans $D$ as a $\mathbb{Q}_{p^2}$-vector space. The fact that $\varphi$ commutes with the action of $\bar{\tau}$ is then equivalent to the fact that the restriction of $\varphi$ to $D_0$ is a $\mathbb{Q}_p$-linear automorphism of $D_0$.

The group $I\Gamma_2$ is cyclic of order $p^2 - 1$ and acts linearly on $D$ and, if we want $D$ not to be semi-stable, the inertia group of $F_2/F_1$, which is the subgroup of index $p + 1$, cannot act trivially. Therefore $I\Gamma_2$ acts on $D$ through characters and at least one of them is of order not dividing $p - 1$. This means that one can find a line $D'$ of $D$ and an integer $i$ satisfying $0 \leq i < p^2 - 1$ and $i$ not divisible by $p+1$ such that $gx = \hat{\xi}_2^i(g) \cdot x$ if $g \in I\Gamma_2$ and $x \in D'$. Then, $g(\varphi x) = \varphi(gx) = \varphi(\hat{\xi}_2^i(g) \cdot x) = \hat{\xi}_2^{pi}(g) \cdot \varphi x$; but $i$ not divisible by $p+1$ means $pi \neq i \pmod{p^2 - 1}$ which implies that $\varphi x \in D'$; we see that $D = D' \oplus D''$ with $gy = \hat{\xi}_2^{pi}(g) \cdot y$ if $g \in |\Gamma_2$ and $y \in D''$. Permuting $i$ and $pi$ mod $p^2 - 1$ if necessary, we can assume that $i = i_1 + pi_2$ and $pi \equiv i_2 + pi_1 \pmod{p^2 - 1}$ with integers $i_1, i_2$ satisfying $0 \leq i_1 < i_2 \leq p - 1$.

Now if $x \in D'$, we have $g\bar{\tau}x = \bar{\tau}g^p x = \bar{\tau}(\hat{\xi}_2^{pi}(g) \cdot x) = \hat{\xi}_2^i(g) \cdot \bar{\tau}x$, hence $\bar{\tau}$ leaves $D'$ fixed; similarly it leaves also $D''$ fixed. If $x \in D_0$ and if $x = x' + x''$, with $x' \in D'$ and $x'' \in D''$, we then see that $\bar{\tau}x' = x'$ and $\bar{\tau}x'' = x''$, hence we can write $D_0 = D'_0 \oplus D''_0$ with $D'_0 = D' \cap D_0$ and $D''_0 = D'' \cap D_0$. If $x' \in D'_0$, we must have $g\varphi x' = \varphi g x' = \varphi(\hat{\xi}_2^i(g) \cdot x') = \hat{\xi}_2^{pi}(g) \cdot \varphi x'$, hence $\varphi x' \in D''_0$. Similarly $\varphi^2 x' \in D'_0$. Summarizing, we see that we can find $d \in \mathbb{Q}_p$ and a basis $e_1, e_2$ of $D$ such that

$$\bar{\tau}e_1 = e_1, \bar{\tau}e_2 = e_2, ge_1 = \hat{\xi}_2^{i_1+pi_2}(g)e_1 \text{ and } ge_2 = \hat{\xi}_2^{i_2+pi_1}(g)e_2$$

for $g \in I\Gamma_2$, $\varphi e_1 = e_2$, $\varphi e_2 = -de_1$.

As $N$ is nilpotent, the fact that it commutes with the action of $I\Gamma_2$ implies $N = 0$. Now we have actions of $\varphi, N$ and $\text{Gal}(F_2/\mathbb{Q}_p)$ satisfying the required properties. If we want $D$ to be weakly admissible, the fact that the Hodge polygon lies below the Newton polygon and that they end up at the same point amounts to requiring that $v_p(d) = s$.

If $s = 0$, we then have a weakly admissible module. There is no line stable under $\varphi$ and $\Gamma_2$, hence $D$ is irreducible and we get $D \simeq D_{IV}(0,0;d;i_1,i_2)$.

If $s \geq 1$, $\Delta$ can be any line of $D_{F_2}$ stable under the action of $\Gamma_2$. One sees immediately that there is a unique $\alpha \in \mathbb{P}^1(\mathbb{Q}_p)$ such that this is the line generated by $\pi_2^{(p-1)i_1} \otimes e_1 + \alpha \cdot \pi_1^{(p-1)i_2} \otimes e_2$. Again because there is no line of $D$ stable under $\varphi$ and $\Gamma_2$, this $D$ is weakly admissible and does not contain any proper weakly admissible subobject. Hence we have proved that *the $D$'s of type IV listed in theorem A are irreducible weakly admissible and that they exhaust the possibilities of $D$'s which are $F_2$-semi-stable but not $F_1$-semi-stable.*

(d) The only thing which is left to prove is the fact that, if $p \geq 5$, *any irreducible 2-dimensional weakly admissible pst-module is $F_2$-semi-stable.* : One can assume that this module is $F$-semi-stable, for $F$ a finite Galois extension of $\mathbb{Q}_p$ containing $F_2$ and contained in $\overline{\mathbb{Q}_p}$. The inertia group $\text{In}(F/\mathbb{Q}_p)$ of the extension $F/\mathbb{Q}_p$ acts linearly on the corresponding two dimensional $F_0$-vector space $D$. If $g \in \text{In}(F/\mathbb{Q}_p)$, the fact that the action of $g$ commutes with the action of $\varphi$ implies that the characteristic polynomial of $g$ has coefficients in $\mathbb{Q}_p$. Because $p \geq 5$, $[\mathbb{Q}_p(\sqrt[p]{1}) : \mathbb{Q}_p] > 2$ and any element of $\text{In}(F/\mathbb{Q}_p)$ of order a power of $p$ acts trivially; so the p-Sylow group $P$ of $\text{In}(F/\mathbb{Q}_p)$ acts trivially. Replacing $F$ by $F^P$, one can assume $F/\mathbb{Q}_p$ tame, i.e. that $\text{In}(F/\mathbb{Q}_p)$ is cyclic of order prime to $p$; if $g$ is a generator of this group, the roots of the polynomial characteristic of $g$ acting on $D$ are in $\mathbb{Q}_{p^2}$, hence of order dividing $p^2 - 1$. Therefore $g^{p^2-1}$ acts trivially and the result follows from the fact that this element generates the inertia group of $F/F_1$. $\square$

## §B : Proof of theorems B1 and B2.

(a) *Proof of theorem* B1: By twisting, it is enough to prove the theorem for $r = 0$ and $i = 0$. In this case we observe that $D_I(0, s; a, d; 0)$ can be viewed as a weakly admissible filtered $\varphi$-module over $\mathbb{Q}_p$ ($N = 0$ and the Galois group action is trivial) which satisfies $\mathrm{Fil}^0 D = D$ and $\mathrm{Fil}^p D = 0$ hence is admissible ([15],[30]).

To prove (i) we are reduced to checking that $V_I^*(0, 0; a, d; 0)$ is isomorphic to a one dimensional $\mathbb{Q}_p(\lambda)$-vector space on which $G_p$ acts through the unramified character $\eta_\lambda$. But $\mathrm{Fil}^0 B_{\mathrm{cris}}$ and $\mathrm{Fil}^0 B_{\mathrm{st}}$ contain the completion $\hat{\mathbb{Q}}_p^{nr}$ of the maximal unramified extension of $\mathbb{Q}_p$ contained in $\overline{\mathbb{Q}}_p$ and hence we have an injective map $\mathrm{Hom}_{\mathbb{Q}_p[\varphi]}(D_I(0, 0; a, d; 0), \hat{\mathbb{Q}}_p^{nr}) \to V_I^*(0, 0; a, d; 0)$, which is an isomorphism for reasons of dimensions. The assertion is then an easy exercise.

(ii) is a consequence of theorem B2 which will be prove below. □

(b) *The category* $\underline{MF}^f_{]-p+1,0]}$ *and the functor* $\overline{V}_{cr}$: Let $\underline{\mathrm{Rep}}_{\mathbb{Q}_p}(G_p)_{\mathrm{cr},[0,p-1[}$ be the full subcategory of $\underline{\mathrm{Rep}}_{\mathbb{Q}_p}(G_p)_{\mathrm{cr},[0,p-1]}$ consisting of those $V$ which have no non-trivial subobject $V'$ such that $V'(-p+1)$ is unramified. Similarly denote by $\underline{\mathrm{Rep}}^f_{\mathbb{Z}_p}(G_p)_{\mathrm{cr},[0,p-1[}$ the full subcategory of $\underline{\mathrm{Rep}}^f_{\mathbb{Z}_p}(G_p)_{\mathrm{cr},[0,p-1]}$ consisting of those $T$ which are isomorphic to a subquotient of an object of $\underline{\mathrm{Rep}}_{\mathbb{Q}_p}(G_p)_{\mathrm{cr},[0,p-1[}$. It is easy to see that, for any $V$ in $\underline{\mathrm{Rep}}_{\mathbb{Q}_p}(G_p)_{\mathrm{cr},[0,p-1]}$, one can find a short exact sequence
$$0 \to V' \to V \to V'' \to 0$$
such that $V'(-p+1)$ is unramified and $V''$ is in $\underline{\mathrm{Rep}}_{\mathbb{Q}_p}(G_p)_{\mathrm{cr},[0,p-1[}$.

Let $\overline{V}$ be a 2-dimensional $\mathbb{F}_p$-vector space on which the action of the inertia group is irreducible. We see that, for any local artinian $\mathbb{Z}_p$-algebra $A$ of residue field $\mathbb{F}_p$, any $A$-deformation of $\overline{V}$ which lies in $\underline{\mathrm{Rep}}^f_{\mathbb{Z}_p}(G_p)_{\mathrm{cr},[0,p-1]}$ actually lies in $\underline{\mathrm{Rep}}_{\mathbb{Q}_p}(G_p)_{\mathrm{cr},[0,p-1[}$.

For $a, b \in \mathbb{Z}$ satisfying $a \leq b$, let $\underline{MF}^f_{[a,b]}$ be the following category:
- the objects are $\mathbb{Z}_p$-modules $M$ of finite length, equipped with
i) a filtration of $M$ by sub-$\mathbb{Z}_p$-modules
$$M = \mathrm{Fil}^a M \supset \mathrm{Fil}^{a+1} M \supset \cdots \supset \mathrm{Fil}^j M \supset \cdots \supset \mathrm{Fil}^b M \supset \mathrm{Fil}^{b+1} M = 0;$$
ii) for each $j$, a $\mathbb{Z}_p$-linear map $\varphi^j : \mathrm{Fil}^j M \to M$ such that $\varphi^{j+1}(x) = p\varphi^j(x)$ if $x \in \mathrm{Fil}^{j+1} M$, and such that $M = \Sigma_{a \leq j \leq b} \varphi^j(\mathrm{Fil}^j M)$.
- the morphisms are $\mathbb{Z}_p$-linear maps compatible with all the structures.

Recall [16] that this is an abelian category and that for any object $M$ the $\mathrm{Fil}^j M$'s are direct summands (as $\mathbb{Z}_p$-modules). If $a < b$, we denote by $\underline{MF}_{]a,b]}$ the full subcategory of $\underline{MF}_{[a,b]}$ consisting of those $M$ which have no non-trivial subobject $N$ such that $\mathrm{Fil}^{a+1} N = 0$.

Let $\overline{k}$ be the residue field of $\overline{\mathbb{Q}}_p$ and $\sigma$ the absolute Frobenius acting on $\overline{k}$ (via $x \mapsto x^p$) and on the ring $W(\overline{k})$ of Witt vectors with coefficients in $\overline{k}$ by functoriality. Let $A_{\mathrm{cris}}$ be the ring constructed in [11]. Recall that this is a $W(\overline{k})$-algebra which is a domain equipped with i) an action of $G_p$ compatible with the ring structure and with the obvious action on $W(\overline{k})$, ii) a Frobenius $\varphi : A_{\mathrm{cris}} \to A_{\mathrm{cris}}$ compatible with the ring structure, commuting with the action of $G_p$ and $\sigma$-semi-linear, iii) a decreasing filtration ($\text{Fil}^j A_{\text{cris}})_{j\in\mathbb{N}}$ by ideals, which are direct summands as $W(\overline{k})$-modules, stable under $G_p$.

Moreover, for $0 \leq j \leq p-1$, we have $\varphi(\text{Fil}^j M) \subset p^j \text{Fil}^j M$, therefore, one can define $\varphi^j : \text{Fil}^j A_{\text{cris}} \to A_{\text{cris}}$ by $\varphi^j x = p^{-j}\varphi x$.

For any object $M$ of $\underline{MF}^f_{[0,p-1]}$ and $0 \leq j \leq p-1$ we can use the natural map $\text{Fil}^j A_{\text{cris}} \otimes_{\mathbb{Z}_p} \text{Fil}^{-j} M \to M$, which is injective, to identify $\text{Fil}^j A_{\text{cris}} \otimes_{\mathbb{Z}_p} \text{Fil}^{-j} M$ with a sub-$A_{\text{cris}}$-module of $A_{\text{cris}} \otimes_{\mathbb{Z}_p} M$ and define $\text{Fil}^0 (A_{\text{cris}} \otimes_{\mathbb{Z}_p} M) = \Sigma_{0 \leq j \leq p-1} \text{Fil}^j A_{\text{cris}} \otimes_{\mathbb{Z}_p} \text{Fil}^{-j} M$. It is easy to see that there is a unique $\sigma$-semi-linear map

$$\varphi : \text{Fil}^0(A_{\text{cris}} \otimes_{\mathbb{Z}_p} M) \to A_{\text{cris}} \otimes_{\mathbb{Z}_p} M$$

such that, for $0 \leq j \leq p-1$, $\varphi(\lambda \otimes x) = \varphi^j \lambda \otimes \varphi^{-j} x$ if $\lambda \in \text{Fil}^j A_{\text{cris}}$ and $x \in \text{Fil}^j M$.

Recall ([16], see also [30]) that if for $M$ in $\underline{MF}_{[-p+1,0]}$, we define

$$\underline{V}_{\text{cr}}(M) = \{v \in \text{Fil}^0(A_{\text{cris}} \otimes_{\mathbb{Z}_p} M) | \varphi v = v\},$$

then $\underline{V}_{\text{cr}}(M)$ is a finite $\mathbb{Z}_p$-representation of $G_p$. If $M$ is in $\underline{MF}_{]-p+1,0]}$, this is an object of $\underline{\text{Rep}}^f_{\mathbb{Z}_p}(G_p)_{\text{cr},[0,p-1[}$, which has the same length as $M$ as a $\mathbb{Z}_p$-module. The functor

$$\underline{V}_{\text{cr}} : \underline{MF}^f_{]-p+1,0]} \to \underline{\text{Rep}}^f_{\mathbb{Z}_p}(G_p)_{\text{cr},[0,p-1[}$$

defined in this way induces an equivalence between those two categories[9]. We denote by $\underline{D}_{\text{cr}}$ a quasi-inverse.

(c) *The mod p representation*: Recall that $r, s \in \mathbb{Z}$ with $0 \leq r < s \leq p-1$ and $\varepsilon \in \mathbb{F}_p^*$. We give to the two dimensional $\mathbb{F}_p$-vector space $(\mathbb{F}_p)^2$ (with $\overline{u}_1, \overline{u}_2$ the canonical basis) the structure of an object $\overline{M} = \overline{M}(r, s; \varepsilon)$ of $\underline{\text{Rep}}^f_{\mathbb{Z}_p}(G_p)_{\text{cr},[0,p-1[}$ by defining

$$\text{Fil}^{-s}\overline{M} = \overline{M}, \text{Fil}^{-s+1}\overline{M} = \text{Fil}^{-r}\overline{M} = \mathbb{F}_p \overline{u}_1, \text{Fil}^{-r+1}\overline{M} = 0,$$

$$\varphi^{-r}\overline{u}_1 = \overline{u}_2, \varphi^{-s}\overline{u}_2 = -\varepsilon^{-1} \cdot \overline{u}_1.$$

It is not hard to check (compare with [16]): i) that one can identify $\underline{V}_{\text{cr}}(\overline{M})$ with the residue field $\mathbb{F}_{p^2}$ of $\mathbb{Q}_{p^2}$ in such a way that the inertia subgroup $I_p$ acts via the character $\xi_2^{r+ps}$; ii) that the determinant of the action of $G_p$ is the character $\eta_\varepsilon \chi^{r+ps}$; all together this proves that $\underline{V}_{\text{cr}}(\overline{M}) \simeq \underline{V}_{i,\varepsilon}$ and this proves the first part of the assertion (ii) of Theorem B1.

(d) Let's prove the following lemma:

**Lemma.** *Let $A$ be a local artinian ring, $\mathfrak{S}$ an abelian $A$-linear category, $F, G$ two covariant functors from $\mathfrak{S}$ to the category of $A$-modules of finite length which are $A$-linear, exact, faithful and such that, for any object $N$ of $\mathfrak{S}$, the $A$-modules $F(N)$ and $G(N)$ have the same length. Then, if $d \in \mathbb{N}$ and $M$ is an object of $\mathfrak{S}$, $F(M)$ is free of rank $d$ over $A$ if and only $G(M)$ is.*

Indeed, for $a \in A$ and $N$ an object of $\mathfrak{S}$, call $[a]_N$ the endomorphism of $N$ which is the action of $a$. Let $m_A$ be the maximal ideal of $A$ and $k = A/m_A$. The

---

[9]actually, what is defined in [16] is a contravariant version of this construction but the passage from one construction to the other is straightforward.

faithfulness of $F$ implies that, for any object $N$ of $\mathfrak{S}$, one has $[a]_N = 0$ for all $a \in m_A$ if and only if $F(N)$ is a $k$-vector space. Now let $\overline{M}$ be the biggest quotient of $M$ such that $[a]_{\overline{M}} = 0$ for all $a \in m_A$. It is clear that $F(\overline{M}) = F(M)/m_A F(M)$ and $G(\overline{M}) = G(M)/m_A G(M)$. Assume that $F(M)$ is free of rank $d$. Then $\dim_k F(\overline{M}) = d$. Hence we have also $\dim_k G(\overline{M}) = d$ and if $e_1, e_2, \cdots, e_d$ are lifting in $G(M)$ of a basis of $G(\overline{M})$ over $k$, the $e_j$ generate $G(M)$. But they are linearly independent over $A$, for otherwise, $\text{length}_A G(M) < d \cdot \text{length}_A A = \text{length}_A F(M)$.

(e) *Deformations*: Let $A$ be a local artinian $\mathbb{Z}_p$-algebra. Any $A$-representation $T$ of $V$ can be viewed as a $\mathbb{Z}_p$-representation together with an imbedding of $A$ into the ring of the endomorphisms of $T$. Therefore, if $T$ is an object of $\underline{\text{Rep}}^f_{\mathbb{Z}_p}(G_p)_{\text{cr},[0,p-1[}$, $\underline{D}_{\text{cr}}(T)$ is an **$A$-object of** $\underline{MF}^f_{]-p+1,0]}$; that is, there is a natural structure of $A$-module on it, the filtration is given by sub-$A$-modules and the maps $\varphi^r$ are $A$-linear. Conversely, given any $A$-object $M$ of $\underline{MF}^f_{]-p+1,0]}$, $\underline{V}_{\text{cr}}(M)$ is an $A$-representation of $G_p$. Therefore, for any $A$, the functor $\underline{V}_{\text{cr}}$ induces an equivalence between the category of $A$-objects of $\underline{MF}^f_{]-p+1,0]}$ and the category of $A$-objects in $\underline{\text{Rep}}^f_{\mathbb{Z}_p}(G_p)_{\text{cr},[0,p-1[}$. Applying the lemma to the first of those two categories and to the functors which associate to $M$ respectively the underlying $A$-module and the $A$-module underlying $\underline{V}_{\text{cr}}(M)$, we see, that for any $A$-object $M$ of $\underline{MF}^f_{]-p+1,0]}$, $\underline{V}_{\text{cr}}(M)$ is flat as an $A$-module if and only if $M$ is.

Now let $V$ be an $A$-deformation of $\overline{V}$ and $M = \underline{D}_{\text{cr}}(V)$; if $m_A$ is the maximal ideal of $A$, one can identify $M/m_A M$ with $\overline{M}$; as $M$ must be a free-$A$-module, any couple $\{u_1, u_2\}$ of elements of $M$ lifting $\{\overline{u}_1, \overline{u}_2\}$ is a basis of $M$ over $A$; one knows [16] that any morphism in $\underline{MF}^f_{]-p+1,0]}$ is strictly compatible with the filtrations. This implies that $\text{Fil}^{-s} M = M$ and that one can choose $u_1 \in \text{Fil}^{-r} M$; as $\varphi^{-r} \overline{u}_1 = \overline{u}_2$, one can choose $u_2 = \varphi^{-r} u_1$; then we must have $\varphi^{-s} u_2 = -[\varepsilon^{-1}](1 + x_1) u_1 + x_2 u_2$ with $x_1, x_2 \in m_A$.

Now $\text{Fil}^{-r} M$ contains the free-$A$-module of rank one spanned by $u_1$. It can't be bigger, for otherwise one could find a nonzero $\lambda \in A$ such that $\lambda u_2 \in \text{Fil}^{-r} M$; multiplying $\lambda$ by a suitable power of $p$, one could assume $p\lambda = 0$; but then $p^{s-r} \cdot \varphi^{-r}(\lambda u_2) = \varphi^{-s}(\lambda u_2) = \lambda(-[\varepsilon^{-1}](1+x_1)u_1 + x_2 u_2) \neq 0$ because $\lambda$ is a unit in $A$; on the other hand, $p^{s-r} \varphi^{-r}(\lambda u_2) = p^{s-r-1} \varphi^{-r}(p\lambda \cdot u_2) = 0$ giving us a contradiction. So we see that

$$\text{Fil}^{-s} M = M, \ \text{Fil}^{-s+1} M = \text{Fil}^{-r} M = Au_1, \ \text{Fil}^{-r+1} M = 0,$$

$$\varphi^{-r} u_1 = u_2, \quad \varphi^{-s} u_2 = -[\varepsilon^{-1}](1+x_1) u_1 + x_2 u_2.$$

Now if we change the lifting of $u_1$, it must be to the element of the form $\lambda u_1$ with $\lambda$ a unit in $A$; then $\varphi^{-r}(\lambda u_1) = \lambda u_2$ and $\varphi^{-s} u_2 = -[\varepsilon^{-1}](1 + x_1) \cdot \lambda u_1 + x_2 \cdot \lambda u_2$ and we see that $x_1$ and $x_2$ depends only on the isomorphism class of $M$.

Conversely it is clear that if we now identify $\{u_1, u_2\}$ with the canonical basis of $A^2$ we have equipped $A^2$, with the structure of an $A$-object $M = M_A(r, s, \varepsilon; x_1, x_2)$ of $\underline{MF}^f_{]-p+1,0]}$ and that

$$V_A(r, s, \varepsilon; x_{,1}, x_2) = \underline{V}_{\text{cr}}(M(r, s, \varepsilon; x_{,1}, x_2))$$

is an $A$-deformation of $\overline{V}$ and that for any $A$-deformation $V$ of $\overline{V}$, there is a unique $(x_1, x_2) \in m_A$ such that $V \simeq V(r, s, \varepsilon; x_{,1}, x_2)$.

Now let $\mathfrak{a}$ be the set of artinian quotients of the ring $\mathbb{Z}_p[\![X_1, X_2]\!]$. For any $A \in \mathfrak{a}$, consider $M_A = M_A(r, s, \varepsilon; x_{,1}, x_2)$ and $V_A = V_A(r, s, \varepsilon; x_1, x_2)$ with $x_i = $ the image of $x_i$ in $A$. Then the $M_A$'s form a projective system of objects of $\underline{MF}^f_{]p+1,0]}$ and therefore the $V_A$'s form a projective system of finite representations of $G_p$ which are in $\underline{\operatorname{Rep}}^f_{\mathbb{Z}_p}(G_p)_{\operatorname{cr},[0,p-1[}$. Then it is clear that $\lim\cdot\operatorname{proj}\cdot_{A \in \mathfrak{a}} V_A$ is a free $\mathbb{Z}_p[\![X_1, X_2]\!]$-module of rank two equipped with an action of $G_p$ which gives an identification of $\mathbb{Z}_p[\![X_1, X_2]\!]$ with $R_{i,\varepsilon}(\operatorname{cr}, [0, p-1])$. Now if we put $Y_1 = (1+x)^{-1} - 1$ and $Y_2 = [\varepsilon](1 + x_1)^{-1}x_2$, we have $\mathbb{Z}_p[\![X_1, X_2]\!] = \mathbb{Z}_p[\![Y_1, Y_2]\!]$.

(f) *End of the proof of* (ii): It is easy to see that if we send $\mathbb{Z}_p[\![X_1, X_2]\!]$ to $\mathbb{Z}_p$ via $X_1 \mapsto x_1, X_2 \mapsto x_2$ if $y_1 = (1 + x_1)^{-1} - 1$ and $y_2 = -[\varepsilon](1 + x_1)^{-1}x_2$ and if $T$ and $V$ are as in the statement of the theorem, then $V$ is crystalline and that the dual $D$ of $D^* = \underline{D}^*_{\operatorname{cris}}(V) = \underline{D}_{\operatorname{st}, \mathbb{Q}_p}(V)$ is the two dimensional $\mathbb{Q}_p$-vector space $D = \mathbb{Q}_p u_1 + \mathbb{Q}_p u_2$ with

$$\operatorname{Fil}^{-s} D = D, \quad \operatorname{Fil}^{-s+1} D = \operatorname{Fil}^{-r} D = \mathbb{Q}_p u_1, \quad \operatorname{Fil}^{-r+1} D = 0,$$

and $\varphi u_1 = p^{-r} u_2, \quad \varphi u_2 = p^{-s}(-[\varepsilon^{-1}](1 + x_1)u_1 + x_2 u_2)$.

Hence $D$ is of Hodge-Tate type $(-s, -r)$ and the characteristic polynomial of $\varphi$ acting on $D$ is $X^2 - p^{-s}x_2 \cdot X + p^{-r-s}[\varepsilon]^{-1}(1 + x_1)$. Therefore, $D^*$ is of Hodge-Tate type $(r, s)$ and the characteristic polynomial of $\varphi$ acting on $D^*$ is $X^2 - p^r x_2[\varepsilon](1 + x_1)^{-1} \cdot X + p^{r+s}[\varepsilon](1 + x_1)^{-1}$; hence $D^* \simeq D_I(r, s; p^r x_2[\varepsilon](1 + x_1)^{-1}, p^{r+s}[\varepsilon](1 + x_1)^{-1} = D_I(r, s; p^r y_2, p^{r+s}[\varepsilon](1 + y_1) \cdot 0)$.

## §C : About the proof of theorems C1, C2 and C3:

a) *Proof of theorem C1 for $s = r$*: Twisting if necessary, we are reduced to case $r = s = 0$ and this is an easy exercise on representations of $G_p$ on which the action of $I_p$ is finite. □

b) *About theorem C2* : Let $V$ be as in theorem C2. The inplications i) $\Rightarrow$ ii) $\Rightarrow$ iii) are well known. If $V$ is potentially semi-stable, there is a $D$ listed in theorem A which is of Hodge-Tate type $(0, 1)$ and admissible such that $V \simeq V^*_{\operatorname{st}}(D)$. Looking at the list, we see that $D$ is either one of the modules $D_I(0, 1; a, d; i)$ or one of the modules $D_{IV}(0, 1; d, i_1, i_2, \alpha)$, hence iii) $\Rightarrow$ iv).

**Remark** : Before giving a sketch of proof of the other statements, let us explain how one could easily deduce theorem C2 and the admissibility statement of theorem C1 from results of Laffaille : We already know (th.B1) that any $D = D_I(0, 1; a, d; i)$ is admissible; moreover $V^*_I(0, 1; a, d; i)(\hat{\xi}^i_1) \simeq V^*_I(0, 1; a, d; 0)$ is a crystalline representation of $G_p$ of Hodge-Tate type $(0, 1)$ and [21] gives that this representation comes from a Barsotti-Tate group defined over $\mathbb{Z}_p$; using the fact that the $p$-adic valuation of each root of the characteristic polynomial of $\varphi$ acting on $D$, which is $X^2 - aX + d$ is $1/2$, it is easy to see that $\Gamma$ is strictly of slope $1/2$, hence we get *the assertion (a) of theorem C2*.

Similarly, let $D$ be any of the $D_{IV}(0,1;d,i_1,i_2,\alpha)$'s of theorem A. Denote by $D_0$ the sub-$\mathbb{Q}_p$-vector space fixed under $\overline{\tau}$. Observe that the filtration on $D_{F_2}$ is actually defined on $E$ and that $D_0$ equipped with the action of $\varphi$ and the filtration on $E \otimes D_0$ can be viewed as a "weakly admissible filtered $\varphi$-module over $E$" of dimension 2 and of Hodge-Tate type $(0,1)$. Laffaille's [20] gives that this module is admissible and that there is a Barsotti-Tate group $J$ defined over $\mathcal{O}_E$ such that the associated $p$-adic representation of $\text{Gal}(\overline{\mathbb{Q}}_p/\mathbb{Q}_p)$, $\text{Hom}_{\text{filt-}\varphi\text{-mod}}(D_0, B_{\text{cris}})$, is isomorphic to $V_p(J)$. From that, we easily get the admissibility of $D$ and *the assertion (b) of theorem C2*, except for the fact that one can choose $J$ strictly of slope $1/2$; this latter fact can be shown by a direct computation. By twisting, we get the *admissibility of all* $D_{IV}(r, r+1; d, i_1, i_2, \alpha)$'s. From the above discussion we have that all the w.a. modules $D_I(0,1;a,d;I)$ and $D_{IV}(0,1;d,i_1,i_2,\alpha)$ are actually admissible and that the corresponding representations are potentially Barsotti-Tate, hence iv) $\Rightarrow$ i).

*Sketch of the proof of the other statements*:

1. - If $S$ is any scheme, let us call a **$p$-group scheme over** $S$ any inductive system $(J_n)_{n \in \mathbb{N}}$ of finite and flat commutative group schemes over $S$ such that map $J_n \to J_{n+1}$ identifies $J_n$ with the kernel of the multiplication by $p^n$ in $J_{n+1}$. Thus, if the map $J_n \to J_{n+1}$ is also an isomorphism for $n$ big enough, we have a **finite $p$-group scheme**; that is, a finite and flat commutative group scheme killed by a power of $p$. If there is an integer $h$ such that $J_n$ is free of rank $p^{nh}$ for all $n$, we get a **Barsotti-Tate group**.

2. - Any formal group, and therefore also any $p$-group scheme, over $\mathbb{F}_p$ is equipped with two endomorphisms, the Frobenius $F = \varphi$ and the Verschiebung $V$ satisfying $\varphi V = V\varphi = p$.

A **slope 1/2 structure** on a $p$-group scheme $J$ over $\mathbb{F}_p$ is an automorphism $u$ of $J$ such that $\varphi = Vu$. If $J$ is a Barsotti-Tate group over $\mathbb{F}_p$, there is at most one slope 1/2 structure $u$ on $J$ and there is such a $u$ if and only if $J$ is strictly of slope 1/2.

The $p$-group schemes over $\mathbb{F}_p$ with slope 1/2 structure form, in an obvious way an additive category $\underline{pGS}_{\mathbb{F}_p,1/2}$ which turns out to be abelian.

Forgetting $u$ we get an additive functor from $\underline{pGS}_{\mathbb{F}_p,1/2}$ to the category $\underline{pGS}_{\mathbb{F}_p}$ of $p$-group schemes over $\mathbb{F}_p$ which is exact and faithful; if $(J_1, u)$ and $(J_2, u)$ are two objects of $\underline{pGS}_{\mathbb{F}_p,1/2}$, the cokernel of the injective map

$$\text{Hom}_{\underline{pGS}_{\mathbb{F}_p,1/2}}((J_1,u)(J_2,u)) \to \text{Hom}_{\underline{pGS}_{\mathbb{F}_p}}(J_1, J_2)$$

is killed by $p$.

3. - Let $CW$ be the formal group of covectors over $\mathbb{F}_p$ ([8], chap.II, §4). Let us recall that, for any finite $\mathbb{F}_p$-algebra $A$, $CW(A)$ consists of covectors $a = (\cdots, a_{-n}, \cdots, a_{-1}, a_0)$ with $a_{-n} \in A$ for all $n$ and $a_{-n} \in r_A$ for almost all $n$, where $r_A$ denotes the radical of $A$; with the obvious notation, we have $a + b = c$, with

$$c_{-n} = S_m(a_{-n-m}, \cdots, a_{-n-1}, a_{-n}; b_{-n-m}, \cdots, b_{-n-1}, b_{-n}) \text{ for } m >> 0,$$

where the $S_m$'s are the polynomials which define the addition on Witt vectors. We have also

$$\varphi a = (\cdots, (a_{-n})^p, \cdots, (a_{-1})^p, (a_0)^p) \text{ and } V_a = (\cdots, a_{-n-1}, \cdots, a_{-2}, a_{-1}).$$

To any $p$-group scheme $J$ over $\mathbb{F}_p$, one can associate its "contravariant Dieudonné-module" $\underline{M}(J) = \text{Hom}(J, CW)$. One can view $\underline{M}$ as a contravariant additive functor from the category of $p$-group schemes over $\mathbb{F}_p$ to the category of $\mathbb{Z}_p[\varphi, V]$-modules which are $\mathbb{Z}_p$-modules of finite type. This is an anti-equivalence of categories. A quasi-inverse to $\underline{M}$ is given by associating to such a $\mathbb{Z}_p[\varphi, V]$-module $M$ the $p$-group scheme $\underline{J}(M)$ defined by $\underline{J}(M)(A) = \text{Hom}_{\mathbb{Z}_p[\varphi, V]}(M, CW(A))$ for all finite $\mathbb{F}_p$-algebras $A$ ([8], chap.III).

Denote by $\mathcal{O}_{1/2}$ the ring $\mathbb{Z}_p[u, u^{-1}, \varphi]/(\varphi^2 - pu)$. Setting $V = \varphi u^{-1}$, we can view $\mathbb{Z}_p[F, V]$ as a subring of $\mathcal{O}_{1/2}$. If $(J, u)$ is an object of $\underline{pGS}_{\mathbb{F}_p, 1/2}$, then $\underline{M}(J)$ has a natural structure of $\mathcal{O}_{1/2}$-module; we get in this way an anti-equivalence between $\underline{pGS}_{\mathbb{F}_p, 1/2}$ and the category of $\mathcal{O}_{1/2}$-modules which are $\mathbb{Z}_p$-modules of finite type.

4. - For any finite $\mathbb{F}_p$-algebra $A$, denote $BW_{1/2}(A)$ the set of the $a = (a_n)_{n \in \mathbb{Z}}$ with $a_n \in r_A$ for all $n$. With obvious the notation, one sees that, for all fixed $n \in \mathbb{Z}$, if $a, b \in BW_{1/2}(A)$, the sequence

$$S_m((a_{n-m})^{p^m}, \cdots, (a_{n-1})^p, a_n; (b_{n-m})^{p^m}, \cdots, (b_{n-1})^p, b_n) \text{ for } m \in \mathbb{N}$$

is stationary; if we denote by $c_n$ its limit and if we set

$$a + b = c = (c_n)_{n \in \mathbb{Z}},$$

$BW_{1/2}(A)$ becomes an abelian group and $BW_{1/2}$ may be viewed as a commutative formal group over $\mathbb{F}_p$ which is equipped with an automorphism $u$ defined by

$$u((a_n)_{n \in \mathbb{Z}}) = (a_{n+1})_{n \in \mathbb{Z}}.$$

We have $\varphi(a_{n \in \mathbb{Z}}) = ((a_n)^p)_{n \in \mathbb{Z}}$ and $V = \varphi u^{-1}$.

We have also a natural morphism from $BW_{1/2}$ to $CW$ sending $(a_n)_{n \in \mathbb{Z}}$ to $(\cdots, (a_{-n})^{p^n}, \cdots, (a_{-1})^p, a_0)$. If $(J, u)$ is an object of $\underline{pGS}_{\mathbb{F}_p, 1/2}$, this map induces an isomorphism from $\text{Hom}((J, u), BW_{1/2})$ to the $\mathcal{O}_{1/2}$-module $\underline{M}(J)$ and we use it to identify these two modules.

5. - We denote by $\underline{pGS}_{\mathcal{O}_E}$ the additive category of $p$-group schemes over $\mathcal{O}_E$. The couples $(J, u)$ consisting of a $p$-group scheme $J$ over $\mathcal{O}_E$ and a slope $1/2$ structure $u$ on the special fiber $J_{\mathbb{F}_p}$ are in an obvious way the objects of an additive category $\underline{pGS}_{\mathcal{O}_E, 1/2}$. We denote by $\underline{pGS}_{\mathcal{O}_E, 1/2, \text{strict}}$ the full subcategory of $\underline{pGS}_{\mathcal{O}_E, 1/2}$ consisting of those $(J, u)$, for which one can find two Barsotti-Tate groups $J'$ and $J''$ over $\mathcal{O}_E$, strictly of slope $1/2$ and morphisms $J \to J', J' \to J''$ of $p$-group schemes such that the sequence

$$0 \to J \to J' \to J'' \to 0$$

is exact and the diagram

$$\begin{array}{ccccccccc} 0 & \to & J_{\mathbb{F}_p} & \to & J'_{\mathbb{F}_p} & \to & J''_{\mathbb{F}_p} & \to & 0 \\ & & \downarrow u & & \downarrow u' & & \downarrow u'' & & \\ 0 & \to & J_{\mathbb{F}_p} & \to & J'_{\mathbb{F}_p} & \to & J''_{\mathbb{F}_p} & \to & 0 \end{array}$$

(where $u'$ (resp. $u''$) is the unique slope 1/2 structure which exists on $J'_{\mathbb{F}_p}$ (resp. $J''_{\mathbb{F}_p}$)) is commutative.

6. - Set $\mathcal{O}_{E,1/2} = \mathcal{O}_E[v, v^{-1}]$. The ring $\mathcal{O}_E \otimes_{\mathbb{Z}_p} \mathcal{O}_{1/2}$ is a domain and, setting $1 \otimes \varphi = \pi^{(p+1)/2} \cdot v$ and $1 \otimes u = -v^2$, we identify $\mathcal{O}_{E,1/2}$ with the normalization of $\mathcal{O}_E \otimes_{\mathbb{Z}_p} \mathcal{O}_{1/2}$ in its field of fractions. Also $\mathcal{O}_{E,1/2}$ is a faithfully flat $\mathcal{O}_{1/2}$-algebra.

Define the **category** $\underline{MF}_{E,1/2}$ as follows:
- an object is a couple $(M, \wedge)$ where $M$ is an $\mathcal{O}_{1/2}$-module, which is of finite type as a $\mathbb{Z}_p$-module, and $\wedge$ is a sub $\mathcal{O}_E$-module of $\mathcal{O}_{E,1/2} \otimes_{\mathcal{O}_{1/2}} M$ such that the map from $\wedge \oplus \wedge$ to $\mathcal{O}_{E,1/2} \otimes_{\mathcal{O}_{1/2}} M$ sending $(x, y)$ to $x + vy$, is an isomorphism;
- a morphism $(M, \wedge) \to (M', \wedge')$ is an $\mathcal{O}_{1/2}$-linear map from $M$ to $M'$ such that the $\mathcal{O}_{E,1/2}$-linear map induced by scalar extension sends $\wedge$ into $\wedge'$.

7. - Now, for any finite and flat $\mathcal{O}_E$-algebra $\mathfrak{a}$, there is a unique $\mathcal{O}_E$-linear map

$$\lambda_{\mathfrak{a}} : \mathcal{O}_{E,1/2} \otimes_{\mathcal{O}_{1/2}} BW_{1/2}(\mathfrak{a}/\pi\mathfrak{a}) \to \mathfrak{a}[1/p]/\pi\mathfrak{a}$$

such that, if $a = (a_n)_{n \in \mathbb{Z}} \in BW_{1/2}(\mathfrak{a}/\pi\mathfrak{a})$ and if $\hat{a}_n$ is a lifting of $a_n$ in $\mathfrak{a}$, then $\lambda_{\mathfrak{a}}(1 \otimes a) = \Sigma_{m \in \mathbb{N}} p^{-m}(\hat{a}_{-m})^{p^{2m}} (\bmod \pi\mathfrak{a})$ and $\lambda_{\mathfrak{a}}(v \otimes a) = \pi^{-(p+1)/2} \cdot \Sigma_{m \in \mathbb{N}} p^{-m}(\hat{a}_{-m})^{p^{2m+1}} (\bmod \pi\mathfrak{a})$.

Let $(M, \wedge)$ be an object of $\underline{MF}_{E,1/2}$. Then $M$ defines an object $(J_{\mathbb{F}_p}, u)$ of $\underline{pGS}_{\mathbb{F}_p,1/2}$ : for any finite $\mathbb{F}_p$-algebra $A$,

$$J_{\mathbb{F}_p}(A) = \mathrm{Hom}_{\mathcal{O}_{1/2}}(M, BW_{1/2}(A)).$$

If for any finite and flat $\mathcal{O}_E$-algebra $\mathfrak{a}$, we define

$$J(\mathfrak{a}) = \{\alpha \in \mathrm{Hom}_{\mathcal{O}_{1/2}}(M, BW_{1/2}(\mathfrak{a}/\pi\mathfrak{a})) | \wedge \subset \mathrm{Ker}\ \lambda_{\mathfrak{a}} \circ (id \otimes \alpha)\},$$

we get a functor $J$ from the category of finite and flat $\mathcal{O}_E$-algebras to abelian groups; one can check that $J$ is actually a $p$-group scheme over $\mathcal{O}_E$. Moreover $(J, u)$ is an object of $\underline{pGS}_{\mathcal{O}_E,1/2,\mathrm{strict}}$.

The correspondence $(M, \wedge) \mapsto (J = \underline{J}(M, \wedge), u = \underline{u}(M))$ defines a contravariant additive functor

$$\underline{Ju} : \underline{MF}_{E,1/2} \to \underline{pGS}_{\mathcal{O}_E,1/2,\mathrm{strict}},$$

which turns out to be an anti-equivalence of categories. If the $\mathbb{Z}_p$-module underlying $M$ is free, $J$ is a Barsotti-Tate group.

8. - Let $\mathcal{O}'_{1/2}$ be the noncommutative ring generated over the commutative ring $\mathbb{Z}_{p^2}[u, u^{-1}]$ by an element $\varphi$ with relations $\varphi^2 = pu, \varphi u = u\varphi$ and $\varphi a = \sigma a \cdot \varphi$, if $a \in \mathbb{Z}_{p^2}$ (where $\sigma$ is the Frobenius). This ring contains $\mathbb{Z}_{p^2}$ and $\mathcal{O}_{1/2}$ as commutative subrings and can be identified as a $\mathbb{Z}_p$-module with

$\mathbb{Z}_{p^2} \otimes_{\mathbb{Z}_p} \mathcal{O}_{1/2}$. For any finite $\mathbb{F}_{p^2}$-algebra $A$, there is unique structure of $\mathbb{Z}_{p^2}$-module on $BW_{1/2}(A)$ such that $[\varepsilon] \cdot (a_n)_{n \in \mathbb{Z}} = (\varepsilon a_n)_{n \in \mathbb{Z}}$ if $\varepsilon \in \mathbb{F}_{p^2}$ and $(a_n)_{n \in \mathbb{Z}} \in BW_{1/2}(A)$. Together with the structure of $\mathcal{O}_{1/2}$-module, $BW_{1/2}(A)$ becomes an $\mathcal{O}'_{1/2}$-module. Moreover, for any $\mathcal{O}_{1/2}$-module $M$, $\mathbb{Z}_{p^2} \otimes_{\mathbb{Z}_p} M$ has a natural structure of $\mathcal{O}'_{1/2}$-module and the obvious map

$$\mathrm{Hom}_{\mathcal{O}_{1/2}}(M, BW_{1/2}(A)) \to \mathrm{Hom}_{\mathcal{O}'_{1/2}}(\mathbb{Z}_{p^2} \otimes_{\mathbb{Z}_p} M, BW_{1/2}(A))$$

is an isomorphism: we use it to identify these two groups.

9. - Let $F = \mathbb{Q}_{p^2}(\pi)$ be the Galois closure of $E$ in $\overline{\mathbb{Q}}_p$; this is a subfield of $F_2 = \mathbb{Q}_{p^2}(\pi_2)$ and $\Gamma_2 = \mathrm{Gal}(F/\mathbb{Q}_p)$ acts on it. Recall (§11) that we have defined an isomorphism $\hat{\xi}$ from the inertia subgroup $I\Gamma$, of $\Gamma_2$ onto the group $\mu_{p^2-1}(\mathbb{Q}_{p^2})$ and called $\overline{\tau}$ the only nontrivial element of $\mathrm{Gal}(F_2/\mathbb{Q}_p(\pi_2))$.

Consider now a couple $(M, gr)$ where $M$ is an $\mathcal{O}_{1/2}$-module which is a $\mathbb{Z}_p$-module of finite type, and $gr$ is a gradation on $M$ indexed by $\mathbb{Z}/(p+1)\mathbb{Z}$,

$$M = \bigoplus_{s \in \mathbb{Z}/(p^2-1)\mathbb{Z}} gr^s M,$$

by sub-$\mathbb{Z}_p[u, u^{-1}]$-modules, such that $\varphi(gr^s M) \subset gr^{ps} M$ for all $s$. The group $\Gamma_2$ acts naturally on $\mathcal{O}'_{1/2} = \mathbb{Z}_{p^2} \otimes_{\mathbb{Z}_p} \mathcal{O}_{1/2}$ (via $g(a \otimes b) = a \otimes b$ if $g \in I\Gamma_2$ and $\overline{\tau}(a \otimes b) = \sigma a \otimes b$). We can define a semi-linear action of $\Gamma_2$ on the $\mathcal{O}'_{1/2}$-module $\mathbb{Z}_{p^2} \otimes_{\mathbb{Z}_p} M$ by setting, if $a \in \mathbb{Z}_{p^2}$ and $x \in gr^s M$,

$$\overline{\tau}(a \otimes x) = \sigma a \otimes x \text{ and } g(a \otimes x) = (\hat{\xi}_2(g))^s \cdot a \otimes x \text{ for all } g \in I\Gamma_2.$$

Let $\overline{\mathbb{Z}}_p$ be the integral closure of $\mathbb{Z}_p$ in the chosen algebraic closure $\overline{\mathbb{Q}}_p$ of $\mathbb{Q}_p$. Define $BW_{1/2}(\overline{\mathbb{Z}}_p/\pi)$ to be the inductive limit of the $BW_{1/2}(\mathcal{O}_{F'}/\pi\mathcal{O}_{F'})$ for $F'$ running through finite Galois extensions of $\mathbb{Q}_p$ contained in $\overline{\mathbb{Q}}_p$ and containing $F$. Set

$$J_{\mathbb{F}_p}(\overline{\mathbb{Z}}_p/\pi) = \mathrm{Hom}_{\mathcal{O}_{1/2}}(M, BW_{1/2}(\overline{\mathbb{Z}}_p/\pi)) =$$
$$\mathrm{Hom}_{\mathcal{O}'_{1/2}}(\mathbb{Z}_{p^2} \otimes_{\mathbb{Z}_p} M, BW_{1/2}(\overline{\mathbb{Z}}_p/\pi)).$$

This abelian group is equipped with an action of $G_p$: if

$$u : \mathbb{Z}_{p^2} \otimes_{\mathbb{Z}_p} M \to BW_{1/2}(\overline{\mathbb{Z}}_p/\pi)$$

is an $\mathcal{O}'_{1/2}$-linear map and if $\gamma \in G_p$, then $\gamma(u) = \gamma \circ u \circ \gamma^{-1}$.

10. - The group $\Gamma_2$ acts naturally on $\mathcal{O}_F$ and on $\mathbb{Z}_{p^2} \otimes_{\mathbb{Z}_p} \mathcal{O}_{E,1/2} = \mathbb{Z}_{p^2} \otimes_{\mathbb{Z}_p} \mathcal{O}_E[v, v^{-1}] = \mathcal{O}_F[v, v^{-1}]$ (with $gv = v$ if $g \in I\Gamma_2$ and $\overline{\tau}v = -v$). If $(M, gr)$ is as above, the action of $\Gamma_2$ on $\mathbb{Z}_{p^2} \otimes_{\mathbb{Z}_p} M$ extends uniquely to a semi-linear action on $\mathbb{Z}_{p^2} \otimes_{\mathbb{Z}_p} (\mathcal{O}_{E,1/2} \otimes_{\mathcal{O}_{1/2}} M)$. Define the category $\underline{MF}_{E/\mathbb{Q}_p, 1/2}$ as follows:

- an object is a triple $(M, \wedge, gr)$ with $(M, \wedge)$ an object of $\underline{MF}_{E/\mathbb{Q}_p, 1/2}$ and $(M, gr)$ as above such that $\mathbb{Z}_{p^2} \otimes_{\mathbb{Z}_p} \wedge \subset \mathbb{Z}_{p^2} \otimes_{\mathbb{Z}_p} (\mathcal{O}_{E,1/2} \otimes_{\mathcal{O}_{1/2}} M)$ is stable under $\Gamma_2$;

- a morphism is a morphism of the underlying objects of $\underline{MF}_{E,1/2}$ which is compatible with the gradations.

This category is abelian.

If $(M, \wedge, gr)$ is an object of $\underline{MF}_{E/\mathbb{Q}_p, 1/2}$, and if $J$ is the $p$-group scheme over $\mathcal{O}_E$ associated to $(M, \wedge)$, one can view $J(\overline{\mathbb{Z}}_p)$ ( = the inductive limit of the $J(\mathcal{O}_{F'})$ for $F'$ describing the finite Galois extensions of $\mathbb{Q}_p$ contained in $\overline{\mathbb{Q}}_p$ and containing $F$) as a subgroup of the group $J_{\mathbb{F}_p}(\overline{\mathbb{Z}}_p/\pi)$ defined above and the condition that $\wedge$ is stable under $\Gamma$ implies that $J(\overline{\mathbb{Z}}_p)$ is stable under $G_p$.

If $\underline{\text{Rep}}_{p-\text{tor}}(G_p)$ denotes the category of $p$-torsion abelian groups $V$ such that the kernel of multiplication by $p$ is finite, equipped with a linear and continuous action of $G_p$, we can see $J(\overline{\mathbb{Z}}_p)$ as an object of this category.

The correspondence $(M, \wedge, gr) \mapsto J(\overline{\mathbb{Z}}_p)$ can be viewed as a contravariant additive functor

$$\underline{J}_{1/2} : \underline{MF}_{E/\mathbb{Q}_p, 1/2} \to \underline{\text{Rep}}_{p-\text{tor}}(G_p).$$

11. - The function $\underline{J}_{1/2}$ is exact and faithful. Moreover, if $(M, \wedge, gr)$ and $(M', \wedge', gr')$ are two objects of $\underline{MF}_{E/\mathbb{Q}_p, 1/2}$, the cokernel of the map

$$\text{Hom}((M, \wedge, gr), (M', \wedge', gr')) \to \text{Hom}(\underline{J}_{1/2}(M, \wedge, gr), \underline{J}_{1/2}(M', \wedge', gr'))$$

is killed by $p$.

If the $\mathbb{Z}_p$-module underlying $M$ is free and if $J = \underline{J}(M, \wedge)$ is the corresponding Barsotti-Tate group over $\mathcal{O}_E$, $\underline{J}_{1/2}(M, \wedge, gr) = J(\overline{\mathbb{Z}}_p)$, when viewed as a $\text{Gal}(\overline{\mathbb{Q}}_p/F_2)$-module. Therefore,

$$V = V_p((\underline{J}_{1/2})(M, \wedge, gr)) = \mathbb{Q}_p \otimes_{\mathbb{Z}_p} \text{lim.proj} \cdot \underline{J}_{1/2}(M, \wedge, gr)_{p^n}$$

is a potentially Barsotti-Tate $p$-adic representation of $G_p$, hence a fortiori is potentially crystalline. It is $F_2$-semi-stable; the corresponding admissible filtered $(\varphi, N, \Gamma_2)$-module $D = \underline{D}^*_{\text{st}, F_2}(V)$ can be identified with $\underline{D}(M, \wedge, gr)$ defined as follows: the underlying $\mathbb{Q}_{p^2}$-vector space is $\mathbb{Q}_{p^2} \otimes_{\mathbb{Z}_p} M = \mathbb{Q}_p \otimes_{\mathbb{Z}_p} (\mathbb{Z}_{p^2} \otimes_{\mathbb{Z}_p} M)$ with the given action of $\varphi$ and of $\Gamma_2$ and with $N = 0$; the filtration on

$$D_{F_2} = F_2 \otimes_{\mathbb{Q}_{p^2}} D = F_2 \otimes_E (E \otimes_{\mathbb{Q}_{p^2}} D) = F_2 \otimes_{\mathcal{O}_E} (\mathcal{O}_{E, 1/2} \otimes_{\mathcal{O}_{1/2}} M)$$

is given by $\text{Fil}^0 D_{F_2} = D_{F_2}$, $\text{Fil}^1 D_{F_2} = F_2 \otimes_{\mathcal{O}_E} \wedge$, $\text{Fil}^2 D_{F_2} = 0$.

12. - With these results in mind, the proof of the theorems becomes an exercise in the category $\underline{MF}_{E/\mathbb{Q}_p, 1/2}$:

a) For each w.a. pst-module $D = D_I(0, 1; a, d, i)$ or $D = D_{IV}(0, 1; d, i_1, i_2; \alpha)$ one exhibits an object $(M, \wedge, gr)$ of $\underline{MF}_{E/\mathbb{Q}_p, 1/2}$ such that $D \simeq \underline{D}(M, \wedge, gr)$; this gives us statement a) and b) of the theorem C2, hence also, as we already explained in the above remark, the admissibility statement of theorem C1 and the implication iv) $\Rightarrow$ i) of theorem C2, whose proof is completed. □

b) To this $(M, \wedge, gr)$, we can associate

$$T_p((\underline{J}_{1/2})(M, \wedge, gr)) = \text{lim.proj} \cdot \underline{J}_{1/2}(M, \wedge, gr)_{p^n}$$

which can be identified with a lattice $T$ of $V = \underline{V}^*_{\text{st}}(D)$ stable under $G_p$. Reducing $(M, \wedge, gr)$ mod $p$ we can compute explicitly the two dimensional $\mathbb{F}_p$-representation $\overline{V} = T/pT = \underline{J}_{1/2}((M, \wedge, gr) \text{ mod } p)$ of $G_p$ and check the assertions of theorem C1 for $(r, s) = (0, 1)$. By twisting we deduce the assertions for $s - r = 1$ and the proof of theorem C1 is completed. □

c) To prove the theorem C3, we consider the category $\mathfrak{M}$ opposite to the full subcategory of $\underline{MF}_{E/\mathbb{Q}_p,1/2}$ whose objects are torsion objects. We can view an object of $\mathfrak{M}$ as a $\mathbb{Z}_p$-module $N$ of finite length and an object $(M, \wedge, gr)$ of $\underline{MF}_{E/\mathbb{Q}_p,1/2}$ such that the $\mathbb{Z}_p$-module underlying $M$ is $N^\wedge = \mathrm{Hom}_{\mathbb{Z}_p}(N, \mathbb{Q}_p/\mathbb{Z}_p)$ the Pontrjagin dual of $N$. Now, $\underline{J}_{1/2}$ can be viewed as a *covariant* additive exact functor from $\mathfrak{M}$ to $\underline{\mathrm{Rep}}^f_{\mathbb{Z}_p}(G_p)$. The proof then consists of playing the same game on $\mathfrak{M}$ that we played with the category $\underline{MF}^f_{]-p+1,0]}$ to prove the theorem B2 (see Appendix, §B). □

*References*

[1] V.A. Abrashkin. *Modular representations of the Galois group of a local field and generalization of Shafarevich conjecture*, Math. USSR Izv., 35 (1990), pp. 469-518.

[2] N. Boston. *Some Cases of the Fontaine-Mazur Conjecture*, J.Number Theory 42 (1992), pp. 285-291.

[3] P. Deligne.*Les constantes de l'équation fonctionnelle des fonctions L*, pp. 501-595 in *Modular functions of one variable II*, Lecture Notes in Math. 349, Springer-Verlag, Berlin (1973).

[4] P. Deligne. *Motifs et groupes de Taniyama.* pp. 261-279 in *Hodge cycles, motives and Shimura varieties*, Lecture Notes in Math. 900, Spring-Verlag, Berlin (1982).

[5] G. Faltings. *Endlichkeitssätze für abelsche Varietäten über Zahlkörpern*, Inv. Math. 73 (1983), pp. 349-366.

[6] G. Faltings. *Crystalline cohomology and p-adic Galois representations*, pp. 25-80 in *Algebraic Analysis, Geometry and Number Theory*, the Johns Hopkins University Press (1989).

[7] J.-M. Fontaine. *Sur la décomposition des algèbres de groupe*, Annales Scient. E.N.S., 4 (1971), pp. 121-180.

[8] J.-M. Fontaine. *Groupes p-divisibles sur les corps locaux*, Astérisque 47-48, Soc. Math. de France, Paris (1977).

[9] J.-M. Fontaine. *Modules galoisiens, modules filtrés et anneaux de Barsotti-Tate*, pp. 3-80 in *Journées de géométrie algébrique de Rennes (III)*, Astérisque 65, Soc. Math. de France, Paris (1979).

[10] J.-M. Fontaine. *Arbeitstagung report 1988,* Geometric l-adic Galois representations.

[11] J.-M. Fontaine. *Le corps des périodes p-adiques*, pp. 59-101 in *Périodes p-adiques*, Astérisque 223, Soc. Math. de France, Paris (1994).

[12] J.-M. Fontaine. *Représentations p-adiques semi-stables*, pp. 113-184, in *Périodes p-adiques*, Astérisque 223, Soc. Math. de France, Paris (1994).

[13] J.-M. Fontaine. *Représentations l-adiques potentiellement semi-stables*, pp.321-347, in *Périodes p-adiques*, Astérisque 223, Soc. Math. de France, Paris (1994).

[14] J.-M. Fontaine. *Schémas propres et lisses sur $\mathbb{Z}$*, pp. 43-56, in Proceedings of the Indo-French Conference on Geometry, Hindustan Book Agency, New Dehli (1993).

[15]   J.-M. Fontaine, L. Illusie. *p-adic periods: a survey*, pp. 57-93, in Proceedings of the Indo-French Conference on Geometry, Hindustan Book Agency, New Dehli (1993).

[16]   J.-M. Fontaine, G. Laffaille. *Construction de représentations p-adiques*, Ann. Scient. E.N.S., 15 (1982), pp. 547-608.

[17]   J.-M. Fontaine, B. Perrin-Riou. *Autour des conjectures de Bloch et Kato: cohomologie galoisienne et valeurs de fonctions L*, pp. 599-706 in *Motives*, Proceedings of Symposia in Pure Maths. 55, part I (1994).

[18]   F. Hajir. *On the growth of p-class groups in p-class field towers*, preprint, Calif. Inst. Tech. (1995).

[19]   L. Illusie. *Cohomologie de de Rham et cohomologie étale p-adique*, Séminaire Bourbaki, exp. 726, juin 1990.

[20]   G. Laffaille. *Construction de groupes p-divisibles: le cas de dimension 1*, pp.103-123 in Journées de géométrie algébrique de Rennes (III), Astérisque 65, Soc. Math. de France, Paris (1979).

[21]   G. Laffaille. *Groupes p-divisibles et modules filtrés: le cas peu ramifié*, Bull. Soc. Math. 108 (1980), pp. 187-206.

[22]   B. Mazur. *Deforming Galois representations* pp. 385-437 in *Galois group over* $\mathbb{Q}$, M.S.R.I. Publications, Spring-Verlag, Berlin 1989.

[23]   R. Ramakrishna. *On a variation of Mazur's deformation functor*, Compositio Math., 87 (1993), pp. 269-286.

[24]   S. Sen. *Lie algebras of Galois groups arising from Hodge-Tate modules*, Ann. of Math. (1973), pp. 160-170.

[25]   S. Sen *Continuous cohomology and p-adic galois representtions*, Inv. Math. 82 (1980), pp. 89-116.

[26]   J.-P. Serre. *Représentations l-adiques*, pp. 177-193 in Kyoto Int. Symposium on Algebraic Number Theory, Japan Soc. for the Promotion of Science (1977) (oeuvres, t.III, pp. 384-400).

[27]   J.-P. Serre. *Abelian l-adic representations and elliptic curves* (2° éd.) Addison-Wesley, Redwood City (1989).

[28]   J.-P. Serre. *Sur les représentations modulaires de degré 2 de* $Gal(\overline{\mathbb{Q}}/\mathbb{Q})$. Duke Math. J. 54 (1987) pp. 179-230.

[29]   R.Taylor, A. Wiles. *Ring Theoretic Properties of Certain Hecke Algebras*, Preprint 1994.

[30]   N.Wach. *Représentations p-adiques cristallines du groupe de Galois d'un corps local*, thèse, univ. Paris-Sud, Orsay (1994).

[31]   A. Wiles. *Modular Elliptic Curves and Fermat's Last Theorem*, Preprint 1994.

Conference on Elliptic Curves and Modular Forms
Hong Kong, December 18-21, 1993
Copyright ©1995 International Press

# On elliptic curves with isomorphic torsion structures and corresponding curves of genus 2

GERHARD FREY
INSTITUTE FOR EXPERIMENTAL MATHEMATICS
UNIVERSITY OF ESSEN
ELLERNSTRASSE 29 D-45326 ESSEN, GERMANY
*E-mail address*: FREY@EXP-MATH.UNI-ESSEN.DE

After stating conjectures about elliptic curves $E_1$ and $E_2$ which have Galois isomorphic submodules in their torsion groups and discussing relations with conjectures about solutions of equations of Fermat type we describe properties of curves of genus 2 which are related to such isomorphisms. These discussions include the case that $E_1$ is isogenous to $E_2$, and they are used to find a possibility to write down explicit equations for twisted modular curves of level $p$ where $p$ is an odd prime.

**Introduction.** Solutions of equations of ternary diophantine equations are closely related to questions about arithmetical properties of elliptic curves and corresponding Galois representations induced by torsion points of such curves. Especially it is of interest whether different elliptic curves can give rise to equivalent representations.

In the case that the ground field is $\mathbb{Q}$ it is well known that this leads to very interesting questions about congruence primes of modular forms, and the celebrated theorem of Ribet [9] shows how closely for instance Fermat's conjecture is related to Taniyama's conjecture or Serre's conjecture (cf. [1]), and for general ground fields $K$ the height conjecture for elliptic curves would answer many questions about solutions of ternary diophantine equations. In the first section we formulate some conjectures arising in this area and discuss their relations.

Among these conjectures the height conjecture for elliptic curves plays a prominent role. It is well known that it would follow from a conjecture of Parshin-Bogomolov-Miyaoka-Yau type about the self intersection of the relative canonical sheaf of arithmetic surfaces with generic fiber equal to a curve of genus $\geq 2$ (cf. [8]). In [2] it is discussed how curves of genus two with elliptic differentials can possibly be used to come closer to the height conjecture.

The aim of this paper is to discuss such curves $C$ under the additional assumption that they are already defined over $K$. This is equivalent to the condition that there are two elliptic curves $E_1$ and $E_2/K$ and a number $n$ such that $E_{1,n}$ (which is the $G_K$-module of the points of order dividing $n$ in $E(\bar{K})$) is isomorphic to $E_{2,n}$ under an "anti-isometric" isomorphism (definition in §1) satisfying certain conditions. In section 2 we examine ramification properties of $C/E$ and study its Galois closure. This enables us to translate conjectures of "asymptotic Fermat type" into conjectures about geometric fundamental groups of curves of genus 2 over global fields (Corollary 2.6).

In section 3 we treat the special case that the Jacobian of $C$ is isogenous to $E \times E$; a special result is that there are curves of genus 2 over finite fields whose geometric fundamental group is infinite.

It is easily seen that covers of elliptic curves $E$ by curves of genus 2 induce related covers of projective lines with very restricted ramification patterns. This is used in section 4 to determine systems of equations for twisted modular curves, they follow from the result of theorem 4.1 which characterizes such covers with the help of solutions of twists by polynomials of degree 3 of an elliptic curve defined over $K(U)$ where $U$ is an indeterminate.

It should be remarked that in order to simplify the exposition we did not formulate the conjectures and results in the greatest possible generality. For instance nearly everywhere the condition "$p$ is an odd prime" could be replaced by "$p$ is an odd number".

The following paper is strongly influenced by common work with E. Kani, and many suggestions from and many discussions with him were of great value for me.

Parts of the results were presented in the lecture I gave during the conference on "Elliptic Curves and Modular Forms"; it is a great pleasure for me to thank the organizers and the Chinese University of Hong Kong for their generous support, for the stimulating scientific atmosphere they succeeded to create and for their warm hospitality which made my visit to Hong Kong an unforgettable experience.

**1 Conjectures.** Let $K$ be a field with absolute Galois group $G_K$. For simplicity assume that char $(K) \neq 2, 3$. Throughout the whole paper we shall assume that $p$ is a prime not dividing $2 \cdot$ char $(K)$.

Let $E/K$ be an elliptic curve. With $E_p$ we denote the group of points of $E(\bar{K})$ whose order divides $p$. Let $\rho_{E,p}$ be the representation of $G_K$ induced by its action on $E_p$.

Now fix an elliptic curve $E_0$ and denote by $\mathcal{R}_K(E_0)$ the following set:

$$\mathcal{R}_K(E_0) = \{E/K; \exists p \in \mathbb{P} \backslash \{2, 3, 5, \text{char } K\} \text{such that } E_p \cong$$
$$E_{0,p} \text{ as } G_K - \text{module and that } E \text{ is semistable at all } \mathfrak{p}|p\}.[1]$$

$K$ is called a global field if either $K$ is a finite number field or $K$ is a function field of one variable over a perfect field $K_0$ as field of constants. Let $h(E)$ be the Faltings height of elliptic curves over $K$.

Now we can state

**Conjecture** $(R_{E_0,K})$: *There is a number $M$ only depending on $K$ and on the conductor of $E_0$ such that for $E \in \mathcal{R}_K(E_0)$ we have:*

$$h(E) < M.$$

*In particular, $\mathcal{R}_K(E_0)$ is finite if $K$ is a number field, or $K_0$ is finite.*

---
[1]The condition that $E$ is semistable at divisors of $p$ should not be necessary for the conjecture stated below.

Let $S_0$ be a finite set of primes (including the archimedean primes) such that $E_0$ has good reduction outside of $S_0$. Assume that $E/K$ is an elliptic curve such that for a prime $p > 3$ we have an $G_K$-isomorphism

$$\alpha_p : E_{0,p} \to E_p.$$

It follows that for places $\mathfrak{q} \notin S_0 \cup \{\mathfrak{p}|p\}$ the curve $E$ is semistable at $\mathfrak{q}$ and $\min\{0, v_\mathfrak{q}(j_E)\} \equiv 0 \bmod p$.

Now assume in addition that $E \in \mathcal{R}_K(E_0)$. Let $j_E \in K$ be the absolute invariant of $E$ and $h(j_E)$ its height. Since $E$ is semistable outside of $S_0$, we get

$$h(j_E) = h_{S_0}(j_E) + \tilde{h}(j_E) \geq h_{S_0}(j_E) + p\deg(\tilde{N}_E)$$

where $h_{S_0}$ is the contribution of places in $S_0$, $\tilde{h}$ is the contribution of places not in $S_0$ and $\tilde{N}_E$ is the part of the conductor $N_E$ of $E$ which is prime to $S_0$.

So

$$h(j_E) \geq \sum_{\substack{\mathfrak{q} \in S_0 \\ \mathfrak{q}|N_E}} \deg \mathfrak{q} + p \sum_{\substack{\mathfrak{q} \notin S_0 \\ \mathfrak{q}|N_E}} \deg \mathfrak{q},$$

which implies a related formula for the Faltings height of $E$ (cf. [1]).

Let us recall that the **height conjecture** $(H_E)$ for elliptic curves (cf. [1]) states:

Let $S_0$ be a finite set of places of $K$. Then there exist constants $c, d$ depending on $S_0$ and on $K$ such that for all elliptic curves $E/K$ which are semistable outside of $S_0$ one has

$$h(E) \leq d \deg N_E + c.$$

Assuming the truth of this conjecture we get that

$$\sum_{\substack{\mathfrak{q} \in S_0 \\ \mathfrak{q}|N_E}} \deg \mathfrak{q} + p \sum_{\substack{\mathfrak{q} \notin S_0 \\ \mathfrak{q}|N_E}} \deg \mathfrak{q} \leq d \deg N_E + c \leq \tilde{d}(\sum_{\mathfrak{q}|N_E} \deg \mathfrak{q}) + c$$

with $\tilde{d}$ depending on $K$ and $d$ only.

Hence for almost all $p$ the support of $N_E$ must lie inside of $S_0$, and so $E$ has to have bounded height. Since for $p > 5$ the modular curve $X(p)$ has genus $\geq 2$ there are only finitely many elliptic curves corresponding to $X(p)(K)$, and so we get

**Proposition 1.1.** *The height conjecture* $(H_E)$ *implies* $(R_{E_0,K})$ *for all* $E_0/K$.

**Corollary 1.2.** *If $K$ is a function field then* $(R_{E_0,K})$ *is true.*

*Proof.* The height conjecture for elliptic curves over function fields is true (cf. [1]). □

Next we weaken Conjecture $(R_{E_0,K})$ to

**Conjecture** $(R_{E_0,K})(2)$: *There is a number $M$ depending on $K$ and the conductor of $E_0$ such that for all $E \in \mathcal{R}_K(E_0)$ with the additional property that $E_{0,2} \simeq E_2$ we have*

$$h(E) \leq M.$$

Let us discuss this conjecture.

Assume that $E \in \mathcal{R}_K(E_0)$ and that $E_2 \cong E_{0,2}$. Without loss of generality we can assume that $E_{0,2} \subset E_0(K)$, and so $E$ can be given by an equation

$$E : Y^2 = X(X - A)(X - B)$$

with $A, B \in K$. Moreover there is a finite set $S_0$ of places of $K$ depending on $E_0$ and on $K$ only, such that the ring of $S_0$-integers $O_{S_0}$ is a principal ideal domain and such that $A, B$ are relatively prime elements in $O_{S_0}$ (as seen above $E$ has to be semistable outside a finite set of places depending on $E_0$).

So for $\mathfrak{p} \notin S_0$ we get: $v_\mathfrak{p}(A^2 B^2 (A-B)^2) = v_\mathfrak{p}(j_E)$ if $v_\mathfrak{p}(j_E) < 0$, and hence we have for $\mathfrak{p} \notin S_0$:

$$v_\mathfrak{p}(A^2 B^2 (A - B)^2) \equiv 0 \mod p$$

if $E \in \mathcal{R}_K(E_0)$ and $\alpha_p : E_{0,p} \to E_p$ is a $G_K$-isomorphism for $p \geq 5$.

So we can find relatively prime elements $z_1, z_2, z_3 \in O_{S_0} \setminus \{0\}$, with $A = az_1^p$, $B = bz_2^p$, $A - B = cz_3^p$ and with $a, b, c \in O_{S_0}^*$. Hence the equation:

$$aZ_1^p - bZ_2^p = cZ_3^p$$

has the solution $(z_1, z_2, z_3)$ in $O_{S_0}$.

Next we state the

**Asymptotic Fermat conjecture** $\underline{F_{as}}$: *Let $S_0$ be a finite set of primes of $K$. Define*

$$L_{S_0} = \{(z_1, z_2, z_3) \in O_{S_0}, \exists a, b, c \in O_{S_0}^*$$

$$and\, p \geq 5 \text{ such that } az_1^p + bz_2^p = cz_3^p\}.$$

*Then there exists a number $M$ depending on $K$ and $S_0$ such that for all $(z_1, z_2, z_3) \in L_{S_0}$ the projective height*

$$h((z_1, z_2, z_3)) \leq M.$$

**Proposition 1.3.** *The conjecture $\underline{(F_{as})}$ implies $\underline{(R_{E_0,K}(2))}$.*

REMARK. It is obvious that $\underline{(F_{as})}$ follows from the height conjecture for elliptic curves as well (cf [1]).

*Proof.* We assume that $E$ is given by

$$Y^2 = X(X - az_1^p)(X - bz_2^p) \text{ for } p \geq 5.$$

Conjecture $\underline{(F_{as})}$ implies that there are only finitely many places dividing $\Delta_E = (abc)^2 (z_1 z_2 z_3)^{2p}$, and so $E$ has good reduction outside a finite set of places $S_1$ depending on $K$ and $S_0$ only, and the proposition follows. □

The main concern of the paper is to investigate another weakened version of $\underline{(R_{E_0,K})}$.

But before doing this we want to make some remarks concerning the special case that $K = \mathbb{Q}$.

We <u>assume</u> that $E_0$ is a modular elliptic curve, i.e. there is a $\mathbb{Q}$-rational non constant morphism
$$\varphi : X_0(N_E) \to E_0.$$
Let $\mathcal{R}_{\text{mod}}(E_0)$ be the set of elliptic curves in $\mathcal{R}_\mathbb{Q}(E_0)$ which are modular curves and $\mathcal{R}_{\text{mod}}(E_0)(2)$ the set of elliptic curves $E$ in $\mathcal{R}_{\text{mod}}(E_0)$ with $E_2 \cong E_{0,2}$.

For a finite set $S_0$ of primes define

$L_{S_0,\text{mod}} = \{(z_1, z_2, z_3) \in \mathbb{Z}_{S_0}; \exists a,b,c \in \mathbb{Z}^*_{S_0} \text{ and } p \geq 5 \text{ with } az_1^p - bz_2^p = cz_3^p \text{ and } E_{\underline{z}} : Y^2 = X(X - az_1^p)(X - bz_2^p) \text{ is modular}\}.$

Then we get

**Proposition 1.4.** *$\mathcal{R}_{mod}(E_0)(2)$ is finite for all elliptic curves $E_0$ whose conductor is in $\mathbb{Z}^*_{S_0}$ if and only if $L_{S_0,mod}$ is finite.*

*Proof.* If $\mathcal{R}_{\text{mod}}(E_0)(2)$ is infinite for $E_0$ modular and $N_{E_0} \in \mathbb{Z}^*_{S_0}$ then $L_{S_0,\text{mod}}$ is shown to be infinite as in the proof of Proposition 1.3. Conversely, assume that there are infinitely many $(z_1, z_2, z_3) \in \mathbb{Z}_{S_0}$ and primes $p_i \geq 5$ with $a_1 z_1^{p_i} - b z_2^{p_i} = c z_3^{p_i}$ and $E_{\underline{z}}$ modular. Using Ribet's theorem about minimal levels of cusp forms related to $\rho_{E,p}$, an easy argument due to Mazur (cf. [1]) proves that there is a modular elliptic curve $E_0$ with $E_{0,2} \subset E_0(\mathbb{Q})$ and with conductor $N_{E_0} \in O^*_{S_0}$ such that $\rho_{E_0,p_i} = \rho_{E_{\underline{z}},p_i}$ and hence $\mathcal{R}_{\text{mod}}(E_0)(2)$ is infinite. □

$\mathcal{R}_{\text{mod}}(E_0)$ is closely related to the question how often there are elliptic quotients of the modular curves $X_0(N)$ with equivalent Galois representation induced by submodules of the group of torsion points. On the other hand it is known (cf. [1]) that the height conjecture for modular elliptic curves is equivalent to the following

**Conjecture:** *Let $S_0$ be a finite set of primes. There exist $c, d \in \mathbb{R}$ such that the following holds:*

*Let $E$ be a modular elliptic curve which is semistable outside of $S_0$ with conductor $N_E$. Let $E^*$ be the unique elliptic curve in $J_0(N_E)$ which is isogenous to $E$, and $A$ its complementary abelian variety in $J_0(N_E)$. Then $|(E^* \cap A)| \leq c \cdot N_E^{2d}$.*

So it becomes obvious from the representation theoretical point of view as well, how closely the question of congruence primes and solutions of equations of Fermat type are related if we are working over $\mathbb{Q}$ and if we assume Taniyama's conjecture.

The conjecture $(R_{E_0,K})$ is rather coarse in the sense that it doesn't care about the symplectic structure of points of order $p$ of elliptic curves induced by the Weil pairing $e_p$.

DEFINITION. *Let $E_0$ and $E$ be two elliptic curves, $\alpha_p : E_{0,p} \to E_p$ a $G_K$-isomorphism. $\alpha_p$ is <u>anti-isometric</u> if for $P, Q \in E_{0,p}$ we have $e_p(P,Q) = e_p(\alpha_p(P), \alpha_p(Q))^{-1}$.*

*In other words:* $\alpha_p$ *is anti-isometric if and only if* $\Delta_{\alpha_p} := \{(P, \alpha_p P)\} \subset E_{0,p} \times E_p$ *is (with respect to the Weil pairings) an isotropic* $G_K$*-submodule.*

Let $\overline{\mathcal{R}_K(E_0)}$ be the set of elliptic curves $E \in \mathcal{R}_K(E_0)$ for which there exists an anti-isometric $G_K$-isomorphism from $E_{0,p}$ to $E_p$ for some prime $p > 5$. The main concern of the rest of the paper is to study properties of elements in $\overline{\mathcal{R}_K(E_0)}$.

We state an obvious weakening of conjecture $(R_{E_0, K})$:

CONJECTURE. $\overline{R_{(E_0),K}}$: *There is a number $M$ depending on $K$ and on the conductor of $E_0$ such that for all $E \in \overline{\mathcal{R}_K(E_0)}$ we have: $h(E) \le M$.*

First we want to discuss the relation between $\mathcal{R}_K(E_0)$ and $\overline{\mathcal{R}_K(E_0)}$.
So assume that $\alpha_p : E_{0,p} \to E_p$ is a $G_K$-isomorphism.
Let $\alpha'_p$ be another isomorphism.
Then $\alpha'^{-1}_p \circ \alpha_p = \beta_p \in \mathrm{Aut}_{G_K}(E_{0,p})$. So $\beta_p$ is in the centralizer of $\mathrm{Im}\,(\rho_{E_0,p})$ in $\mathrm{Aut}\,(E_{0,p})$. Let $(P_1, P_2)$ be a base of $E_{0,p}$ and use it to map $\mathrm{Aut}\,(E_{0,p})$ into $\mathrm{Gl}\,(2, \mathbb{Z}/p\mathbb{Z})$. Since $e_p(\beta_p P_1, \beta_p P_2) = e_p(P_1, P_2)^{\det(\beta_p)}$ we get:

**Proposition 1.5.** *Let $\alpha : E_{0,p} \to E_p$ be a $G_K$-isomorphism. Assume that $e_p(\alpha P_1, \alpha P_2) = e_p(P_1, P_2)^{k_\alpha}$. Then there are $k$ anti-isometric isomorphisms*

$$\alpha' : E_{0,p} \to E_p$$

*with* $k = |\{\beta \in Z_{\mathrm{Gl}(2,\mathbb{Z}/p\mathbb{Z})}(\mathrm{Im}(\rho_{E,p})); \det(\beta) = -k_\alpha^{-1}\}|.$

**Corollary 1.6.** *Assume that $Z_{\mathrm{Gl}(2,\mathbb{Z}/p\mathbb{Z})}(\mathrm{Im}\rho_{E,p})) = (\mathbb{Z}/p\mathbb{Z})^* \cdot id$ and that $\alpha \in \mathrm{Iso}_{G_K}(E_{0,p}, E_p)$ with $e_p(\alpha P_1, \alpha P_2) = e_p(P_1, P_2)^{k_\alpha}$. Then there is an antiisometric element $\alpha'$ in $\mathrm{Iso}_{G_K}(E_{0,p}, E_p)$ if and only if $-k_\alpha \in (\mathbb{Z}/p\mathbb{Z})^{*2}$, and in this case $\alpha'$ is, up to sign, uniquely determined.*

**Corollary 1.7.** *Assume that $\alpha \in \mathrm{Iso}_{G_K}(E_{0,p}, E_p)$, and that $\eta : E_0 \to E'_0$ is a $K$-rational isogeny of prime degree $l$ with $\left(\frac{l}{p}\right) = -1$. Then there is an anti-isometric element either in $\mathrm{Iso}_{G_K}(E_{0,p}, E_p)$ or in $\mathrm{Iso}_{G_K}(E'_{0,p}, E_p)$.*

*Proof.* If $e_p(\alpha P_1, \alpha P_2) = e_p(P_1, P_2)^{k_\alpha}$ then $e_p(\eta P_1, \eta P_2) = e_p(P_1, P_2)^l = e_p(\alpha P_1, \alpha P_2)^{l \cdot k_\alpha}$ and either $-k_\alpha$ or $-k_{\alpha'} l$ is a square modulo $p$. $\square$

EXAMPLE. Assume that $E_0$ contains a $K$-rational point $P$ of order 2. Put $E'_0 = E_0/\langle P \rangle$. Let $\mathcal{R}'_K(E_0)$ be the set of curves $E \in \mathcal{R}_K(E_0)$ with $\alpha_p \in \mathrm{Iso}_{G_K}(E_{0,p}, E_p)$ and $p \not\equiv \pm 1 \bmod 8$. Then $E \in \mathcal{R}'_K(E_0)$ is either in $\overline{\mathcal{R}_K(E_0)}$ or in $\overline{\mathcal{R}_K(E'_0)}$.

**Corollary 1.8.** *Assume that $\mathrm{Im}(\rho_{E_0,p})$ is contained in a Cartan subgroup $A$ of $\mathrm{Gl}(2, \mathbb{Z}/p\mathbb{Z})$ e.g. assume that $E_0$ has complex multiplication. Let*

$\mathrm{Iso}_{G_K}(E_{0,p}, E_p) \neq \emptyset$. Then there are at least $p-1$ anti-isometric elements $\alpha \in \mathrm{Iso}_{G_K}(E_{0,p}, E_p)$ if $A$ is split, and $p+1$ such elements if $A$ is non split.

*Proof.* The centralizer of $\mathrm{Im}(\rho_{E_0,p})$ contains $A$, and $|A \cap \mathrm{Sl}(2, \mathbb{Z}/p\mathbb{Z})| = p-1$ if $A$ is split, and $p+1$ if $A$ is not split. Hence $A/C \cap \mathrm{Sl}(2, \mathbb{Z}/p\mathbb{Z})$ has $p-1$ elements, and so $\det : C \to \mathbb{F}_p^*$ is surjective, and the assertion follows. $\square$

**2 Curves of genus 2 covering elliptic curves.** In this section we shall relate curves of genus 2 to elements in $\mathrm{Iso}_K(E_{0,p}, E_p)$ which are anti-isometric, and we shall show that Conjecture $\overline{(R_{E_0,K})}$ is closely connected to a conjecture about unramified geometric Galois coverings of such curves.
We begin by recalling some facts which can be found in [**2**] and [**4**].

Assume that $E_0$ and $E_1$ are elliptic curves over $K$ and that for an odd prime $p$ different from $\mathrm{char}(K)$ there is a $G_K$-isomorphism $\alpha : E_{0,p} \to E_p$ with $e_p(P,Q) = e_p(\alpha P, \alpha Q)^{-1}$ for all $P,Q \in E_{0,p}$ i.e., $\alpha$ is anti-isometric and $\Delta_\alpha = \{(P, \alpha P)\} \subset E_{0,p} \times E_{1,p}$ is isotropic in $E_{0,p} \times E_{1,p}$.
Then
$$J_\alpha := E_0 \times E_1 / \Delta_\alpha$$
is a principally polarized abelian surface over $K$ (cf. [**2**]).

$J_\alpha$ is the Jacobian of a curve $C_\alpha/K$ of genus 2 if and only if there is $\underline{\mathrm{no}}$ isogeny
$$\eta : E_0 \to E_1$$
with $\deg \eta = k(p-k)$ for some $1 \leq k \leq p-1$ such that
$$k \circ \alpha = \eta | E_{0,p}$$
(cf. [**6**]).
We assume that this is the case.

Then there is a map
$$\pi_\alpha : C_\alpha \to E_0$$
with $\deg(\pi_\alpha) = p$, and $wC_\alpha | E_0 = -\mathrm{id}$ where $w$ is the hyperelliptic involution of $C_\alpha$.
Let $\overline{K}$ be the algebraic closure of $K$. Hurwitz genus formula yields:
Either there are two points $P_1, P_2$ on $E_0(\overline{K})$ with $P_1 = -P_2 \neq P_2$ and $Q_i \in C_\alpha(\overline{K})$ with $\pi_\alpha Q_i = P_i (i=1,2)$, the ramification order of $Q_i/P_i$ is equal to two and all points $Q$ of $C$ different from $Q_1, Q_2$ are unramified in $C_\alpha/E_0$. (**Type I**) (which is the generic case)
or, as degeneration there is a Weierstrass point $Q$ of $C_\alpha$ which ramifies of order 3 over $E_0$ and $C_\alpha \setminus \{Q\}$ is unramified, or there is a Weierstrass point of which $E_0$ has two ramified extensions which ramify of order 2. (**Type II**).

NOTATION. *For curves $X$ defined over $K$ let $F(X)$ be its function field, and $\bar{F}(X)$ the composite of $F(X)$ with $\bar{K}$.*

Let $\tilde{F}_\alpha$ be the Galois closure of $F(C_\alpha)/F(E_0)$.

**Proposition 2.1.** *Assume that $C_\alpha$ is of type I. Then $\tilde{F}_\alpha/\bar{F}(E_0)$ is a geometric extension with Galois group $S_p$. Hence $\tilde{F}_\alpha$ is the function field of a curve $\tilde{C}_\alpha/K$ which is a Galois cover of $E_0$ with Galois group $S_p$.*

*Proof.* To prove the proposition it is enough to show that the Galois group of $\tilde{\bar{F}}_\alpha|\bar{F}(E_0)$ is equal to $S_p$. Let $\mathfrak{p}_1$ be one of the places of $\bar{F}(E_0)$ which ramify in $\tilde{\bar{F}}_\alpha$.

Then conorm $_{\bar{F}(E_0)/\bar{F}(C_\alpha)}(\mathfrak{p}_1) = \mathfrak{q}_1^2 \cdot \mathfrak{q}_1 \ldots \mathfrak{q}_{p-1}$ where $\mathfrak{q}_1, \ldots \mathfrak{q}_{p-1}$ are different places. It follows that the inertia group of an extension of $\mathfrak{q}_1$ to $\tilde{\bar{F}}_\alpha$ is generated by a transposition. But this implies already that $G(\tilde{\bar{F}}_\alpha/\bar{F}(E_0))$ is equal to $S_p$ (cf. [13]). □

Now assume that $C_\alpha$ is of type II and that there is a place $\mathfrak{p} \in F(E_0)$ such that
conorm $_{\bar{F}(E_0)/\bar{F}(C_\alpha)}(\mathfrak{p}) = \mathfrak{q}^3 \cdot \mathfrak{q}_1 \ldots \mathfrak{q}_{p-3}$ where $\mathfrak{q}_1, \ldots, \mathfrak{q}_{p-3}$ are different places. So we have a 3-cycle inside of $G(\tilde{\bar{F}}_\alpha/\bar{F}(E_0))$ and hence this group contains $A_p$. Since the hyperelliptic involution $w$ of $C_\alpha$ maps $E_0$ into itself the function field of $C_\alpha/w$ (which is a rational field $K(U)$) is isomorphic to $K(X)[Y]/(f(Y))$ where $K(X)$ is the function field of $E_0/w$.

We can describe the ramification of $K(U)/K(X)$ (see §4) and get:
The discriminant of $K(U)/K(X)$ is either equal to
i) $\frac{p-1}{2}(\mathfrak{p}'_1 + \mathfrak{p}'_2) + (\frac{p-3}{2})(\mathfrak{p}'_3 + \mathfrak{p}'_4) + 2\mathfrak{p}_4$
or equal to
ii) $\frac{p-1}{2}(\mathfrak{p}'_1 + \mathfrak{p}'_2 + \mathfrak{p}'_3 + \mathfrak{p}'_4)$
where $\mathfrak{p}'_1, \ldots, \mathfrak{p}'_4$ correspond to the places of $K(X)$ which ramify in $F(E_0)/K(X)$.

If i) occurs then $\Delta_f$ is not a square in $\bar{K}(X)$ and $\bar{K}(X)(\sqrt{\Delta_f})$ is a function field of genus 0. Hence $\Delta_f$ is not a square in the function field of $E_0$ and so $G(\tilde{\bar{F}}_\alpha/\bar{F}(E_0))$ is equal to $S_p$.

If case ii) occurs then $\tilde{\bar{F}}_\alpha/\bar{F}(E_0)$ has Galois group $A_p$, and the algebraic closure of $K$ in $\tilde{F}_\alpha$ is an extension of $K$ of degree $\leq 2$.

The last possibility is that a point of order 2 of $E_0$ has two extensions to $C_\alpha$ which ramify of order 2. (case II iii) ). Hence $G(\tilde{\bar{F}}_\alpha/\bar{F}(E_0))$ contains a product of two transpositions and so it contains $A_p$ if $p > M$,[2] and the algebraic closure of $K$ in $\tilde{F}_\alpha$ is an extension of degree $\leq 2$.

**Proposition 2.2.** *If $\pi_\alpha : C_\alpha \to E_0$ is of type II of degree $\geq M$ then the Galois group of $\tilde{F}_\alpha/F(E_0)$ contains $A_p$. $\tilde{F}_\alpha$ is geometric over $K$ with Galois group equal to $S_p$ if case i) occurs.*

*If case ii) occurs $\tilde{F}_\alpha$ is geometric over $K$ if and only if $G(\tilde{\bar{F}}_\alpha/\bar{F}(E_0))$ is equal to $A_p$.*

Define $F'_\alpha = \tilde{F}_\alpha^{A_p}$, and let $C'_\alpha$ be the curve defined over $\bar{K}$ with function field $\bar{F}'_\alpha$.

---
[2]The author has to thank W. Lempken for pointing out this result.

**Corollary 2.3.** $\tilde{F}_\alpha$ *is a geometric extension of* $F'_\alpha$. *If* $C_\alpha$ *is of type I then* $C'_\alpha$ *is a curve of genus 2 defined over* $K$ *and* $\tilde{C}_\alpha/C'_\alpha$ *is unramified. If* $C_\alpha$ *is of type II then* $C'_\alpha$ *is a curve of genus 1 which is* $\tilde{K}$-*isogenous to* $E_0$.

In the type II-case $C_\alpha/E_0$ is only ramified in the point $P$; it follows that $P$ is $K$-rational and so we can assume that $P$ is the ramification point of $E_0/(E_0/w)$ which corresponds to the place $\mathfrak{p}_\infty$ of the function field $K(X)$ of $E_0/w$.

To simplify the exposition we shall assume from now on that $E_0(K)_2 = \mathbb{Z}/2 \times \mathbb{Z}/2$ if type II occurs. It will be shown in section 4 that there is up to isomorphism exactly one curve $C_3$ of genus 2 covering $E_0$ of degree 3 which is ramified of order 3 at $P$ (cf. proposition 4.2). If case II iii) occurs we can find a curve $C_4$ of genus 2 covering $E_0$ of degree 4 which ramifies at $P$ of order 2, and so the composite of its function field with $\tilde{F}_\alpha$ is unramified over $F(C_4)$. It follows that for $p > M$ we have:

**Corollary 2.4.** $\tilde{F}_\alpha \cdot F(C_3)/F(C_3)$ *resp.* $\tilde{F}_\alpha F(C_4)/F(C_4)$ *is an unramified Galois extension whose Galois group contains* $A_p$ *and the algebraic closure of* $K$ *in* $\tilde{F}_\alpha \cdot F(C_3)$ *is of degree* $\leq 2$ *over* $K$. *In other words: For every elliptic curve* $E_0$ *with* $E_{0,2} \subset E_0(K)$ *there exists an extension* $K_1$ *of* $K$ *with* $[K_1 : K] \leq 2$ *and a curve* $C'$ *of genus 2 covering* $E_0$ *of degree 3 or 4 such that every covering of degree* $p > M$ *of type II of* $E_0$ *by a curve of genus 2 gives rise to an unramified geometric Galois cover of* $C'$ *defined over* $K_1$ *whose Galois group contains* $A_p$.

Now assume conversely that $\tilde{C}/E_0$ is a geometric Galois cover with group $S_p$ defined over $K$. Assume moreover that there are exactly 2 points $P_1$ and $P_2$ of $E_0(\bar{K})$ which ramify in $\tilde{C}$ with ramification order 2, or equivalently, assume that $g(\tilde{C}) = \frac{p!}{2} + 1$, and that the inertia group of $P_i$ is generated by a transposition.

Let $C'$ be equal to $\tilde{C}/A_p$.

It follows that $C'/E_0$ is ramified in $P_1$ and $P_2$ and that $\tilde{C}/C'$ is unramified.

Let $Q_1$ be an extension of $P_1$ to $\tilde{C}$ with inertia field $K_{Q_1}$ which has index 2 in the function field of $\tilde{C}$ and whose Galois group is generated by a transposition $\tau$. Hence there are $p-2$ copies of $S_{p-1}$ inside of $S_p$ which contain $\tau$, and so $Q_1$ is ramified in exactly two extensions of degree $p$ over the function field of $E_0$. It follows that in each of the fields of degree $p$ over $E_0$ exactly $(p-1)!$ extensions of $P_1$ to $\tilde{C}$ ramify, or, in other words, there is exactly one extension of $P_1$ to $C = \tilde{C}/S_{p-1}$ (take some embedding of $S_{p-1}$ into $S_p$) which is ramified. Since for $P_2$ the same result holds we get that $C$ is a curve of genus 2 covering $E_0$ over $K$ of degree $p$ which is of type I.

By using analogous arguments one can characterize coverings of type II: Assume that $\tilde{C}/E_0$ is a Galois cover whose Galois group is contained in $S_p$ and contains $A_p$, and that exactly one point $P$ of $E_0$ is ramified in $\tilde{C}$, assume furthermore that the inertia group of an extension of $P$ to $\tilde{C}$ is generated by a 3-cycle resp. by a product of two transpositions. Let $C$ be a subcover of $E_0$ of degree $p$. Then $C$ is a curve of genus 2 defined over $K$ covering $E_0$ of degree $p$ which is of type II.

Having $\pi : C \to E_0$ it follows that there is an elliptic curve $E_1/K$ such that the Jacobian $J_C$ of $C$ is equal to $(E_0 \times E)/\Delta_\alpha$ where $\alpha$ is an anti-isometric element $\alpha$ in $\mathrm{Iso}_{G_K}(E_{0,p}, E_p)$.

Summarizing our results we get

**Theorem 2.5.** *Assume that is a prime $p > u$ not dividing $2 \cdot char(K)$. Let $\alpha : E_{0,p} \to E_p$ be an anti-isometric $G_K$-isomorphism between the points of order $p$ of the elliptic curves $E_0$ and $E$. Then either there is a geometric Galois cover $\tilde{C}/K$ of $E_0$ with Galois group $S_p$ with ramifies of order 2 in two points of $E_0$ and which is unramified over the curve $\tilde{C}/A_p$ (which has genus 2), or there is a Galois cover $\tilde{C}/E_0$ in which exactly one point $P$ of $E_0$ is ramified of order 2 or 3, the Galois group of $\tilde{C}/E_0$ contains $A_p$ and is contained in $S_p$, $\tilde{C}/E_0$ is geometric if and only if $(\tilde{C}/A_p)/E_0$ is geometric, and there is a curve $C'$ of genus 2 covering $E_0$ over $K(E_{0,2})$ such that $\tilde{C} \times_{E_0} C'/C'$ is unramified.*

*Conversely to every Galois cover $\pi : \tilde{C} \to E_0$ whose Galois group is contained in $S_p$ and contains $A_p$, which is ramified either in 2 points of $E_0$ of order 2 or in one point of $E_0$ of order 2 or 3 such that the ramification groups are generated by 2- or resp. 3-cycles resp. a product of two two-cycles there corresponds an anti-canonical $G_K$-isomorphism $\alpha : E_{0,p} \to E_p$, where $E/K$ is an elliptic curve such that a subextension $C$ of $E_0$ in $\tilde{C}$ of degree $p$ is a curve $C$ of genus 2 with $J_C = E_0 \times E_1/\Delta_\alpha$.*

Type II forces us to formulate Theorem 2.5 in a rather complicated way. By using arguments from [2] one can show that if type II occurs and if $K$ is a number field then the height of the corresponding curve $C$ is bounded. Using this we can reformulate conjecture $(\overline{R_{E_0,K}})$ in terms which involve arithmetical properties of the "$K$-rational fundamental group" of curves of genus 2 with elliptic differentials.

**Corollary 2.6.** *Let $K$ be a number field. The conjecture $(\overline{R_{E_0,K}})$ is true if the following holds:*
*There is a constant $M$ depending on $K$ and the conductor of $N_{E_0}$ such that for all curves $C'$ of genus 2 with the properties:*
*i) $C'$ covers $E_0$ of degree 2,*
*ii) there is a prime $p > 5$ such that $C'$ has a geometric $K$-rational unramified Galois cover $\tilde{C}$ with Galois group $A_p$ which is normal over $E_0$ with Galois group $S_p$ and for all embeddings of $S_{p-1}$ to $G(\tilde{C}/E_0)$ we have:*
*The modular height of $\tilde{C}/S_{p-1}$ is bounded by $M$.*

## 3 Covers induced by isogenies between elliptic curves.

The height conjectures stated in the previous sections don't take care of the case that $E_0$ is isogenous to $E_1$. In this section we want to discuss this situation more closely.
So let
$$\eta : E_0 \to E$$
be a $K$-isogeny of degree $l_0$.

Let $p$ be a prime not dividing $2 \cdot l_0 \cdot char(K)$. The restriction of $\eta$ to $E_{0,p}$ induces a $K$-rational isomorphism
$$\alpha_{0,p} : E_{0,p} \to E_p,$$

and so $E_1 \in \mathcal{R}_K(E_0)$.

Can we use $\alpha_{0,p}$ to construct an anti-isometric isomorphism $\alpha$ such that $(E_0 \times E_1/\Delta_\alpha)$ is the Jacobian of a curve $C_\alpha$ covering $E_0$ of degree $p$?

**Proposition 3.1.** *Let $\rho_{E_0,p}$ be the representation of $G_K$ induced by the action of $G_K$ on $E_{0,p}$. We assume that $E_0$ has no complex multiplication and that the center of $Im(\rho_{E_0,p}) \in Gl(E_{0,p})$ is equal to $(\mathbb{Z}/p\mathbb{Z})^* \cdot id$.*

*Then there is a curve $C/K$ of genus 2 covering $E_0$ of degree $p$ with $J_C$ isogenous to $E_0 \times E_1$ if and only if $p$ is split in $\mathbb{Q}(\sqrt{-l_0})$ but for all divisors $l'$ of $l_0$ we get: $l' \cdot p$ is not norm of an element in $\mathbb{Z} + \mathbb{Z} \cdot \sqrt{-l'_0}$ where $l'_0$ is the square free part of $l_0$.*

*Proof.* In order to find an anti-isometric isomorphism $\alpha$ from $E_{0,p}$ to $E_p$ we have to find an $l \in \mathbb{Z}$ with

$$\deg(l \cdot \eta) \equiv -1 \bmod p$$

or:

$$l^2 \cdot l_0 \equiv -1 \bmod p.$$

So $-l_0$ has to be a square mod $p$, and hence $p$ is split in $\mathbb{Q}(\sqrt{-l_0})$.

Now assume that $p$ is split in $\mathbb{Q}(\sqrt{-l_0})$, and that $l^2 l_0 \equiv -1 \bmod p$, and define $\alpha := l \cdot \eta_{|E_0,p}$.

We know that $E_0 \times E_1/\Delta_\alpha$ is principally polarized and that this polarization is an irreducible curve $C_\alpha$ of genus 2 if and only if there is no isogeny $\eta' : E_0 \to E_1$ with $\deg \eta' = k(p-k)$ for $1 \leq k \leq p-1$ and $k \cdot \alpha = \eta'_{|E_{0,p}}$.

Let $\eta'$ be any isogeny from $E_0$ to $E_1$. Let $\hat{\eta}$ be the dual isogeny of $\eta$ from $E_1$ to $E_0$.

Then $\eta' \circ \hat{\eta}$ is an endomorphism of $E_1$, and since $E_1$ has no complex multiplication there is $n \in \mathbb{Z}$ with $\eta' \circ \hat{\eta} = n \cdot id_{E_1}$.

It follows that
$$l_0 \cdot \eta' = n \cdot \eta, \quad \text{and so}$$
$$\deg \eta' = \frac{n^2}{l_0}.$$

So if
$$\deg \eta' = k(p-k) \quad \text{with} \quad 1 \leq k \leq p-1$$

we have:
$$\frac{n^2}{l_0} = k(p-k).$$

So
$$k = \frac{n_1^2}{l_1} \text{ and } (p-k) = \frac{n_2^2}{l_2}$$

with $l_1 \cdot l_2 = l_0$.
Hence
$$p = \frac{n_1^2}{l_1} + \frac{n_2^2}{l_2} = \frac{l_2 n_1^2 + l_1 n_2^2}{l_0}.$$

Since $\frac{n_1^2}{l_1}$ and $\frac{n_2^2}{l_2}$ are integers we have:

$$\frac{n_i^2}{l_i} = l_i' \cdot (n_i')^2$$

where $l_i'$ is the square free part of $l_i$, and so

$$p = l_1' n_1'^2 + l_2' n_2'^2$$

with $l_1' \cdot l_2' = l_0'$ the square free part of $l_0$.
By multiplying with $l_1'$ we see that $l_1' \cdot p$ is norm of an element of $\mathbb{Z} + \mathbb{Z}\sqrt{-l_0'}$, and so one direction of the proposition is proved.

Conversely assume that for a square free divisor $l'$ of $l_0$ we have:

$$l'p = n_1^2 + l_0' n_2^2.$$

Since $l'$ divides $n_1$ we get: $k := \frac{n_1^2}{l'} \in \mathbb{N}$, and $p = k + \frac{(n_2')^2}{l_2}$ with $l_2 = \frac{l_0}{l'}$.
Then

$$\eta' := \frac{(n_1 n_2')}{l_0} \cdot \eta$$

is an isogeny of $E_0$ to $E_1$ of degree $k(p-k)$ with $k = \frac{n_1^2}{l'}$.

It induces an isomorphism $\alpha' \in \text{Iso}_{G_K}(E_{0,p}, E_p)$ with $e_p(\alpha' P, \alpha' Q) = e_p(P, Q)^{-k^2}$.
Since $k \cdot \alpha$ has the same property and the center of $\text{Im}(\rho_{E_{0,p}}) \in \text{Gl}(E_{0,p})$ is equal to $(\mathbb{Z}/p\mathbb{Z})^* \cdot \text{id}$ we must have: $\alpha' = \pm \alpha$, and by replacing $\eta'$ by $-\eta'$ if necessary we get equality and hence $E_0 \times E_1/\Delta_\alpha$ is not a Jacobian of a curve of genus 2. □

EXAMPLE. Let $E_0$ be an elliptic curve over a number field $K$ without complex multiplication but with a $K$-rational isogeny of degree 5.

We can apply the result of prop. 3.1 to construct covers of degree $p$ of $E_0$ by curves of genus 2 whose Jacobian is equal to $(E_0 \times \eta E_0)/\Delta_\alpha$ for all primes $p$ which are split in $\mathbb{Q}(\sqrt{-5})$ into factors which are not principal ideals.

Next let us assume that $E_0$ has complex multiplication over $K$ and that $O_{-d}$ the ring of integers in $\mathbb{Q}(\sqrt{-d})$, is equal to its ring of endomorphisms.

Let $p$ be a prime not dividing $d$.
$\text{Im}(\rho_{E_{0,p}})$ is contained in a Cartan subgroup $A$ of $\text{Gl}(2, \mathbb{Z}/p\mathbb{Z})$, and so its center contains $A$.
Using Corollary 1.8 we get:

To each isogeny $\eta : E_0 \to E$ defined over $K$ of degree $l_0$ prime to $p$ there exist $p+1$ (if $A$ is non split) resp. $p-1$ (if $A$ is split) anti-isometric isomorphism $\alpha_p$ of the form

$$\eta_{|E_{0,p}} \circ \beta$$

with $\beta \in A$.

We recall: $\alpha_p$ does not belong to a cover $\pi_p : C_{\alpha p} \to E_0$ with $\deg \pi_p = p$ if and only if there is an isogeny $\eta' : E_0 \to E_1$ of degree $k(p-k)$ $(1 \le k \le p-1)$ with $\eta'_{|E_{0,p}} = k \circ \alpha_p$.

By applying the dual map $\hat{\eta}$ of $\eta$ we get an endomorphism $\tilde{\eta} = \hat{\eta} \circ \eta'$ of $E_0$ with degree $(\hat{\eta}) = k(p-k)l'_0$ with $l'_0|l_0$, and in order to be sure that there is an anti-isometric isomorphism $\alpha : E_{0,p} \to E_{1,p}$ inducing a cover $\pi_p : C_\alpha \to E_0$ of degree $p$ and with $g(C_\alpha) = 2$ we have to bound the number of endomorphisms of $E_0$ with degree $l'_0 k(p-k)$ by a number smaller than $p-1$.

Let us assume that $E_0 = E$ and hence $l'_0 = 1$. I. Kiming has proved the following result:

**Proposition 3.2.** *Let $D_0$ be a square free natural number and let $D$ be the discriminant of $\mathbb{Q}(\sqrt{-D_0})$. Let $h(D_0)$ be the class number of $\mathbb{Q}(\sqrt{-D_0})$ and $w$ be the number of roots of unity in $\mathbb{Q}(\sqrt{-D_0})$. Define $\alpha(D_0) = \prod\limits_{\substack{l \in \mathbb{P}\setminus\{2\} \\ l | D_0}} \left(l + \left(\frac{-1}{l}\right)\right)$.*

*For a prime $p$ let $\beta(p)$ be the number of elements $\alpha$ in the ring of integers of $\mathbb{Q}(\sqrt{-D_0})$ with $N(\alpha) = k(p-k)$ for some $1 \leq k \leq p-1$.*

*Define*
$$\epsilon(D_0) = \begin{cases} 24 & \text{if } D_0 \equiv 1 \bmod 4 \\ 8 & \text{if } D_0 \equiv 3 \bmod 4 \\ 4 & \text{if } D_0 \equiv 2 \bmod 4. \end{cases}$$
*Then*
$$\beta(p) = \frac{\epsilon(D_0)h(D_0)}{w \cdot \alpha(D_0)}(p+1) + O(p^{1/2}).$$

For the proof of this proposition we refer to [**7**]. In the same paper one finds: $h(D_0)/\alpha(D_0) = O(\log D_0 \log \log D_0/\sqrt{D_0})$. Using this we get the

**Corollary 3.3.** *Assume that $D_0$ is a square free number such that $\frac{\epsilon(D_0)h(D_0)}{w \cdot \alpha(D_0)} < 1$, for instance assume that $D_0$ is sufficiently large. Assume moreover that $E/K$ is an elliptic curve with $\mathrm{End}_{\bar{K}}(E) = O_{-D_0}$, the ring of integers of $\mathbb{Q}(\sqrt{-D_0})$. Then for $p$ large enough there are anti-isometric $G_K$-isomorphisms*
$$\alpha_p : E_p \to E_p$$
*such that $E \times E/\Delta_{\alpha_p}$ is the Jacobian of a curve $C$ which covers $E$ of degree $p$.*

*Proof.* By proposition 3.2 it follows that for large $p$ the union of the set of all endomorphisms of $E$ of degree $k(p-k)$ for some $1 \leq k \leq p-1$ has less than $p-1$ elements. □

This result can be used to prove the existence of curves of genus 2 defined over finite fields $k$ whose geometric fundamental group over $k$ is infinite:

**Corollary 3.4.** *Let $k_0$ be a finite field and let $E_0/k_0$ be an elliptic curve with $E_{0,2} \subset E_0(k_0)$ and $\mathrm{End}_{\bar{k}_0}(E_0) = O_{-D_0}$ where $D_0$ is a square free natural number with $\frac{\epsilon(D_0)h(D_0)}{w\alpha(D_0)} < 1$. Let $k_1$ be the quadratic extension of $k_0$.*
*Then there is (at least) one curve $C$ of genus 2 defined over $k_0$ such that for infinitely many primes $p_i$ there exists an unramified Galois cover $\tilde{C}_i$ of $C$ defined*

over $k_0$ whose Galois group contains $A_{p_i}$. Hence the composite $F$ of the function fields $F_i = F(C_i)$ is an unramified Galois extension of $F(C)$ with infinite Galois group and the algebraic closure of $k_0$ in $F$ is contained in $k_1$.

The proof of this corollary follows immediately from Corollary 3.3 and theorem 2.5 and the fact that there are only finitely many curves of genus 2 defined over $k_0$.

**4 Parametrization of anti-isometric level structures.** In Section 2 we discussed the relation between curves of genus 2 covering elliptic curves and anti-isometric isomorphisms between groups of torsion points of elliptic curves. In this section we'll use the ramification pattern of these coverings in order to "parametrize" them.

To simplify the exposition we shall assume that the field $K$ is quadratically closed, i.e. $K^{*2} = K^*$. (Recall that we always assume that char$(K) \neq 2, 3$.)

Next we fix an elliptic curve $E_0/K$ and again we simplify by assuming that $E_{0,2} \subset E(K)$.

Let
$$\pi : C \to E_0$$
be a cover of degree $p$ with $g(C) = 2$ and let $w$ be the hyperelliptic involution which acts as -id on $E_0$.

Then $K(U) := F(C)^w$ is a rational function field which contains $K(X) = F(E_0)^w$, and $\deg[K(U) : K(X)] = p$. We'll describe the ramification of $K(U)/K(X)$.

The Weierstrass points of $C$ are the six places of $\bar{K}(U)$ which ramify in $\bar{F}(C)/\bar{K}(U)$, these places are extensions of the four places of $\bar{K}(X)$ which ramify in $\bar{F}(E_0)/\bar{K}(X)$ and which correspond to points in $E_{0,2}$. Since the projection under $\pi$ of the Weierstrass points of $C$ maps three of these points to one point in $E_{0,2}$ and the remaining three Weierstrass points to different points in $E_{0,2}$ (cf. [2]) we can choose $X$ such that the places $\mathfrak{p}_0(X = 0), \mathfrak{p}_1(X = 1), \mathfrak{p}_\infty(\frac{1}{X} = 0)$ and $\mathfrak{p}_a(X = a)$ for some $a \in K^* \setminus \{1\}$ ramify in $F(E_0)/K(X)$ and such that $\mathfrak{p}_0, \mathfrak{p}_1, \mathfrak{p}_\infty$ have exactly one extension whose ramification order is prime to 2, and that $\mathfrak{p}_a$ has three such extensions.

If the cover $C \to E_0$ is of type I then only ramification of order two occurs, and so we get: $\mathfrak{p}_0, \mathfrak{p}_1, \mathfrak{p}_\infty$ have exactly one unramified extension and $\frac{p-1}{2}$ ramified extensions with ramification order 2, and $\mathfrak{p}_a$ has three unramified extensions and $\frac{p-3}{2}$ ramified extensions with ramification order 2. Moreover there is exactly one place $\mathfrak{p}$ of $\bar{K}(X)$ different from $\mathfrak{p}_0, \mathfrak{p}_1, \mathfrak{p}_\infty, \mathfrak{p}_a$ which has one extension ramified of order 2, and $\bar{K}(U)/\bar{K}(X)$ is unramified outside of $\{\mathfrak{p}_0, \mathfrak{p}_1, \mathfrak{p}_\infty, \mathfrak{p}_a, \mathfrak{p}\}$.

If $C \to E_0$ is of type II then one of the extensions of $\mathfrak{p}_0, \mathfrak{p}_1, \mathfrak{p}_\infty, \mathfrak{p}_a$ is ramified of order 3 or 4, and $\bar{K}(U)/\bar{K}(X)$ is unramified outside of $\{\mathfrak{p}_0, \mathfrak{p}_1, \mathfrak{p}_\infty, \mathfrak{p}_a\}$.

We can assume that $\mathfrak{p}_0$ and $\mathfrak{p}_1$ have only extensions with ramification order dividing 2, and so there are $\frac{p-1}{2}$ ramified extensions of $\mathfrak{p}_0$ and $\mathfrak{p}_1$ in $\bar{K}(U)/\bar{K}(X)$.

If $\mathfrak{p}_\infty$ has a ramified extension with ramification order 3 then $\mathfrak{p}_\infty$ and $\mathfrak{p}_a$ have $\frac{p-3}{2}$ extensions with ramification order 2, we get case i) of the discussion in Section 2.

In case II ii) $\mathfrak{p}_\infty$ has $\frac{p-1}{2}$ extensions and $\mathfrak{p}_a$ has $\frac{p-5}{2}$ extensions which ramify of order 2, $\mathfrak{p}_\infty$ has one extension which is unramified, and $\mathfrak{p}_a$ has two unramified extensions.

In case II iii) one of the places $\mathfrak{p}_\infty$ and $\mathfrak{p}_a$ has an extension which is ramified of order 4.

Now choose $U$ such that $\mathfrak{q}_0(U = 0), \mathfrak{q}_1(U = 1)$ and $\mathfrak{q}_\infty(\frac{1}{U} = 0)$ are the extensions of $\mathfrak{p}_0, \mathfrak{p}_1$ and $\mathfrak{p}_\infty$ whose ramification orders are odd. We note that $U$ is completely determined by this choice, and that $\mathfrak{q}_0$, $\mathfrak{q}_1$ and $\mathfrak{q}_\infty$ are of degree 1 over $K$.

The curve $C$ is given by the equation

$$Z^2 = U(U-1)f_3(U)$$

where $f_3$ is a polynomial of degree 3 in $K(U)$ whose roots correspond to the places of $K(U)$ extending $\mathfrak{p}_a$ with odd ramification order.

Conversely we can begin with $K(X) \subset K(U)$ with $[K(U) : K(X)] = p$ and the ramification pattern of $K(U)/K(X)$ as described above. Then $C : Z^2 = U(U-1)f_3(U)$ is a curve of genus 2 covering

$$E : Y^2 = X(X-1)(X-a)$$

of degree $p$.

Hence to determine all elliptic curves $E/K$ such that there exists $K$-rational anti-isometric isomorphisms

$$\alpha_p : E_{0,p} \to E_p$$

is equivalent with the determination of all subfields $K(X)$ of $K(U)$ with the ramification pattern as above.

So assume that

$$X = \frac{P(U)}{Q(U)}$$

with relatively prime $P, Q \in K(U)$ and $\deg(X) = \max(\deg(P), \deg Q) = p$.

The ramification conditions at $\mathfrak{p}_0$ resp. $\mathfrak{p}_\infty$ imply:

$$X = \frac{P_1(U)^2 U}{P_2(U)^2}$$

with

$$\deg P_1 = \frac{p-1}{2},$$

$$\deg P_2 = \begin{cases} \frac{p-3}{2} & \text{if type II case i occurs} \\ \frac{p-1}{2} & \text{else} \end{cases}$$

(Note that $K^{*2} = K^*$ and so constants are absorbed by $P_1$.)

Now look at the place $\mathfrak{p}_1(X = 1)$.

We get:

$$P_1(U)^2 U - P_2(U)^2 = S_1(U)^2(U-1)$$

with $S_1 \in K[U]$, $\deg S_1 = \frac{p-1}{2}$ and $S_1(1) \neq 0$.
So $Q_1 = \frac{P_1}{S_1}$ and $Q_2 = \frac{P_2}{S_1}$ are elements in $K(U)$ which solve

$$Q_2^2 - Q_1^2 U = (1-U)$$

a norm equation with respect to the extension $K(\sqrt{U})$ of $K(U)$ with $Q_i(1) \neq 0$.

We know one solution of this equation, namely

$$Q_1^0 = Q_2^0 = 1, \quad \text{and hence} \quad \frac{Q_2 + Q_1\sqrt{U}}{1+\sqrt{U}}$$

has norm 1. By Hilbert's theorem so it follows:

$$\frac{Q_2 + Q_1\sqrt{U}}{1+\sqrt{U}} = \frac{f_1(U) - f_2(U)\sqrt{U}}{f_1(U) + f_2(U)\sqrt{U}} \quad \text{with} \quad f_i \in K(U).$$

Hence

$$Q_2 + Q_1(\sqrt{U}) = \frac{f_1^2 + Uf_2^2 - 2f_1f_2 U + \sqrt{U}(f_1^2 + Uf_2^2 - 2f_1f_2)}{f_1^2 - Uf_2^2}.$$

So

$$Q_2 = \frac{f_1^2 + Uf_2^2 - 2f_1f_2 U}{f_1^2 - Uf_2^2}$$

and

$$Q_1 = \frac{f_1^2 + Uf_2^2 - 2f_1f_2}{f_1^2 - Uf_2^2}.$$

Because of homogeneity we can assume that $f_i \in K[U]$ and $f_1$ and $f_2$ are relatively prime with $f_1(1) \neq f_2(1)$. Hence $f_1^2 + Uf_2^2 - 2f_1f_2$ is prime to $f_1^2 - Uf_2^2$ and because of

$$\frac{P_1}{S_1} = \frac{f_1^2 + Uf_2^2 - 2f_1f_2}{f_1^2 - Uf_2^2}, \quad \frac{P_2}{S_1} = \frac{f_1^2 + Uf_2^2 - 2f_1f_2 U}{f_1^2 - Uf_2^2}$$

we get

$$P_1 = f_1^2 + Uf_2^2 - 2f_1f_2, \quad P_2 = f_1^2 + Uf_2^2 - 2f_1f_2 U, \quad S_1 = f_1^2 - Uf_2^2.$$

It remains to exploit the ramification pattern at $\mathfrak{p}_a$:
We must have:

$$P_1^2 U - aP_2^2 = S_2^2 f_3(U)$$

where $f_3 \in K[U]$ is a polynomial with three distinct zeros $\lambda_1, \lambda_2, \lambda_3$ in $\bar{K}$ different from 0 and 1. Hence

$$U(f_1^4 + U^2 f_2^4 + 4f_1^2 f_2^2 + 2Uf_1^2 f_2^2 - 4Uf_1 f_2^3 - 4f_1^3 f_2)$$
$$-a(f_1^4 + U^2 f_2^4 + 4f_1^2 f_2^2 U^2 + 2Uf_1^2 f_2^2$$
$$-4U^2 f_1 f_2^3 - 4Uf_1^3 f_2) = S_2^2 f_3(U).$$

Let us compute the degree of these polynomials:
Since $\deg(P_1) = \frac{p-1}{2}$ we get: If $p \equiv 1 \mod 4$:

$\deg f_1 = \frac{p-1}{4}$ and $\deg f_2 < \deg f_1$ for type I-covers
or for type II-covers, case ii

$\deg f_1 = \frac{p-1}{4}$ and $\deg f_2 = \frac{p-5}{4}$ else

If $p \equiv 3 \mod 4$:

$\deg f_2 = \frac{p-3}{4}$ and $\deg f_1 \leq \deg f_2$ for type I-covers or
type II-covers, case ii

$\deg f_2 = \frac{p-3}{4} = \deg f_1$ else.

Now put $w = \frac{f_1}{f_2}$:
Then

$$(U-a)w^4 + 4U(a-1)w^3 + (4U + 2U^2 - 4aU^2 - 2aU)w^2$$
$$+(4aU^2 - 4U^2)w + (U^3 - aU^2) = S_2^2 f_3(U).$$

Hence $(w, \frac{S_2}{f_2^2})$ is a solution of $E_{a,f_3(U)}$, the twist by $f_3(U)$ of the curve $E_a$ of genus 1 defined over $K(U)$ and given by the equation

$$E_a : (U-a)W^2 + 4U(a-1)W^3 + 2U(U(1-2a) + 2 - a)W^2$$
$$+4U^2(a-1)W + U^2(U-a) = Z^2$$

This solution has the special property that the degree of its $W$-coordinate is equal to $\frac{p-1}{4}$ if $p \equiv 1 \mod 4$, and equal to $\frac{p-3}{4}$, if $p \equiv 3 \mod 4$.

Conversely if $(w, z)$ is a solution of $E_{a,f_3(U)}$ where $f_3 \in K[U]$ is a separable polynomial of degree 3 prime to $U(U-1)$ and $w = \frac{f_1}{f_2}$; $f_i \in K[U]$ which satisfy the degree conditions from above with $f_i(0) \neq 0, f_1(0) \neq 2f_2(0)$ and $f_1(1) \neq f_2(1)$ then define,

$$P_1 := f_1^2 + f_2^2 U - 2f_1 f_2, \quad P_2 := f_1^2 + f_2^2 U - 2f_1 f_2 U$$

and

$$X := \frac{P_1^2 U}{P_2^2}.$$

Then $K(X) \subset K(U)$ has index $p$ with ramification pattern as required.
So we get

**Theorem 4.1.** *For $a \in K^* \setminus \{0, 1\}$ let $E_a$ be the curve of genus 1 defined over $K(U)$ by the equation*

$$E_a : Z^2 = (U-a)W^4 + 4U(a-1)W^3 + 2U(U(1-2a) + 2 - a)W^2$$
$$+4U^2(a-1)W + U^2(U-a).$$

*Let $f_3(U)$ be a separable polynomial in $K(U)$ with zeros different from 0 and 1. Then the curve*

$$C : V^2 = U(U-1)f_3(U)$$

is a curve of genus 2 covering the elliptic curve
$$Y^2 = X(X-1)(X-a)$$
of degree $p$ if and only if the quadratic twist of $E_a$ by $f_3$ has a $K(U)$-rational point $\left(\frac{f_1}{f_2}, z\right)$ with $f_i \in K[U]$, relatively prime, and

$$\deg f_1 \equiv \tfrac{p-1}{4}, \deg f_2 \leq \tfrac{p-5}{4} \quad \text{if } p \equiv 1 \bmod 4$$
$$\deg f_2 = \tfrac{p-3}{4}, \deg f_1 \leq \deg f_2 \quad \text{if } p \equiv 3 \bmod 4$$

and $f_1(1) \neq f_2(1), 2f_2(0) \neq f_1(0)$.

REMARK. 1) If $E$ is not isogenous to $Y^2 = f_3(U)$ then $\left(\frac{f_1}{f_2}, z\right)$ is a solution of minimal degree of $E_{a, f_3(U)}$.

2) If $K$ is a function field we know that Conjecture $(R_{E,K})$ is true for all elliptic curves $E/K$. Hence there is a number $M$ such that for $f_3(U) \in K[U]$ with height $(f_3) > M$ the equation $E_{a,f_3}$ has no solution in $K[U]$.

We end by discussing the cases $p = 3$ and $p = 5$.

First assume that $p = 3$.

In this case the degrees of $f_1$ and $f_2$ have to be zero. So we have to determine all $f_3(U) \in K[U]$ of the form

$$f_3(U) = (U-a)w^4 + 4U(a-1)w^3 + 2U(U(1-2a) + 2-a)w^2$$
$$+ 4U^2(a-1)w + U^2(U-a)$$

with $w \in K^* \setminus \{1, 2\}$.

For $f_3(U) = c(U^3 + \alpha U^2 + \beta U + \gamma)$ this implies:

$$c = 1$$
$$\gamma = -aw^4$$
$$\beta = w^4 + 4(a-1)w^3 + (4-2a)w^2$$
$$\alpha = (2-4a)w^2 + 4(a-1)w - a$$

Type II case i) occurs if $\deg P_2 = 0$ which is equivalent with $f_2^2 - 2f_1 f_2 = 0$, or: $w = \tfrac{1}{2}$.

The cases II ii) and II iii) cannot occur for $p = 3$.

We can interpret this result as follows:

**Proposition 4.2.** *Assume that $K^* = K^{*2}$ and that $E$ is an elliptic curve given by the equation:*
$$Y^2 = X(X-1)(X-a).$$
*Then the pairs {elliptic curves $E'$ defined over $K$, anti-isometric isomorphisms*
$$\alpha_3 : E_3 \to E'_3\}$$
*are parametrized by the rational family of curves of genus 2 given by the equation*

$$C_w : Y^2 = X(X-1)X^3 + [(4a-2)w^2 + (4a-4)w + a]X^2$$
$$+ [w^4 + (4a-4)w^3 + (4-2a)w^2]X$$
$$+ [(2-4a)w^2 + 4(a-1)w - a]$$

where $w$ runs over $K^*\backslash\{1,2\}$. The value $w = \frac{1}{2}$ gives rise to the unique covering of $E_0$ of degree 3 by a curve of genus 2 which is of type II.

REMARKS.

1) In [10] one finds a parametrization of elliptic curves $E'$ with <u>isometric</u> isomorphisms from $E$ to $E'_3$. We recall that $E$ has a point $P$ of order 2 by assumption, and if $\alpha : E_3 \to E'_3$ is isometric then $\alpha' : E/<P> \to E \overset{\alpha}{\to} E'$ is anti-isometric.

Hence Proposition 4.2 as well as the equations of [R-S] can be used to describe all elliptic curves $E'$ with isomorphisms from $E_3$ to $E'_3$.

2) Having $C_w$ it is possible to determine the complementary elliptic curve $E'$ of $E$ in $J_{C_w}$. If we want to determine its $j$-invariant we can assume that $E'_2 \subset E'(K)$ and so we have to find a rational function $Z = Z(U)$ of degree 3 in $K(U)$ with the following properties: Let $\lambda_1, \lambda_2, \lambda_3$ be the zeros of $U^3 + \alpha(w)U^2 + \beta(w)U + \gamma(w)$. Then the place $U = \lambda_1$ is the only unramified extension of $Z = 0$, $U = \lambda_2$ is the only unramified extension of $Z = 1$, $U = \lambda_3$ is the only unramified extension of $\frac{1}{Z} = 0$, and the place $Z = \beta$ has the extensions $U = 0$, $U = 1$ and $U = \infty$. Then $E'$ is given by $Y^2 = X(X-1)(X-\beta)$.

To determine $Z$ and $\beta$ one can proceed completely analogous as in the proof of proposition 4.2. Since Rubin and Silverberg give a very nice description of the curves $E'$ (up to a 2-isogeny) we don't carry out this procedure explicitely here.

Next we discuss the case $p = 5$.

We get: $\deg f_1 = 1$ and $\deg f_2 = 0$.

Take $w = w_1 U + w_2$ with $w_i \in K^*$, $w_1 + w_2 \neq 1$, $w_2 \neq 2$

We have to solve:

$(U-a)(w_1 U + w_2)^4 + (4U(a-1)(w_1 U + w_2)^2 + 2U(U + 2 - 2a - 4aU)(w_1 U + w_2)^2 + (4U^2 a - 4U^2)(w_1 U + w_2) + U(U-a) = (s_1 U + s_2)^2 f_3(U)$

with $f_3(U) = U^3 + \alpha U^2 + \beta U + \gamma$.

By comparing coefficients one gets the following system of equations:

(1) $\qquad -s_1^2 + w_1^4 = 0$

(2) $\qquad -(\alpha s_1^2) - 2s_1 s_2 + 2w_1^2 - 8aw_1^2 - aw_1^4 + 4w_1^3 w_2 = 0$

(3) $\qquad -(\beta s_1^2) - 2\alpha s_1 s_2 - s_2^2 - 4w_1 + 4aw_1$
$\qquad +4w_1 w_2 - 16aw_1 w_2 - 4aw_1^3 w_2 + 6w_1^2 w_2^2 = 0$

(4) $\qquad 1 - \gamma s_1^2 - 2\beta s_1 s_2 - \alpha s_2^2 - 4w_2 + 4aw_2$
$\qquad +2w_2^2 - 8aw_2^2 - 6aw_1^2 w_2^2 + 4w_1 w_2^3 = 0$

(5) $\qquad -a - 2\gamma s_1 s_2 \beta s_2^2 - 4aw_1 w_2^3 + w_2^4 = 0$

(6) $\qquad \gamma s_2^2 + aw_2^4 = 0$

Let us discuss type II. In case i) $P_2$ has to have degree 1, and so $w_1^2 U^2 + U - 2w_1 U^2$ has degree 1 which means that $w_1 = \frac{1}{2}$. Hence $\alpha_1$ has to be $\pm \frac{1}{4}$, and the remaining 5 inequalities determine $f_3(U)$ uniquely.

In case ii) $s_1 U - s_2$ has to divide $f_3(U)$, and so $\frac{s_2}{s_1}$ is one of the zeros of $f_3(U)$. Again we see that there are only finitely many possibilities to $f_3(U)$.

REMARKS.

1. It would be nice to "solve" this system of equations, i.e. to find a rational parametrization of the solutions.

2. These two examples illustrate the general situation, too. For each $p$ we'll get a system of $p+1$ equations with $(p+2)$ variables which determine all possible polynomials $f_3(U)$. Since for fixed $p$ and fixed $E$, the set of curves of genus 2 which cover $E$ of degree $p$, is parametrized by a curve $\tilde{X}(p)$ which is a twist of the modular curve $X(p)$, $X(p)$ is birationally equivalent over $\bar{K}$ to the variety defined by these equations. The occurence of type II implies one more condition either for the leading coefficients of $f_1$ and $f_2$ or for the zeros of $f_3(U)$, hence for fixed $p$ and $E$ this occurs only finitely often.

*References*

[1]  G. Frey, Links between solutions of $A - B = C$ and elliptic curves; in: Number Theory, Ulm 1987, Springer LNM 1380 (1989), (31-62).

[2]  G. Frey, E. Kani, Curves of genus 2 covering elliptic curves and an arithmetical application; in: Arithmetic Algebraic Geometry; Progr. Math. 89, 1991, Boston- Basel-Berlin, (153-175).

[3]  T. Hayashida, M. Nishi, Existence of curves of genus 2 on a product of two elliptic curves; J. Math. Soc. Japan (17) (1965) (1-16).

[4]  E. Kani, Curves of genus 2 with elliptic differentials and the height conjecture for elliptic curves; Proc. Conf. Numb. Theory and Arithm. Geom.; Preprint 18 (1991); Inst. f. Exp. Math. Essen; (30-39).

[5]  E. Kani, Curves with elliptic differentials; preprint.

[6]  E. Kani, The number of curves of genus two with elliptic differentials; preprint.

[7]  I. Kiming, On certain problems in the analytical arithmetic of quadratic forms arising from the theory of curves of genus 2 with elliptic differentials.

[8]  A.N. Parshin, The Bogomolov-Miyaoka-Yau inequality for arithmetical surfaces and its applications; Sem. Théorie des Nombres Paris 1986/87; Progr. Math. 75, Boston-Basel-Berlin (1989), 299-312.

[9]  K. Ribet, On modular representations of $G(\bar{\mathbb{Q}}/\mathbb{Q})$ arising from modular forms; Inv. Math. 100 (1990), 431-476.

[10] K. Rubin, A. Silverberg, Families of elliptic curves with constant mod $p$ representations, this volume.

[11] J.P. Serre, Sur les représentations modulaires de degré 2 de $G(\bar{\mathbb{Q}}/\mathbb{Q})$; Duke Math. J. 54 (1987), 179-230.

[12] J.P. Serre, Propriétés galoisiennes des points d'ordre fini des courbes elliptiques; Inv. Math. 15 (1972), 259-331.

[13] H. Wielandt, Finite permutation groups, sec. ed.; New York-London, 1968.

Conference on Elliptic Curves and Modular Forms
Hong Kong, December 18-21, 1993
Copyright ©1995 International Press

# Complete intersections and Gorenstein rings

H.W. Lenstra, Jr.
Department of Mathematics # 3840
University of California, Berkeley, CA 94720–3840 USA
*E-mail address*: HWL@MATH.BERKELEY.EDU

> This paper is devoted to the proof of the following fact from commutative algebra, which is a slight sharpening of a result of Wiles. Let $\mathcal{O}$ be a complete discrete valuation ring, $R$ a complete noetherian local $\mathcal{O}$-algebra, $B$ a finite flat local $\mathcal{O}$-algebra, and $\varphi\colon R \to B$, $\pi\colon B \to \mathcal{O}$ surjective $\mathcal{O}$-algebra homomorphisms. Suppose that the length of the $\mathcal{O}$-module $(\ker \pi\varphi)/(\ker \pi\varphi)^2$ is finite and bounded by the length of $\mathcal{O}/\pi(\mathrm{Ann}_B \ker \pi)$. Then $\varphi$ is an isomorphism and $B$ is a complete intersection.

We prove the following fact from commutative algebra, due to Wiles in the case that $B$ is a Gorenstein ring.

**Theorem.** *Let $\mathcal{O}$ be a complete discrete valuation ring, $R$ a complete noetherian local $\mathcal{O}$-algebra, $B$ a finite flat local $\mathcal{O}$-algebra, and $\varphi\colon R \to B$, $\pi\colon B \to \mathcal{O}$ surjective $\mathcal{O}$-algebra homomorphisms. Then the following are equivalent:*

(i) *the length of the $\mathcal{O}$-module $(\ker \pi\varphi)/(\ker \pi\varphi)^2$ is finite and less than or equal to the length of $\mathcal{O}/\pi(\mathrm{Ann}_B \ker \pi)$;*

(ii) *the length of the $\mathcal{O}$-module $(\ker \pi\varphi)/(\ker \pi\varphi)^2$ is finite and equal to the length of $\mathcal{O}/\pi(\mathrm{Ann}_B \ker \pi)$;*

(iii) *$B$ is a complete intersection, $\pi(\mathrm{Ann}_B \ker \pi) \neq 0$, and $\varphi$ is an isomorphism.*

The terms are explained below.

Rings are supposed to be commutative with 1. For the basic definitions from commutative algebra we refer to [1]. We write $\mathfrak{m}_R$ for the maximal ideal of a local ring $R$. By $\mathcal{O}$ we shall always denote a complete discrete valuation ring; the completeness assumption can be dropped, except where complete intersections are involved (this is mostly due to the naïve nature of our definition of a complete intersection below).

*Finite flat.* We shall call an $\mathcal{O}$-algebra *finite flat* if it is finitely generated and free as an $\mathcal{O}$-module. This is equivalent to it being finitely generated and *flat* as an $\mathcal{O}$-module, which is an easy fact that we shall not need (cf. [1], Exercise 7.16).

---

Key words: one-dimensional local ring, complete intersection, congruence ideal.
1991 Mathematics subject classification: 13H10. The author thanks B. de Smit, B. Mazur, and R. Pink for their assistance and comments. He was supported by NSF under grant No. DMS 92-24205. Part of the work reported in this paper was done while the author was on appointment as a Miller Research Professor in the Miller Institute for Basic Research in Science.

For a finitely generated free $\mathcal{O}$-module $M$ we shall put $M^\dagger = \mathrm{Hom}_\mathcal{O}(M, \mathcal{O})$; this is an autoduality of the category of finitely generated free $\mathcal{O}$-modules.

*Local $\mathcal{O}$-algebras.* A *local $\mathcal{O}$-algebra* is an $\mathcal{O}$-algebra $B$ that is local as a ring and for which the structure map $\mathcal{O} \to B$ maps $\mathfrak{m}_\mathcal{O}$ inside $\mathfrak{m}_B$.

*Gorenstein rings.* A finite flat local $\mathcal{O}$-algebra $B$ is called *Gorenstein* if $B^\dagger$ is free of rank 1 as a $B$-module. This is *not* a relative notion: there is an absolute notion of "Gorenstein ring" that is equivalent to the given one for finite flat local algebras over a discrete valuation ring (see [3], Section 18), and which we will not need.

*Complete intersections.* Let $B$ be a finite flat local $\mathcal{O}$-algebra that has the same residue class field as $\mathcal{O}$. The latter condition means that the natural map $\mathcal{O}/\mathfrak{m}_\mathcal{O} \to B/\mathfrak{m}_B$ is an isomorphism; it is satisfied if there is an $\mathcal{O}$-algebra homomorphism $B \to \mathcal{O}$, which is the case in the Theorem. We call $B$ a *complete intersection* if, for some non-negative integer $n$, there are elements $f_1, \ldots, f_n \in \mathcal{O}[[X_1, \ldots, X_n]]$ (the two $n$'s are the same!) such that $B \cong \mathcal{O}[[X_1, \ldots, X_n]]/(f_1, \ldots, f_n)$ as $\mathcal{O}$-algebras. Again, this is *not* a relative notion: there is an absolute notion of "complete intersection" that is equivalent to the given one for finite flat local algebras over a complete discrete valuation ring with the same residue class field (see [3], Section 21). We will not need this fact. What we do need about complete intersections is summarized in the following lemma.

**Lemma 1.** *Let $n$ be a non-negative integer and $f_1, \ldots, f_n \in \mathcal{O}[[X_1, \ldots, X_n]]$. Suppose that $B = \mathcal{O}[[X_1, \ldots, X_n]]/(f_1, \ldots, f_n)$ is finitely generated and non-zero as an $\mathcal{O}$-module. Then $B$ is a finite flat local $\mathcal{O}$-algebra with the same residue class field as $\mathcal{O}$; it is Gorenstein; and $B^\dagger$ has a $B$-generator $\lambda$ with the property that the trace map $\mathrm{Tr}_{B/\mathcal{O}}: B \to \mathcal{O}$ is given by $\mathrm{Tr}_{B/\mathcal{O}} = d \cdot \lambda$, where $d$ is the image of $\det\left(\dfrac{\partial f_i}{\partial X_j}\right)_{i,j}$ in $B$.*

*Proof. (sketch)* Let $\pi_\mathcal{O}$ be a prime element of $\mathcal{O}$. For the first statement, it suffices to check that $f_1, \ldots, f_n, \pi_\mathcal{O}$ is a "regular" $\mathcal{O}[[X_1, \ldots, X_n]]$-sequence. One way to do this is by means of the "Koszul-complex" ([3], Theorem 16.8). Once one knows about regular sequences and the Koszul complex, one can prove the remaining statements by means of an argument due to Tate ([4], Appendix). □

A more general version of Lemma 1, in which $\mathcal{O}$ is allowed to be any noetherian ring, is proved in [2]. The main tool in the proof is again the Koszul complex.

*The congruence ideal.* Let $B$, $C$ be rings and let $\pi: B \to C$ be a surjective ring homomorphism. Then the *congruence ideal* $\eta_\pi$ of $\pi$ is defined to be the $C$-ideal $\pi(\mathrm{Ann}_B \ker \pi)$, where $\mathrm{Ann}_B \ker \pi$ denotes the annihilator of $\ker \pi$ in $B$.

The terminology is explained by the following example. Let $C$, $D$ be rings with ideals $I$, $J$, and suppose that an isomorphism $C/I \cong D/J$ is given. Put $B = C \times_{C/I} D = \{(x,y) \in C \times D : x \text{ and } y \text{ have the same image in } C/I\}$, and let $\pi: B \to C$ be the first projection. Then $\ker \pi = \{0\} \times J$, and if $\mathrm{Ann}_D J = 0$ then $\mathrm{Ann}_B \ker \pi = I \times \{0\}$ so that $\eta_\pi = I$. Since the elements of $B$ are defined by means of a congruence mod $I$ (more precisely, an equality in $C/I$), the ideal $I$ may indeed be called a "congruence ideal".

The definition can be reformulated as follows. View $C$ as a $B$-algebra via $\pi$. Then there is an isomorphism $\mathrm{Hom}_B(C, B) \cong \mathrm{Ann}_B \ker \pi$ sending $f$ to $f(1)$, so $\eta_\pi$ is just the image of the map $\mathrm{Hom}_B(C, B) \to \mathrm{Hom}_B(C, C) \cong C$ induced by $\pi$.

If $\eta_\pi = C$ then the sequence $0 \to \ker \pi \to B \to C \to 0$ of $B$-modules splits, so that $B$ becomes a product of two rings. Hence if $B$ and $C$ are local, then one has $\eta_\pi = C$ if and only if $\pi$ is an isomorphism. A "relative" version of this statement will, under additional hypotheses, be proved below (Lemma 3).

Suppose now that $B$ is a flat $\mathcal{O}$-algebra and that $\pi\colon B \to \mathcal{O}$ is an $\mathcal{O}$-algebra homomorphism (necessarily surjective) with $\eta_\pi \neq 0$. We prove that

$$(2) \qquad (\ker \pi) \cap (\mathrm{Ann}_B \ker \pi) = 0,$$

so that the surjective map $\mathrm{Ann}_B \ker \pi \to \eta_\pi$ given by $\pi$ is actually an isomorphism. Let $x \in (\ker \pi) \cap (\mathrm{Ann}_B \ker \pi)$, choose $a \in \eta_\pi$, $a \neq 0$, and write $a = \pi(b)$ with $b \in \mathrm{Ann}_B \ker \pi$. Then we have $ax = (a - b)x = 0$, the first equality because $b \in \mathrm{Ann}_B \ker \pi$ and $x \in \ker \pi$, and the second because $a - b \in \ker \pi$ and $x \in \mathrm{Ann}_B \ker \pi$. Since $B$ is flat, multiplication by $a$ is injective, so $x = 0$, as required.

*Gorenstein rings and the congruence ideal.* Suppose that $B$ and $C$ are finite flat local $\mathcal{O}$-algebras that are Gorenstein, and let $\pi\colon B \to C$ be a surjective $\mathcal{O}$-algebra homomorphism. Choosing isomorphisms $B \cong B^\dagger$, $C \cong C^\dagger$ of $B$-modules we find that $\mathrm{Hom}_B(C, B) \cong_B \mathrm{Hom}_B(C^\dagger, B^\dagger)$. The latter module is, by duality, isomorphic to $\mathrm{Hom}_B(B, C)$, which is easily seen to be generated by the map $\pi$. Thus $\eta_\pi$ is a principal $C$-ideal, generated by the image of $\pi$ under the map $\mathrm{Hom}_B(B, C) \cong \mathrm{Hom}_B(C, B) \to C$. This can be used as an alternative definition of the congruence ideal in the Gorenstein situation.

**Lemma 3.** *Let $A$ and $B$ be finite flat local $\mathcal{O}$-algebras, and let $\varphi\colon A \to B$, $\pi\colon B \to \mathcal{O}$ be surjective $\mathcal{O}$-algebra homomorphisms. Suppose that $A$ is Gorenstein and that $\eta_{\pi\varphi} = \eta_\pi \neq 0$. Then $\varphi$ is an isomorphism.*

*Proof.* One easily checks that $\varphi$ induces a map $\mathrm{Ann}_A \ker \pi\varphi \to \mathrm{Ann}_B \ker \pi$. Applying (2) to $\pi$ and to $\pi\varphi$ we find that $\pi$ and $\pi\varphi$ induce isomorphisms $\mathrm{Ann}_B \ker \pi \to \eta_\pi$ and $\mathrm{Ann}_A \ker \pi\varphi \to \eta_{\pi\varphi}$. Thus from $\eta_{\pi\varphi} = \eta_\pi$ it follows that $\mathrm{Ann}_A \ker \pi\varphi \to \mathrm{Ann}_B \ker \pi$ is an isomorphism as well, and that $\varphi \mathrm{Ann}_A \ker \pi\varphi = \mathrm{Ann}_B \ker \pi$. Therefore we have

$$A/(\ker \varphi + \mathrm{Ann}_A \ker \pi\varphi) \cong \varphi A/\varphi \mathrm{Ann}_A \ker \pi\varphi = B/\mathrm{Ann}_B \ker \pi,$$

which is free as an $\mathcal{O}$-module since $B/\mathrm{Ann}_B \ker \pi$ can be viewed as a submodule of $\mathrm{End}_\mathcal{O} \ker \pi$. Also, applying (2) to $\pi\varphi$ we obtain

$$\ker \varphi \cap \mathrm{Ann}_A \ker \pi\varphi \subset \ker \pi\varphi \cap \mathrm{Ann}_A \ker \pi\varphi = 0.$$

We conclude that there is an exact sequence of $A$-modules

$$0 \to \ker \varphi \oplus \mathrm{Ann}_A \ker \pi\varphi \to A \to B/\mathrm{Ann}_B \ker \pi \to 0$$

consisting of finitely generated free $\mathcal{O}$-modules. Dualizing, we obtain an exact sequence of $A$-modules

$$0 \to (B/\operatorname{Ann}_B \ker \pi)^\dagger \to A^\dagger \to (\ker \varphi)^\dagger \oplus (\operatorname{Ann}_A \ker \pi\varphi)^\dagger \to 0.$$

Since $A$ is supposed to be Gorenstein, we have $A^\dagger \cong_A A$. Tensoring with the residue class field $k$ of $A$ we find that $\dim_k(A^\dagger \otimes_A k) = 1$. By the exact sequence, this implies that one of $\dim_k((\ker \varphi)^\dagger \otimes_A k)$ and $\dim_k((\operatorname{Ann}_A \ker \pi\varphi)^\dagger \otimes_A k)$ is 0. Hence by Nakayama's lemma and duality one of $\ker \varphi$ and $\operatorname{Ann}_A \ker \pi\varphi$ is 0. But $\operatorname{Ann}_A \ker \pi\varphi \cong \eta_{\pi\varphi} \neq 0$, so $\ker \varphi = 0$ and $\varphi$ is an isomorphism. □

The condition that $A$ be Gorenstein cannot be omitted in Lemma 3. This is shown by the example $A = \{(x,y,z) \in \mathcal{O} \times \mathcal{O} \times \mathcal{O} : x \equiv y \equiv z \bmod \mathfrak{m}_\mathcal{O}\}$, $B = \{(x,y) \in \mathcal{O} \times \mathcal{O} : x \equiv y \bmod \mathfrak{m}_\mathcal{O}\}$, $\varphi((x,y,z)) = (x,y)$, $\pi((x,y)) = x$, in which $\eta_{\pi\varphi} = \eta_\pi = \mathfrak{m}_\mathcal{O}$. The ring $B$ in this example is Gorenstein (even a complete intersection).

*Intermezzo on the Fitting ideal.* Let $B$ be a ring and let $M$ be a finitely generated $B$-module, with generators $m_1, \ldots, m_r$. Let $f: B^r \to M$ map $(b_i)_{i=1}^r$ to $\sum_i b_i m_i$. Then the *Fitting ideal* $F_B M$ is the $B$-ideal generated by all elements of $B$ of the form $\det(v_1, \ldots, v_r)$, with $v_1 \in \ker f, \ldots, v_r \in \ker f$ (viewed as column vectors); evidently, it suffices to let the $v_i$ range over a set of *generators* for $\ker f$. The Fitting ideal is independent of the choice of the generators $m_i$. To see this, let $m_{r+1} = \sum_{i=1}^r c_i m_i$, with $c_i \in B$. One obtains generators for the kernel of $f': B^{r+1} \to M$, $f'((b_i)_{i=1}^{r+1}) = \sum_{i=1}^{r+1} b_i m_i$, by taking generators for $\ker f$ (with a zero coordinate appended) together with the element $(-c_1, \ldots, -c_r, 1)$. The latter element will have to occur in any non-zero determinant built up from these generators of $\ker f'$. It follows that the Fitting ideal does not change if the system of generators $m_1, \ldots, m_r$ is changed into $m_1, \ldots, m_r, m_{r+1}$. Inductively, this implies that any two systems $m_1, \ldots, m_r$ and $m'_1, \ldots, m'_s$ of generators give rise to the same Fitting ideal as their union $m_1, \ldots, m_r, m'_1, \ldots, m'_s$.

We need three properties of the Fitting ideal. The first is

(4) $$F_B M \subset \operatorname{Ann}_B M.$$

Namely, if $\sum_{j=1}^r v_{ij} m_j = 0$ for $1 \leq i \leq r$, then "multiplying by the adjoint" we see that $\det(v_{ij})$ annihilates each $m_j$ and therefore $M$. Secondly, we have

(5) $$F_C(M \otimes_B C) = \pi(F_B M)$$

when $\pi: B \to C$ is a surjective ring homomorphism. This is because $M \otimes_B C$ is, as a $C$-module, defined by the 'same' relations as those that define $M$ as a $B$-module.

Thirdly, for $B = \mathcal{O}$ the Fitting ideal just measures the length: if $M$ is a finitely generated $\mathcal{O}$-module, then $F_\mathcal{O} M = \mathfrak{m}_\mathcal{O}^{\operatorname{length} M}$ (if $M$ does not have finite length, interpret the right side to be 0). To prove this, one writes $M$ as a direct sum of cyclic modules $\mathcal{O}/I_i$ and checks that $F_\mathcal{O} M$ is the product of the ideals $I_i$.

*The congruence ideal and* $(\ker \pi)/(\ker \pi)^2$. Let $B$ and $C$ be rings and let $\pi: B \to C$ be a surjective ring homomorphism for which $\ker \pi$ is finitely generated. Then $(\ker \pi)/(\ker \pi)^2$ is a $C$-module, and one has

(6) $$F_C((\ker \pi)/(\ker \pi)^2) \subset \eta_\pi \subset \operatorname{Ann}_C((\ker \pi)/(\ker \pi)^2).$$

Namely, we have $F_B(\ker \pi) \subset \mathrm{Ann}_B \ker \pi \subset \mathrm{Ann}_B((\ker \pi)/(\ker \pi)^2)$, the first inclusion by (4) and the second one trivially. Now apply $\pi$. By (5) and $\mathrm{Ann}_B((\ker \pi)/(\ker \pi)^2) = \pi^{-1} \mathrm{Ann}_C((\ker \pi)/(\ker \pi)^2)$ this gives (6).

In the case that is of interest to us, the first inclusion of (6) can be found in [5], Proposition 6.2, with a rather more complicated proof.

*Complete intersections and the congruence ideal.* Let $B$ be a finite flat local $\mathcal{O}$-algebra, let $\pi\colon B \to \mathcal{O}$ be an $\mathcal{O}$-algebra map, and suppose that $B$ is a complete intersection. Then we have

(7)  $$F_\mathcal{O}((\ker \pi)/(\ker \pi)^2) = \eta_\pi.$$

This is proved by an explicit computation. Let $B = \mathcal{O}[[X_1,\ldots,X_n]]/(f_1,\ldots,f_n)$. The images $b_j$ of $X_j$ in $B$ belong to $\mathfrak{m}_B$, and replacing $X_j$ by $X_j - \pi(b_j)$ we may assume that $b_j \in \ker \pi$. Then $f_i(0) = 0$ for all $i$. To describe the $\mathcal{O}$-module $(\ker \pi)/(\ker \pi)^2$, one considers the ideal $\mathfrak{p}$ of $\mathcal{O}[[X_1,\ldots,X_n]]$ generated by $X_1,\ldots,X_n$; then $\mathfrak{p}/\mathfrak{p}^2$ is $\mathcal{O}$-free of rank $n$, and $(\ker \pi)/(\ker \pi)^2$ is $\mathfrak{p}/\mathfrak{p}^2$ modulo the submodule spanned by the images of $f_1, \ldots, f_n$. The definition of the Fitting ideal now gives

$$F_\mathcal{O}((\ker \pi)/(\ker \pi)^2) = \mathcal{O} \det\left(\frac{\partial f_i}{\partial X_j}(0)\right)_{i,j} = \mathcal{O}\pi(d),$$

with $d$ as in Lemma 1. To prove (7), it suffices, by (6), to prove the inclusion $\supset$. Let $x \in \eta_\pi$, and write $x = \pi(y)$ with $y \in \mathrm{Ann}_B \ker \pi$. By Lemma 1, we can choose $\lambda \in B^\dagger$ with $\mathrm{Tr}_{B/\mathcal{O}} = d\lambda$. From $\pi(d) - d \in \ker \pi$ we see that $(\pi(d) - d)y = 0$, so $\pi(d)\lambda(y) = (d\lambda)(y) = \mathrm{Tr}_{B/\mathcal{O}}(y)$. The trace of $y$ can be computed from the action of $y$ on the exact sequence $0 \to \ker \pi \to B \to \mathcal{O} \to 0$, and one finds that $\mathrm{Tr}_{B/\mathcal{O}}(y) = \pi(y) = x$. Therefore $x = \pi(d)\lambda(y) \in \mathcal{O}\pi(d) = F_\mathcal{O}((\ker \pi)/(\ker \pi)^2)$, as required.

The following result shows that one can recognize isomorphisms to complete intersections by looking at $(\ker \pi)/(\ker \pi)^2$.

**Lemma 8.** *Let $R$ be a complete noetherian local $\mathcal{O}$-algebra, let $B$ be a finite flat local $\mathcal{O}$-algebra, and let $\varphi\colon R \to B$, $\pi\colon B \to \mathcal{O}$ be surjective $\mathcal{O}$-algebra homomorphisms. Suppose that $B$ is a complete intersection, that the map $(\ker \pi\varphi)/(\ker \pi\varphi)^2 \to (\ker \pi)/(\ker \pi)^2$ induced by $\varphi$ is an isomorphism, and that these modules are of finite length over $\mathcal{O}$. Then $\varphi$ is an isomorphism.*

*Proof.* Let $n$, $f_i$ and the elements $b_j \in \ker \pi$ be as above, so that $\sum_j a_{ij} b_j \in (\ker \pi)^2$, where $a_{ij} = \frac{\partial f_i}{\partial X_j}(0)$. Since $(\ker \pi)/(\ker \pi)^2$ is of finite length we have $\det(a_{ij}) \neq 0$. Choose $r_j \in R$ with $\varphi(r_j) = b_j$. The hypothesis of the lemma implies that $r_1, \ldots, r_n$ generate $\ker \pi\varphi$ modulo $(\ker \pi\varphi)^2$, so by Nakayama's lemma they generate $\ker \pi\varphi$. Hence together with $\mathfrak{m}_\mathcal{O}$ they generate $\mathfrak{m}_R$, so that there is a surjective $\mathcal{O}$-algebra map $\psi\colon \mathcal{O}[[X_1,\ldots,X_n]] \to R$ sending $X_j$ to $r_j$. The hypothesis of the lemma implies that $\sum_j a_{ij} r_j \in (\ker \pi\varphi)^2$, so there are $g_i \in \ker \psi$ with $\frac{\partial g_i}{\partial X_j}(0) = a_{ij}$. We have $g_i \in \ker \varphi\psi = (f_1,\ldots,f_n)$ and therefore $g_i = \sum_l h_{il} f_l$, with $h_{il} \in \mathcal{O}[[X_1,\ldots,X_n]]$. From $a_{ij} = \frac{\partial g_i}{\partial X_j}(0) =$

$\sum_l h_{il}(0) \frac{\partial f_l}{\partial X_j}(0) = \sum_l h_{il}(0) a_{lj}$ and $\det(a_{ij}) \neq 0$ we see that $(h_{il}(0))$ is the identity matrix. This implies that the matrix $(h_{il})$ is invertible, so that $f_l \in \ker \psi$. Hence the map $\psi$ factors through $B$ and gives a map $B \to R$ that is inverse to $\varphi$. This proves the lemma. □

We need one more technical result before we can prove the Theorem.

**Lemma 9.** *Let $B$ be a finite flat local $\mathcal{O}$-algebra, and let $\pi \colon B \to \mathcal{O}$ be an $\mathcal{O}$-algebra homomorphism. Then there is a finite flat local $\mathcal{O}$-algebra $A$, together with a surjective $\mathcal{O}$-algebra homomorphism $\varphi \colon A \to B$, such that $A$ is a complete intersection and the map $(\ker \pi\varphi)/(\ker \pi\varphi)^2 \to (\ker \pi)/(\ker \pi)^2$ induced by $\varphi$ is an isomorphism.*

*Proof.* Let $b_1, \ldots, b_n$ generate $\ker \pi$. We first prove that $\mathcal{O}[b_1, \ldots, b_n] = B$. Let $C = \mathcal{O}[b_1, \ldots, b_n]$. Since $B$ is finite over $C$ the ring $C$ is local, and its maximal ideal $\mathfrak{m}_C = \mathfrak{m}_B \cap C$ contains $b_1, \ldots, b_n$. Clearly $C$ is Noetherian. We have $B = \mathcal{O} + \ker \pi = \mathcal{O} + (\sum_j B b_j) \subset C + \mathfrak{m}_C \cdot B$, so Nakayama's lemma implies that $C = B$.

The surjective $B$-linear map $B^n \to \ker \pi$ sending $(c_j)_{j=1}^n$ to $\sum_j c_j b_j$ gives upon tensoring with $\mathcal{O}$ a surjective map $\mathcal{O}^n \to (\ker \pi)/(\ker \pi)^2$. Choose generators $(a_{ij})_{j=1}^n$, $i = 1, \ldots, n$, for the kernel of the latter map. This can be done, since every submodule of $\mathcal{O}^n$ is generated by $n$ elements. For each $i$ we have $\sum_j a_{ij} b_j \in (\ker \pi)^2$, so $g_i(b_1, \ldots, b_n) = 0$ for some polynomial $g_i \in \mathcal{O}[X_1, \ldots, X_n]$ of the form $g_i = (\sum_j a_{ij} X_j) +$ (terms of degree $\geq 2$); here, and below, "degree" means "total degree".

Since $B$ is finite over $\mathcal{O}$, there is a non-negative integer $m$ with the property that the expressions $\prod_j b_j^{m_j}$ of degree $\sum_j m_j \leq m$ span $B$ as an $\mathcal{O}$-module. Enlarging $m$, if necessary, we can achieve that each $g_i$ has degree at most $m+2$. Write $b_i^{m+1} = h_i(b_1, \ldots, b_n)$, where $h_i \in \mathcal{O}[X_1, \ldots, X_n]$ has degree at most $m$. We define

$$f_i = X_i^{m+3} - X_i^2 h_i + g_i \in \mathcal{O}[X_1, \ldots, X_n] \qquad (1 \leq i \leq n).$$

Evidently, we have

$$f_i(b_1, \ldots, b_n) = 0,$$
$$f_i = X_i^{m+3} + \text{(terms of degree} \leq m+2),$$
$$f_i = \sum_j a_{ij} X_j + \text{(terms of degree} \geq 2)$$

for $1 \leq i \leq n$.

Put $D = \mathcal{O}[X_1, \ldots, X_n]/(f_1, \ldots, f_n)$. Then there is a surjective $\mathcal{O}$-algebra map $\psi \colon D \to B$ sending the image of $X_j$ to $b_j$. Each monomial of degree greater than $n(m+2)$ in $X_1, \ldots, X_n$ is divisible by $X_i^{m+3}$ for some $i$, so is modulo $f_i$ congruent to an $\mathcal{O}$-linear combination of monomials of smaller degrees. This implies that the monomials of degree at most $n(m+2)$ span $D$ as an $\mathcal{O}$-module, so that $D$ is finite over $\mathcal{O}$. Since $\mathcal{O}$ is complete, it follows that $D$ is a product of complete local rings: $D = \prod_\mathfrak{n} D_\mathfrak{n}$, where $\mathfrak{n}$ ranges over the maximal ideals of $D$; to see this, write for each positive integer $t$ the Artin ring $D/\mathfrak{m}_\mathcal{O}^t D$ as a product of local rings (see [1], Theorem 8.7), and take the projective limit over $t$. One of these maximal ideals is the image of the maximal ideal $\mathfrak{m} = (\mathfrak{m}_\mathcal{O}, X_1, \ldots, X_n)$ of $\mathcal{O}[X_1, \ldots, X_n]$ in $D$; so we have $D = D' \times D_\mathfrak{m}$, where $\mathfrak{m}D' = D'$ and $D_\mathfrak{m}$ is complete. Thus, if we complete at $\mathfrak{m}$ then the equality $D = \mathcal{O}[X_1, \ldots, X_n]/(f_1, \ldots, f_n)$ turns into $D_\mathfrak{m} = \mathcal{O}[[X_1, \ldots, X_n]]/(f_1, \ldots, f_n)$, and the surjection $\psi \colon D \to B$ turns into a surjection $\varphi \colon D_\mathfrak{m} \to B$. By Lemma 1 it follows that $D_\mathfrak{m}$ is a complete intersection. From $f_i = \sum_j a_{ij} X_j +$ (terms of degree $\geq 2$) we see that the kernel of the surjective map $\mathcal{O}^n \to (\ker \pi\varphi)/(\ker \pi\varphi)^2$ that sends $(c_j)_{j=1}^n$ to the image of $\sum_j c_j X_j$ is generated by the elements $(a_{ij})_{j=1}^n$, $i = 1, \ldots, n$. This implies that the map $(\ker \pi\varphi)/(\ker \pi\varphi)^2 \to (\ker \pi)/(\ker \pi)^2$ induced by $\varphi$ is an isomorphism. The lemma follows, with $A = D_\mathfrak{m}$. $\square$

**Corollary 10.** *Let $B$ be a finite flat local $\mathcal{O}$-algebra, and let $\pi \colon B \to \mathcal{O}$ be an $\mathcal{O}$-algebra homomorphism with $\eta_\pi \neq 0$. Then $B$ is a complete intersection if and only if $F_\mathcal{O}((\ker \pi)/(\ker \pi)^2) = \eta_\pi$.*

*Proof.* "Only if" we know from (7). To prove "if", we choose $\varphi \colon A \to B$ as in Lemma 9. Then we have

$$\eta_{\pi\varphi} = F_\mathcal{O}((\ker \pi\varphi)/(\ker \pi\varphi)^2) = F_\mathcal{O}((\ker \pi)/(\ker \pi)^2) = \eta_\pi,$$

the first equality by (7), the second from Lemma 9, and the last by hypothesis. Lemma 1 asserts that $A$ is Gorenstein. Now apply Lemma 3. $\square$

We prove the Theorem. The implication (iii)$\Rightarrow$(ii) is immediate from (7) and the second inclusion of (6), and (ii)$\Rightarrow$(i) is clear. To prove (i)$\Rightarrow$(iii), we note that

$$\eta_\pi \subset F_\mathcal{O}((\ker \pi\varphi)/(\ker \pi\varphi)^2) \subset F_\mathcal{O}((\ker \pi)/(\ker \pi)^2) \subset \eta_\pi,$$

the first inclusion by the hypothesis in (i), the second because there is a surjective map $(\ker \pi\varphi)/(\ker \pi\varphi)^2 \to (\ker \pi)/(\ker \pi)^2$, and the third by (6). We conclude that we have equality everywhere. The finite length hypothesis in (i) now implies that $\eta_\pi \neq 0$, so Corollary 10 shows that $B$ is a complete intersection. Lemma 8, finally, shows that $\varphi$ is an isomorphism. This completes the proof of the Theorem.

REMARK. R. Pink pointed out that Lemma 1, of which we only sketched the proof, can be bypassed entirely. To do this one verifies the conclusion of Lemma 1 and the equality (7) directly for the only ring to which it needs to be applied, namely the ring $A$ constructed in the proof of Lemma 9. One proceeds as follows.

One starts by proving that the $\mathcal{O}$-algebra $D = \mathcal{O}[X_1, \ldots, X_n]/(f_1, \ldots, f_n)$ constructed in the proof of Lemma 9 has the two following properties: first, $D$ is free of rank $(m + 3)^n$ as an $\mathcal{O}$-module, the images of the monomials $\prod_{i=1}^n X_i^{k_i}$ with $0 \leq k_i \leq m + 2$ forming a basis; and, second, the $D$-module

$D^\dagger = \operatorname{Hom}_\mathcal{O}(D, \mathcal{O})$ is free of rank 1, a basis being formed by the linear map $\lambda \colon D \to \mathcal{O}$ that sends the monomial $\prod_{i=1}^n X_i^{m+2}$ to 1 and the other basis elements to 0. The proof of these properties is a straightforward verification that exploits the shape of the relations $f_i$. It follows that $D$ is Gorenstein. From $D = D' \times A$ it follows that $A$ is Gorenstein as well.

Next one studies $\eta_{\pi\psi}$, where $\psi \colon D \to B$ is as in the proof of Lemma 9. Since $D$ is Gorenstein, one has $\operatorname{Ann}_D(\ker \pi\psi) = \operatorname{Hom}_D(\mathcal{O}, D) \cong \operatorname{Hom}_D(D, \mathcal{O}) = \mathcal{O} \cdot \pi\psi$, which shows that $\operatorname{Ann}_D(\ker \pi\psi)$ is free of rank 1 over $\mathcal{O}$. To exhibit a generator, one writes $f_i = \sum_{j=1}^n f_{ij} X_j$, where the polynomials $f_{ij}$ are such that $f_{ii} - X_i^{m+2}$ and $f_{ij}$ (for $i \neq j$) have degree at most $m + 1$. From (4), with $M = \ker \pi\psi$, one sees that $\det(f_{ij})$ belongs to $\operatorname{Ann}_D(\ker \pi\psi)$, and it is in fact a generator of $\operatorname{Ann}_D(\ker \pi\psi)$ since $\lambda(\det(f_{ij})) = 1$. Applying $\pi\psi$ one finds that $\eta_{\pi\psi}$ is generated by $\det(a_{ij})$. This is the same as saying that $F_\mathcal{O}((\ker \pi\psi)/(\ker \pi\psi)^2) = \eta_{\pi\psi}$. Passing to $A$ one concludes that $F_\mathcal{O}((\ker \pi\varphi)/(\ker \pi\varphi)^2) = \eta_{\pi\varphi}$.

Now that one knows the Gorenstein property and equality (7) for the complete intersections constructed in Lemma 9 one can pass to the more general case of Corollary 10. That is, if $B$ and $\pi$ are as in Corollary 10, then $B$ is a complete intersection if and only if $B$ is a complete intersection and Gorenstein, and if and only if $F_\mathcal{O}((\ker \pi)/(\ker \pi)^2) = \eta_\pi$. To prove this, suppose that $B$ has one of these properties, and let $\varphi \colon A \to B$ be as in Lemma 9. Since $A$ is known to have all three properties, it suffices to show that $\varphi$ is an isomorphism. In the case that $F_\mathcal{O}((\ker \pi)/(\ker \pi)^2) = \eta_\pi$ this is done as in the proof of Corollary 10 given above. In the other cases $B$ is a complete intersection, so $\varphi$ is an isomorphism by Lemma 8; note that $(\ker \pi)/(\ker \pi)^2$ has finite length by the second inclusion of (6).

In all our results we assumed that the finite flat local $\mathcal{O}$-algebra $B$ is provided with an $\mathcal{O}$-algebra homomorphism $\pi \colon B \to \mathcal{O}$. Similar results can be proved for more general finite flat $\mathcal{O}$-algebras $B$. The role of $\pi$ can then be played by the multiplication map $\mu \colon B \otimes_\mathcal{O} B \to B$, which is defined by $\mu(b_1 \otimes b_2) = b_1 b_2$, and the role of the base ring $\mathcal{O}$ is taken over by $B$. As an example, we prove the following proposition, which was suggested by B. Mazur. Recall that the module $\Omega_{B/\mathcal{O}}$ of Kähler differentials is defined to be the $B$-module $(\ker \mu)/(\ker \mu)^2$ (see [**3**], Section 25).

**Proposition.** *Let $B$ be a finite flat local $\mathcal{O}$-algebra that has the same residue class field as $\mathcal{O}$, and denote by $\mu$ the multiplication map $B \otimes_\mathcal{O} B \to B$. Suppose that the $B$-module $\Omega_{B/\mathcal{O}}$ has finite length. Then $B$ is a complete intersection if and only if the congruence ideal $\eta_\mu$ is principal and equal to $F_B(\Omega_{B/\mathcal{O}})$.*

We note that the finite length condition for $\Omega_{B/\mathcal{O}}$ is equivalent to the $K$-algebra $B \otimes_\mathcal{O} K$ being étale, where $K$ is the field of fractions of $\mathcal{O}$; if $K$ has characteristic 0, then it equivalent to the nil-radical of $B$ being zero.

The proof of the Proposition is analogous to the proof of Corollary 10. We go through the changes that need to be made.

For the "only if" part, suppose that $B \cong \mathcal{O}[[X_1, \ldots, X_n]]/(f_1, \ldots, f_n)$ as $\mathcal{O}$-algebras. Since $B$ is finite flat over $\mathcal{O}$, there is a $B$-algebra isomorphism $\mathcal{O}[[X_1, \ldots, X_n]] \otimes_\mathcal{O} B \cong B[[X_1, \ldots, X_n]]$. This implies that we have $B \otimes_\mathcal{O} B \cong$

$B[[X_1, \ldots, X_n]]/(f_1, \ldots, f_n)$ as $B$-algebras, where $B \otimes_{\mathcal{O}} B$ is viewed as a $B$-algebra via the second factor. As in the first half of the proof of (7) one now checks that $F_B(\Omega_{B/\mathcal{O}})$ equals the principal ideal $B \cdot d$, where $d$ is as in Lemma 1. The equality $\text{Tr}_{B/\mathcal{O}} = d\lambda$ from Lemma 1 implies that $\text{Tr}_{B \otimes B/B} = (d \otimes 1)(\lambda \otimes 1)$. This is used to show that $\eta_\mu \subset B \cdot d$, as in the second half of the proof of (7). Hence the inclusion $\eta_\mu \subset F_B(\Omega_{B/\mathcal{O}})$ holds, and by (6) one has equality. This proves the "only if" part.

The proof of the "if" part depends on the following generalization of Lemma 3. Let a finite flat local algebra over a noetherian local ring $C$ be defined in the same way as for $C = \mathcal{O}$.

**Lemma 11.** *Let $C$ be a one-dimensional local noetherian ring, let $A$ and $B$ be finite flat local $C$-algebras, and let $\varphi \colon A \to B$, $\pi \colon B \to C$ be surjective $C$-algebra homomorphisms. Suppose that $\text{Hom}_C(A, C)$ is free of rank 1 as an $A$-module, that $\eta_{\pi\varphi} = \eta_\pi$, and that $\eta_\pi$ is free of rank 1 as a $C$-module. Then $\varphi$ is an isomorphism.*

*Proof.* If $\eta_\pi = C$ then $\pi\varphi$ and $\pi$ are both isomorphisms, so $\varphi$ is an isomorphism as well. For the rest of the proof we assume that $\eta_\pi \neq C$, so that $\eta_\pi \subset \mathfrak{m}_C$. Since $\eta_\pi$ is $C$-free of rank 1, we have $\eta_\pi = Ca$, where $a$ is a non-zero-divisor of $C$. The proof of (2) now carries through. Hence $\pi$ induces an isomorphism $\text{Ann}_B \ker \pi \to \eta_\pi$.

Let $\mathfrak{p}$ be a minimal prime ideal of $C$. Then $C_\mathfrak{p}$ is an Artin ring, and $(\eta_\pi)_\mathfrak{p}$ is a $C_\mathfrak{p}$-ideal that is free of rank 1. Since $C_\mathfrak{p}$ has finite length as a module over itself, this implies that $(\eta_\pi)_\mathfrak{p} = C_\mathfrak{p}$, so $\eta_\pi \not\subset \mathfrak{p}$. Because $C$ is one-dimensional, this implies that the only prime ideal of the ring $C/\eta_\pi$ is $\mathfrak{m}_C/\eta_\pi$. It follows that $C/\eta_\pi$ is a local Artin ring. Therefore there exists $c \in C$, $c \notin \eta_\pi$, such that $c\mathfrak{m}_C \subset \eta_\pi$.

Next we prove that $B/\text{Ann}_B \ker \pi$ is free as a $C$-module. Write $I = \text{Ann}_B \ker \pi$, so that $I \cong \eta_\pi$. Both $B$ and $I$ are $C$-free, so it suffices to prove that a basis for $I$ can be supplemented to a basis for $B$. By Nakayama's lemma, this can be done if the natural map $I/\mathfrak{m}_C I \to B/\mathfrak{m}_C B$ is injective, i.e., if $\mathfrak{m}_C I = I \cap \mathfrak{m}_C B$. Let $x \in I \cap \mathfrak{m}_C B$. Then $cx \in I \cap c\mathfrak{m}_C B \subset I \cap \eta_\pi B = I \cap aB$. Since $a$ acts as a non-zero-divisor on the free $C$-module $B$, it follows from the definition of $I$ that $I \cap aB = aI$, which equals $\eta_\pi I$. Hence $x$ is an element of $I$ with $cx \in \eta_\pi I$. Since $I$ is $C$-free and $c \notin \eta_\pi$, this implies that $x \in \mathfrak{m}_C I$, as required.

Once (2) and the fact that $B/\text{Ann}_B \ker \pi$ is $C$-free are known, the proof that we gave for Lemma 3 generalizes easily to a proof for Lemma 11. $\square$

Let now $B$ and $\mu$ be as in the Proposition, and suppose that the congruence ideal $\eta_\mu$ is principal and equal to $F_B(\Omega_{B/\mathcal{O}})$. We wish to prove that $B$ is a complete intersection.

View $B \otimes_{\mathcal{O}} B$ as a $B$-algebra via the second factor. We start by constructing a finite flat local $B$-algebra $A$ of the form $A = B[[X_1, \ldots, X_n]]/(f_1, \ldots, f_n)$ together with a surjective $B$-algebra homomorphism $\psi \colon A \to B \otimes_{\mathcal{O}} B$ for which the induced map $(\ker \mu\varphi)/(\ker \mu\varphi)^2 \to (\ker \mu)/(\ker \mu)^2$ is an isomorphism. This is done as in the proof of Lemma 9, with $B \otimes_{\mathcal{O}} B$, $\mu$, and $B$ in the roles of $B$, $\pi$, and $\mathcal{O}$. There are two changes.

First, we need a new argument, in the second paragraph, to show that the kernel of any surjective $B$-linear map $f \colon B^n \to (\ker \mu)/(\ker \mu)^2$ is generated by $n$ elements. This depends on the hypothesis that the ideal $F_B((\ker \mu)/(\ker \mu)^2)$ is *principal*, say with generator $a$. Since the module $(\ker \mu)/(\ker \mu)^2$ is supposed to be of finite length, its Fitting ideal contains a power of a prime element of $\mathcal{O}$, and therefore $a$ is not a zero-divisor. This implies that $F_B((\ker \mu)/(\ker \mu)^2)$ is $B$-free of rank 1. By Nakayama's lemma, any set of generators for $F_B((\ker \mu)/(\ker \mu)^2)$ contains a basis, so one can choose the element $a$ to be of the form $\det(v_1, \ldots, v_n)$, where $v_1, \ldots, v_n \in \ker f$. Let $v \in \ker f$, and replace, for some $1 \leq i \leq n$, the $i$th column of the matrix $(v_1, \ldots, v_n)$ by $v$. The determinant of the resulting matrix belongs to $F_B((\ker \mu)/(\ker \mu)^2)$, and is therefore equal to $b_i a$ for some uniquely determined $b_i \in B$. One now verifies in a straightforward way that $v = \sum_{i=1}^{n} b_i v_i$ ("Cramer's rule"), so that $v_1, \ldots, v_n$ span $\ker f$.

Second, we need a new proof that $A$ is finite flat as a $B$-algebra. For this one can apply a version of Lemma 1 that is valid for general base rings (as in [2]), or one uses R. Pink's argument that we sketched above. In the same way one proves that $\mathrm{Hom}_B(A, B)$ is $A$-free of rank 1 and that the analogue of (7) is valid for $A$, that is, $F_B((\ker \mu\varphi)/(\ker \mu\varphi)^2) = \eta_{\mu\varphi}$.

Having constructed $A$, one shows that the map

$$\varphi \colon A = B[[X_1, \ldots, X_n]]/(f_1, \ldots, f_n) \to B \otimes_{\mathcal{O}} B$$

is an isomorphism of $B$-algebras. To do this one simply copies the proof of Corollary 10, replacing Lemma 3 by Lemma 11 (applied to $B \otimes_{\mathcal{O}} B$ and $B$ in the roles of $B$ and $C$).

We now know that $B$ becomes a "relative complete intersection" after base extension with itself. To finish the proof of the Proposition we descend to $\mathcal{O}$.

Let, generally, $C$ be a complete local noetherian ring, and $R$ a complete local noetherian $C$-algebra with the same residue class field $k$ as $C$. Then there exists $m$ such that $C[[X_1, \ldots, X_m]]$ has an ideal $J$ for which $R \cong C[[X_1, \ldots, X_m]]/J$ as $C$-algebras. The minimal number of generators of the ideal $J$ equals $\dim_k J/\mathfrak{m}J$, where $\mathfrak{m}$ denotes the maximal ideal of $C[[X_1, \ldots, X_m]]$. The number $m - \dim_k J/\mathfrak{m}J$ only depends on the $C$-algebra $R$, and not on the presentation $R \cong C[[X_1, \ldots, X_m]]/J$; this is proved by a straightforward argument, which resembles the proof, given above, that the Fitting ideal is well-defined. Write $\epsilon(R, C) = m - \dim_k J/\mathfrak{m}J$. If $D$ is a finite flat local $C$-algebra, then one readily verifies that $\epsilon(R \otimes_C D, D) = \epsilon(R, C)$.

With $C = \mathcal{O}$, $D = R = B$ we now find that $\epsilon(B, \mathcal{O}) = \epsilon(B \otimes_{\mathcal{O}} B, B) \geq 0$, the inequality coming from the isomorphism $B[[X_1, \ldots, X_n]]/(f_1, \ldots, f_n) \cong B \otimes_{\mathcal{O}} B$. It follows that there exist $m$ and $g_1, \ldots, g_m \in \mathcal{O}[[X_1, \ldots, X_m]]$ such that $\mathcal{O}[[X_1, \ldots, X_m]]/(g_1, \ldots, g_m) \cong B$ as $\mathcal{O}$-algebras. Hence $B$ is a complete intersection. (One actually has $\epsilon(B, \mathcal{O}) = 0$, by [3], Theorem 21.1.) This completes the proof of the Proposition. □

REMARK. Under the hypotheses of the Proposition, $B$ is actually a complete intersection if and only if the Fitting ideal $F_B(\Omega_{B/\mathcal{O}})$ is principal. This can be deduced from Theorem 9.5 in E. Kunz, *Kähler differentials* (Vieweg, Braunschweig, 1986).

*References*

[1] M. F. Atiyah, I. G. Macdonald, *Introduction to commutative algebra*, Addison-Wesley, Reading, Mass., 1969.

[2] B. de Smit, H. W. Lenstra, Jr., *Finite complete intersection algebras*, Report 9453/B, Econometric Institute, Erasmus University Rotterdam, The Netherlands, 1994.

[3] H. Matsumura, *Commutative ring theory*, Cambridge University Press, Cambridge, 1986.

[4] B. Mazur, L. Roberts, *Local Euler characteristics*, Invent. Math. **9** (1970), 201–234.

[5] J. Tilouine, *Théorie d'Iwasawa classique et de l'algèbre de Hecke ordinaire*, Compositio Math. **65** (1988), 265–320.

Conference on Elliptic Curves and Modular Forms
Hong Kong, December 18-21, 1993

## Homologie des courbes modulaires affines et paramétrisations modulaires

### L. MEREL

**Introduction.** Soit $E$ une courbe elliptique modulaire définie sur $\mathbb{Q}$ de conducteur $N$. Supposons que $E$ soit une courbe elliptique modulaire forte, et soit $\phi$ une paramétrisation modulaire forte de $E$ : $\phi$ est un morphisme non constant défini sur $\mathbb{Q}$

$$X_0(N) \to E,$$

transformant la pointe $\Gamma_0(N)\infty$ en $0$ et qui ne se factorise par aucune isogénie de courbes elliptiques $E' \to E$ définie sur $\mathbb{Q}$ de degré $> 1$.

**Théorème 1** *a) Il existe un élément non nul $\sum \mu_x x \in \mathbb{Q}^{(\mathbb{P}^1(\mathbb{Z}/N\mathbb{Z}))}$, unique au signe près, satisfaisant les conditions* i$^-$), ii$^-$), iii$^-$) *et* iv$^-$) *ci-dessous. Il existe un élément non nul $\sum \lambda_x x \in \mathbb{Z}^{(\mathbb{P}^1(\mathbb{Z}/N\mathbb{Z}))}$, unique au signe près modulo $R_N$ (où $R_N$ est défini dans* i$^+$)), *satisfaisant les conditions* ii$^+$), iii$^+$) *et* iv$^+$) *énoncées dans la suite de l'introduction.*

*b) Le degré $\deg \phi$ de $\phi$ et son nombre d'enroulement $n_\phi$ (voir [9] et section 3.2) sont donnés par les formules*

$$\deg \phi = \frac{\epsilon}{2} | \sum_{x \in \mathbb{P}^1(\mathbb{Z}/N\mathbb{Z})} \lambda_x \mu_x |$$

*et*

$$n_\phi = \frac{\epsilon}{2} |\mu_{(0,1)}|,$$

*où $\epsilon$ est le nombre de composantes connexes de l'ensemble des points réels de $E$ (c'est-à-dire 2 si le discriminant de $E$ est positif, 1 s'il est négatif).*

Les conditions indiquées dans le théorème 1, a) suffisent à déterminer algorithmiquement les familles $\mu_x$ et $\lambda_x$ à partir du conducteur de la courbe elliptique et du nombre de points de quelques réductions modulo $p$ de la courbe. Le théorème 1 fournit donc un algorithme pour calculer $\deg \phi$ et $n_\phi$. Remarquons que l'expression de $\deg \phi$ et $n_\phi$ ne dépend pas du choix des $\lambda_x$ et des $\mu_x$.

Posons $c = \begin{pmatrix} -1 & 0 \\ 0 & 1 \end{pmatrix}$, $\sigma = \begin{pmatrix} 0 & -1 \\ 1 & 0 \end{pmatrix}$ et $\tau = \begin{pmatrix} 0 & -1 \\ 1 & 1 \end{pmatrix}$.

On fait opérer à droite les matrices $M = \begin{pmatrix} a & b \\ c & d \end{pmatrix} \in M_2(\mathbb{Z})$ de déterminant premier à $N$ sur $\mathbb{P}^1(\mathbb{Z}/N\mathbb{Z})$ en posant $(u,v)M = (au + cv, bu + dv)$.

Pour tout nombre premier $p$, notons $\mathcal{S}_p$ l'ensemble (fini) des matrices $\begin{pmatrix} a & b \\ c & d \end{pmatrix} \in M_2(\mathbb{Z})$ de déterminant $p$ vérifiant $a > b \geq 0$ et $d > c \geq 0$. Posons $a_p =$

$1 + p - |E(\mathbb{F}_p)|$, où $|E(\mathbb{F}_p)|$ est le nombre de points de la réduction modulo $p$ d'un modèle minimal de Weierstrass de $E$ sur $\mathbb{Z}$.

Notons $P_N$ la réunion disjointe des ensembles $(\mathbb{Z}/(\delta, N/\delta)\mathbb{Z})^*$, où $\delta$ parcourt les diviseurs positifs de $N$ et où par convention $(\mathbb{Z}/1\mathbb{Z})^*$ est constitué d'un élément. Les éléments de $P_N$ sont notés sous la forme $[\delta, x]$ où $\delta$ est un diviseur de $N$ et $x$ un élément de $(\mathbb{Z}/(\delta, N/\delta)\mathbb{Z})^*$ (par abus de notation si $x$ est donné dans $\mathbb{Z}/N\mathbb{Z}$, il s'agit de sa réduction modulo $(\delta, N/\delta)$).

Les conditions mentionnées dans l'énoncé du théorème 1 s'énoncent ainsi :

$i^-$) Pour tout $x \in \mathbb{P}^1(\mathbb{Z}/N\mathbb{Z})$ on a
$$\mu_x + \mu_{x\sigma} = \mu_x + \mu_{x\tau} + \mu_{x\tau^2} = 0.$$

$ii^-$) Pour tout $x \in \mathbb{P}^1(\mathbb{Z}/N\mathbb{Z})$ on a
$$\mu_x = \mu_{xc}.$$

$iii^-$) Pour tout nombre premier $p$ ne divisant pas $N$ et tout $x \in \mathbb{P}^1(\mathbb{Z}/N\mathbb{Z})$, on a
$$a_p \mu_x = \sum_{M \in \mathcal{S}_p} \mu_{xM}.$$

$iv^-$) Pour tout élément $\sum_{(u,v)} \nu_{(u,v)}(u,v) \in \mathbb{Z}^{(\mathbb{P}^1(\mathbb{Z}/N\mathbb{Z}))}$ vérifiant
$$\sum_{(u,v) \in \mathbb{P}^1(\mathbb{Z}/N\mathbb{Z})} \nu_{(u,v)}([(u,N), v^{-1}] - [(v,N), -u^{-1}]) = 0$$
dans $\mathbb{Z}^{(P_N)}$ (c.f. notations ci dessus), on a
$$\sum_x \mu_x \nu_x \in \mathbb{Z}.$$

De plus $\sum \mu_x x$ est minimal pour cette propriété (i.e. n'est multiple par aucun entier $n > 1$ d'un élément de $\mathbb{Q}^{(\mathbb{P}^1(\mathbb{Z}/N\mathbb{Z}))}$ vérifiant cette propriété).

$i^+$) Notons $R_N$ le sous-groupe de $\mathbb{Z}^{(\mathbb{P}^1(\mathbb{Z}/N\mathbb{Z}))}$ engendré par les éléments de la forme $x + x\sigma$, $x + x\tau + x\tau^2$ ($x \in \mathbb{P}^1(\mathbb{Z}/N\mathbb{Z})$) et les éléments $x \in \mathbb{P}^1(\mathbb{Z}/N\mathbb{Z})$ vérifiant $x = x\sigma$ ou $x = x\tau$.

$ii^+$) Pour tout $x \in \mathbb{P}^1(\mathbb{Z}/N\mathbb{Z})$ on a
$$\sum_x \lambda_x x \equiv \sum_x \lambda_x xc \pmod{R_N}.$$

$iii^+$) Pour tout nombre premier $p$ ne divisant pas $N$, on a
$$\sum_{M \in \mathcal{S}_p} \sum_x \lambda_x xM \equiv a_p \sum \lambda_x x \pmod{R_N}.$$

$iv^+$) On a dans $\mathbb{Z}^{(P_N)}$ l'égalité
$$\sum_{(u,v) \in \mathbb{P}^1(\mathbb{Z}/N\mathbb{Z})} \lambda_{(u,v)}([(u,N), v^{-1}] - [(v,N), -u^{-1}]) = 0.$$

Il existe $\sum \nu_x x \in \mathbb{Z}^{(\mathbb{P}^1(\mathbb{Z}/N\mathbb{Z}))}$ vérifiant pour tout $x \in \mathbb{P}^1(\mathbb{Z}/N\mathbb{Z})$ l'égalité suivante $\nu_x + \nu_{x\sigma} = \nu_x + \nu_{x\tau} + \nu_{x\tau^2} = 0$ et tel que

$$\sum_x \lambda_x \nu_x = 1.$$

*Remarques.-* Les éléments $\sum_x \mu_x x$ et $\sum_x \lambda_x x$ du théorème 1 satisfont pour $p|N$ les conditions $iii^-)$ et $iii^+)$ respectivement à condition de restreindre la sommation aux matrices $M \in S_p$ pour lesquelles $xM$ est défini (cf. théorème 6).

Il existe un élément $\sum_x \mu'_x x \in \mathbb{Q}^{(\mathbb{P}^1(\mathbb{Z}/N\mathbb{Z}))}$ unique au signe près satisfaisant les conditions $i^-)$, $iii^-)$, $iv^-)$ et $\mu'_x = -\mu'_{xc}$ pour tout $x \in \mathbb{P}^1(\mathbb{Z}/N\mathbb{Z})$. Il existe un élément $\sum_x \lambda'_x x$ de $\mathbb{Z}^{(\mathbb{P}^1(\mathbb{Z}/N\mathbb{Z}))}$ unique au signe près modulo $R_N$ satisfaisant $iii^+)$, $iv^+)$ et $\sum_x \lambda'_x x \equiv -\sum_x \lambda'_x xc \pmod{R_N}$. On a

$$\deg \phi = \frac{\epsilon}{2} | \sum_{x \in \mathbb{P}^1(\mathbb{Z}/N\mathbb{Z})} \lambda'_x \mu'_x |.$$

Cela se déduit de la partie 3 en échangeant les rôles de $\omega^+$ et $\omega^-$.

Le théorème 1 caractérise les courbes elliptiques modulaires en le sens suivant. Soit $E$ une courbe elliptique définie sur $\mathbb{Q}$. Pour que $E$ soit modulaire, il faut et il suffit qu'il existe $\sum_x \mu_x x \in \mathbb{Q}^{(\mathbb{P}^1(\mathbb{Z}/N\mathbb{Z}))}$ satisfaisant les conditions $i^-)$ et $iii^-)$. Cela se déduit aisément des théorèmes 3 et 5.

La preuve du théorème 1 est fondée sur l'étude de l'homologie singulière des courbes modulaires $X_0(N)$. Nous étudierons l'homologie des courbes modulaires définies plus généralement par des sous-groupes d'indice fini de $SL_2(\mathbb{Z})$. Cet article est divisé en trois parties. La première partie donne une présentation (très semblable à celle obtenue dans [14]) de l'homologie des courbes modulaires privées de leurs pointes. Cette homologie est duale de l'homologie relative aux pointes. La dualité est fournie par les produits d'intersection. Une présentation par générateurs et relations de l'homologie relative a été obtenue par Manin [6]. Notre présentation de l'homologie des courbes modulaires affines s'interprète, grâce aux produits d'intersection, comme duale de la présentation de Manin.

Dans la deuxième partie nous nous restreignons aux sous-groupes de congruence pour étudier l'action des opérateurs de Hecke sur l'homologie des courbes modulaires affines. Les formules obtenues généralisent celles de [12].

La troisième partie est consacrée à la preuve du théorème 1. Nous illustrons ce théorème en donnant des coefficients $\mu_x$ et $\lambda_x$ associés aux deux courbes elliptiques modulaires fortes de conducteur 37.

Je tiens à remercier Joseph Oesterlé, mon directeur de thèse, pour de nombreux commentaires très utiles à la rédaction de cet article.

## 1 Présentation des groupes d'homologie.

*1.1 Homologie des courbes modulaires affines.* Soit $\Gamma$ un sous-groupe d'indice fini de $SL_2(\mathbb{Z})$ contenant $\begin{pmatrix} -1 & 0 \\ 0 & -1 \end{pmatrix}$. Il opère par homographies sur le demi-plan de Poincaré $\mathfrak{H}$. Notons $Y_\Gamma$ la surface de Riemann quotient $\Gamma\backslash\mathfrak{H}$, $X_\Gamma$ sa compactifiée et *ptes* l'ensemble des pointes de $X_\Gamma$. Notons $\pi$ la surjection canonique

$$\mathfrak{H} \cup \mathbb{P}^1(\mathbb{Q}) \to X_\Gamma.$$

Si $A$ et $B$ sont deux parties finies et disjointes de $X_\Gamma$, notons $H_A^B$ le groupe d'homologie singulière relative $H_1(X_\Gamma - A, B; \mathbb{Z})$ (l'ensemble vide étant omis par convention).

Posons $\rho = e^{\frac{2\pi i}{3}}$. Notons $\delta$ le chemin géodésique reliant $i$ à $\rho$. Posons $R = \pi(SL_2(\mathbb{Z})\rho)$ et $I = \pi(SL_2(\mathbb{Z})i)$. Ces ensembles sont disjoints. Pour $g \in SL_2(\mathbb{Z})$ notons $[g]_*$ la classe de $\pi(g\delta)$ dans $H_{ptes}^{R \cup I}$. Cette classe ne dépend que de la classe $\Gamma g$ de $g$ dans $\Gamma\backslash SL_2(\mathbb{Z})$. Nous la noterons encore $[\Gamma g]_*$.

**Théorème 2** *L'homomorphisme de groupes*

$$\mathbb{Z}^{(\Gamma\backslash SL_2(\mathbb{Z}))} \to H_{ptes}^{R \cup I}$$

*prolongeant l'application*

$$\Gamma\backslash SL_2(\mathbb{Z}) \to H_{ptes}^{R \cup I}$$
$$\Gamma g \mapsto [g]_*$$

*est un isomorphisme de groupes.*

Le groupe $H_{ptes}$ s'identifie à un sous-groupe de $H_{ptes}^{R \cup I}$. Ce sous-groupe est décrit par le théorème suivant.

**Théorème 3** *Soit $\sum \mu_g g \in \mathbb{Z}^{(\Gamma\backslash SL_2(\mathbb{Z}))}$. L'élément $\sum \mu_g [g]_*$ de $H_{ptes}^{R \cup I}$ appartient à l'image canonique de $H_{ptes}$ si et seulement si on a pour tout $g \in \Gamma\backslash SL_2(\mathbb{Z})$*

$$\mu_g + \mu_{g\sigma} = 0 \text{ et } \mu_g + \mu_{g\tau} + \mu_{g\tau^2} = 0.$$

Ce théorème est dual d'un théorème de Manin ([**6**]). Les théorèmes 2 et 3 sont très proches de ceux démontrés dans [**14**]. Nous emploierons les mêmes techniques pour les démontrer, c'est-à-dire notamment des formules de produits d'intersection.

*1.2 Produits d'intersection entre les groupes $H_{ptes}^{R \cup I}$ et $H_{R \cup I}^{ptes}$.* Soit $z_0 \in \mathfrak{H}$ tel que $|z_0| = 1$ et $-\frac{1}{2} < \Re(z_0) < 0$. Notons $\gamma$ le chemin de $\bar{\mathfrak{H}}$ composé des chemins géodésiques reliant $0$ à $z_0$ et $z_0$ à $\infty$. Pour $g \in SL_2(\mathbb{Z})$ notons $[g]^*$ la classe dans $H_{R \cup I}^{ptes}$ de $\pi(g\gamma)$.

Les produits d'intersection fournissent un accouplement unimodulaire noté •

$$H_{R \cup I}^{ptes} \times H_{ptes}^{R \cup I} \to \mathbb{Z}.$$

**Proposition 1** *Soient $g$ et $h$ deux éléments de $SL_2(\mathbb{Z})$. Le nombre d'intersection de l'élément $[g]^*$ de $H_{R \cup I}^{ptes}$ et de l'élément $[h]_*$ de $H_{ptes}^{R \cup I}$ est donné par la formule*

$$[g]^* \bullet [h]_* = 1 \quad si \quad \Gamma g = \Gamma h,$$
$$[g]^* \bullet [h]_* = 0 \quad sinon.$$

*Démonstration.-* Le triangle hyperbolique $T$ de sommets $\infty$, $0$ et $\rho$ est un domaine fondamental pour $SL_2(\mathbb{Z})$. Les points des chemins $\gamma$ et $\delta$ distincts des extrémités sont intérieurs à $T$. On en déduit que $g\gamma$ et $h\delta$ n'ont pas de point d'intersection si $g \notin \{h, -h\}$. On a donc

$$[g]^* \bullet [h]_* = 0$$

si $\Gamma g \neq \Gamma h$.

Si $g \in \{h, -h\}$, on constate que les chemins $g\gamma$ et $h\delta$ ont un unique point d'intersection, en lequel ils se coupent transversalement avec pour nombre d'intersection $+1$. On a donc

$$[g]^* \bullet [h]_* = 1$$

si $\Gamma g = \Gamma h$. Cela achève la preuve de la proposition.

**Proposition 2** *L'ensemble $\{[g]_*/g \in SL_2(\mathbb{Z})\}$ engendre $H_{ptes}^{R \cup I}$.*

*Démonstration.-* Le groupe $H_{ptes}^{R \cup I}$ est l'image par l'homomorphisme de groupes déduit fonctoriellement de $\pi$ de $H_1(\mathfrak{H}, SL_2(\mathbb{Z})\{\rho, i\}; \mathbb{Z})$. Les translatés du triangle hyperbolique $T$ par les éléments de $SL_2(\mathbb{Z})$ constituent un pavage connexe par bords de $\mathfrak{H}$. Pout tout $g \in SL_2(\mathbb{Z})$, le chemin $g\delta$ relie les points $gi$ et $g\rho$, et ces points sont les seuls de $gT$ appartenant à $SL_2(\mathbb{Z})\{\rho, i\}$. Comme $\mathfrak{H}$ est simplement connexe on en déduit que $H_1(\mathfrak{H}, SL_2(\mathbb{Z})\{\rho, i\}; \mathbb{Z})$ est engendré par les classes de $g\delta$ ($g \in SL_2(\mathbb{Z})$). On en déduit la proposition.

Prouvons maintenant le théorème 2. D'après la proposition 2 l'homomorphisme

$$\mathbb{Z}^{(\Gamma \backslash SL_2(\mathbb{Z}))} \to H_{ptes}^{R \cup I}$$

est surjectif. D'après la proposition 1 il est injectif.

**Corollaire 1** *L'homomorphisme de groupes*

$$\mathbb{Z}^{(\Gamma \backslash SL_2(\mathbb{Z}))} \to H_{R \cup I}^{ptes}$$

*prolongeant l'application*

$$\Gamma \backslash SL_2(\mathbb{Z}) \to H_{ptes}^{R \cup I}$$
$$\Gamma g \mapsto [g]^*$$

*est un isomorphisme de groupes.*

*Démonstration.-* Cela résulte immédiatement du fait que l'accouplement

$$H_{R \cup I}^{ptes} \times H_{ptes}^{R \cup I} \to \mathbb{Z}$$

défini par les produits d'intersection est unimodulaire et de la proposition 1.

*1.3 Application aux groupes $H_{ptes}$ et $H^{ptes}$.* Notons $<\sigma>$ et $<\tau>$ les sous-groupes finis de $SL_2(\mathbb{Z})$ engendrés par $\sigma$ et $\tau$ respectivement. Les applications

$$\Gamma \backslash SL_2(\mathbb{Z})/<\sigma> \to I \quad \text{et} \quad \Gamma \backslash SL_2(\mathbb{Z})/<\tau> \to R$$

déduites par passage aux quotients de $g \mapsto \Gamma g i$ et $g \mapsto \Gamma g \rho$ sont des bijections. On en déduit la série d'isomorphismes canoniques suivante

$$\mathrm{H}_0(R \cup I; \mathbb{Z}) \simeq \mathbb{Z}^{(R \cup I)} \simeq \mathbb{Z}^{(R)} \times \mathbb{Z}^{(I)}$$
$$\simeq \mathbb{Z}^{(\Gamma \backslash SL_2(\mathbb{Z})/<\tau>)} \times \mathbb{Z}^{(\Gamma \backslash SL_2(\mathbb{Z})/<\sigma>)}.$$

Cela permet d'écrire la suite exacte longue d'homologie relative

$$0 \leftarrow \mathrm{H}_0(X_\Gamma - ptes; \mathbb{Z}) \leftarrow \mathrm{H}_0(R \cup I; \mathbb{Z}) \leftarrow \mathrm{H}_{ptes}^{R \cup I} \leftarrow \mathrm{H}_{ptes} \leftarrow 0$$

sous la forme

$$0 \leftarrow \mathbb{Z} \leftarrow \mathbb{Z}^{(\Gamma \backslash SL_2(\mathbb{Z})/<\tau>)} \times \mathbb{Z}^{(\Gamma \backslash SL_2(\mathbb{Z})/<\sigma>)}$$
$$\leftarrow \mathbb{Z}^{(\Gamma \backslash SL_2(\mathbb{Z}))} \leftarrow \mathrm{H}_{ptes} \leftarrow 0.$$

Les produits d'intersection permettent d'en déduire par dualité une suite exacte

$$0 \to \mathbb{Z} \to \mathbb{Z}^{(\Gamma \backslash SL_2(\mathbb{Z})/<\tau>)} \times \mathbb{Z}^{(\Gamma \backslash SL_2(\mathbb{Z})/<\sigma>)}$$
$$\to \mathbb{Z}^{(\Gamma \backslash SL_2(\mathbb{Z}))} \to \mathrm{H}^{ptes} \to 0,$$

dans laquelle la surjection $\mathbb{Z}^{(\Gamma \backslash SL_2(\mathbb{Z}))} \to \mathrm{H}^{ptes}$ s'identifie à la surjection canonique $\mathrm{H}_{R \cup I}^{ptes} \to \mathrm{H}^{ptes}$.

Prouvons maintenant le théorème 3. Soit $\sum \mu_g g \in \mathbb{Z}^{(\Gamma \backslash SL_2(\mathbb{Z}))}$. L'élément $\sum \mu_g [g]_*$ de $\mathrm{H}_{ptes}^{R \cup I}$ est l'image d'un élément de $\mathrm{H}_{ptes}$ par l'injection canonique si et seulement si son bord est nul, c'est-à-dire si on a dans $\mathbb{Z}^{(R \cup I)}$

$$\sum \mu_g (g\rho - gi) = 0.$$

On obtient la condition énoncée dans le théorème 3 lorsqu'on identifie $\mathbb{Z}^{(R \cup I)}$ à $\mathbb{Z}^{(\Gamma \backslash SL_2(\mathbb{Z})/<\tau>)} \times \mathbb{Z}^{(\Gamma \backslash SL_2(\mathbb{Z})/<\sigma>)}$ comme ci-dessus.

Pour $\sum_x \mu_x x \in \mathbb{Z}(\Gamma \backslash SL_2(\mathbb{Z}))$ vérifiant la condition du théorème 3, notons $\xi_0(\sum_x \mu_x x)$ l'élément correspondant de $H_{ptes}$. Pour $g \in SL_2(\mathbb{Z})$ ou $g \in \Gamma \backslash SL_2(\mathbb{Z})$, notons $\xi^0(g)$ l'image dans $\mathrm{H}^{ptes}$ de $[g]^*$ par la surjection canonique

$$\mathrm{H}_{R \cup I}^{ptes} \to \mathrm{H}^{ptes}.$$

On déduit immédiatement des suites exactes longues figurant plus haut le corollaire suivant.

**Corollaire 2** *Les classes $\xi^0(g)$ engendrent $\mathrm{H}^{ptes}$ lorsque $g$ décrit $SL_2(\mathbb{Z})$ et on a les relations*

$$\xi^0(g) + \xi^0(g\sigma) = 0,$$
$$\xi^0(g) + \xi^0(g\tau) + \xi^0(g\tau^2) = 0$$

*pour tout $g \in SL_2(\mathbb{Z})$.*

*Remarque.-* Pour $(\alpha, \beta) \in \mathbb{P}^1(\mathbb{Q})$ notons $\{\alpha, \beta\}$ la classe dans $\mathrm{H}^{ptes}$ de l'image par $\pi$ du chemin géodésique reliant $\alpha$ à $\beta$ dans $\bar{\mathfrak{H}}$. On a $\xi^0(g) = \{g0, g\infty\}$. Signalons que le corollaire 2 est un théorème de Manin ([6]).

On a la formule suivante relative à l'accouplement
$$\mathrm{H}^{ptes} \times \mathrm{H}_{ptes} \to \mathbb{Z}.$$

**Corollaire 3** *Soit $g \in \Gamma \backslash SL_2(\mathbb{Z})$. Soit $\sum_h \mu_h h \in \mathbb{Z}^{(\Gamma \backslash SL_2(\mathbb{Z}))}$ tel que $\sum_h \mu_h[h]_*$ soit dans l'image de $H_{ptes}$ par l'injection canonique. On a*
$$\xi^0(g) \bullet \xi_0(\sum_h \mu_h h) = \mu_g.$$

*Démonstration.-* Cela résulte de l'égalité
$$\xi^0(g) \bullet \xi_0(\sum_h \mu_h h) = [g]^* \bullet (\sum_h \mu_h [h]_*) = \mu_g.$$

*1.4 L'homologie de $X_\Gamma$.* Remarquons qu'on a une surjection canonique
$$\alpha : \mathrm{H}_{ptes} \to H$$
où suivant nos conventions de la section 1.1 $H$ désigne le groupe $\mathrm{H}_1(X_\Gamma; \mathbb{Z})$.

**Proposition 3** *Soit $\sum_x \mu_x x \in \mathbb{Q}^{(\Gamma \backslash SL_2(\mathbb{Z}))}$ vérifiant pour tout $x \in \Gamma \backslash SL_2(\mathbb{Z})$*
$$\mu_x + \mu_{x\sigma} = \mu_x + \mu_{x\tau} + \mu_{x\tau^2} = 0.$$
*Alors l'image de $\xi_0(\sum_x \mu_x x) \in H_{ptes} \otimes \mathbb{Q}$ dans $H \otimes \mathbb{Q}$ par $\alpha \otimes 1_\mathbb{Q}$ appartient à $H$ si et seulement si pour tout élément $\sum_x \nu_x x \in \mathbb{Z}^{(\Gamma \backslash SL_2(\mathbb{Z}))}$ vérifiant*
$$\sum_x \nu_x((x\infty) - (x0)) = 0$$
*dans $\mathbb{Z}^{(\Gamma \backslash \mathbb{P}^1(\mathbb{Q}))}$ on a*
$$\sum_x \nu_x \mu_x \in \mathbb{Z}.$$

*Démonstration.-* On a une injection canonique
$$\beta : H \to \mathrm{H}^{ptes}.$$
adjointe de $\alpha$ pour les produits d'intersection. Observons d'abord que si $\sum_x \nu_x x \in \mathbb{Z}^{(\Gamma \backslash SL_2(\mathbb{Z}))}$, alors $\sum_x \nu_x \xi^0(x)$ appartient à $\beta(H)$ si et seulement si son bord est nul, c'est-à-dire si et seulement si on a
$$\sum_x \nu_x((x\infty) - (x0)) = 0$$
dans $\mathbb{Z}^{(\Gamma \backslash \mathbb{P}^1(\mathbb{Q}))}$.

Soit alors $\sum_x \mu_x x \in \mathbb{Q}^{(\Gamma \backslash SL_2(\mathbb{Z}))}$ tel que $\mu_x + \mu_{x\sigma} = \mu_x + \mu_{x\tau} + \mu_{x\tau^2} = 0$ pour tout $x \in \Gamma \backslash SL_2(\mathbb{Z})$. Notons $y$ l'image de $\xi_0(\sum_x \mu_x x)$ par $\alpha \otimes 1_\mathbb{Q}$. Elle appartient à $H$ si et seulement si on a $y' \bullet y \in \mathbb{Z}$ pour tout $y' \in H$, c'est-à-dire $\beta(y') \bullet \xi_0(\sum_x \mu_x x) \in \mathbb{Z}$ pour tout $y' \in H$, ou encore, d'après ce qui précède, si et seulement si, pour tout élément $\sum_x \nu_x x$ de $\mathbb{Z}^{(\Gamma \backslash SL_2(\mathbb{Z}))}$ vérifiant $\sum_x \nu_x((x\infty) - (x0)) = 0$ dans $\mathbb{Z}^{(\Gamma \backslash \mathbb{P}^1(\mathbb{Q}))}$, le produit d'intersection $\xi^0(\sum_x \nu_x x) \bullet \xi_0(\sum_x \mu_x x)$, qui est égal à $\sum_x \mu_x \nu_x$, appartient à $\mathbb{Z}$.

## 2 Opérateurs de Hecke.

*2.1 L'action des opérateurs de Hecke sur $H^{ptes}$.* Soient $N$ un entier $> 0$ et $K$ un sous-groupe de $GL_2(\mathbb{Z}/N\mathbb{Z})$ tel que l'application déterminant $GL_2(\mathbb{Z}/N\mathbb{Z}) \to (\mathbb{Z}/N\mathbb{Z})^*$ soit surjective. Notons $\phi_N$ la surjection canonique $M_2(\mathbb{Z}) \to M_2(\mathbb{Z}/N\mathbb{Z})$. Supposons qu'on ait

$$\Gamma = \phi_N^{-1}(K) \cap SL_2(\mathbb{Z}).$$

Soit $m$ un entier $> 0$ premier à $N$. Posons

$$A_m = \{g \in M_2(\mathbb{Z})/\det(g) = m\} \text{ et } A_{m,K} = \phi_N^{-1}(K) \cap A_m.$$

Le groupe $\Gamma$ opère à gauche et à droite sur $A_{m,K}$. Soit $R$ un système de représentants de $\Gamma \backslash A_{m,K}$. Notons $T_{m,K}$ la correspondance de Hecke

$$\Gamma z \mapsto \sum_{r \in R} \Gamma r z$$

sur la courbe modulaire $Y_\Gamma$. Elle définit un endomorphisme encore noté $T_{m,K}$ du groupe $H^{ptes}$ qui à l'image par $\pi$ de la classe d'un chemin $c$ de $\mathfrak{H}$ d'extrémités dans $\mathbb{P}^1(\mathbb{Q})$ associe la somme des images par $\pi$ des classes des chemins $rc$, pour $r \in R$.

Soit $\sum_M u_M M \in \mathbb{Z}^{(A_m)}$ vérifiant la condition $(C)$ suivante : pour tout $\alpha \in A_m/SL_2(\mathbb{Z})$ on a dans $\mathbb{Z}^{(\mathbb{P}^1(\mathbb{Q}))}$

$$\sum_{M \in \alpha} u_M((M\infty) - (M0)) = (\infty) - (0).$$

De tels éléments existent ; des exemples en sont donnés dans [**12**], [**10**] et [**11**].

Pour $M = \begin{pmatrix} a & b \\ c & d \end{pmatrix}$ on pose $\tilde{M} = \begin{pmatrix} d & -b \\ -c & a \end{pmatrix}$ (si $M$ est inversible on a $\tilde{M} = M^{-1}\det M$). Notons $\tilde{K}$ l'image de $K$ par $M \mapsto \tilde{M}$.

L'application canonique $\Gamma \backslash SL_2(\mathbb{Z}) \to \tilde{K} \backslash GL_2(\mathbb{Z}/N\mathbb{Z})$ est bijective. Cela permet d'associer à tout élément $x \in \tilde{K} \backslash GL_2(\mathbb{Z}/N\mathbb{Z})$ un élément $[x]_*^{\tilde{K}}$ de $H^{R \cup I}_{ptes}$. Pour $\sum_x \mu_x x \in \mathbb{Z}^{(\tilde{K} \backslash GL_2(\mathbb{Z}/N\mathbb{Z}))}$ tel que $\sum_x \mu_x [x]_*^{\tilde{K}}$ soit dans l'image de $H_{ptes}$ par l'injection canonique, notons $\xi_0^{\tilde{K}}(\sum_x \mu_x x)$ l'élément correspondant de $H_{ptes}$.

Pour $x \in \tilde{K} \backslash GL_2(\mathbb{Z}/N\mathbb{Z})$, notons encore $\xi_{\tilde{K}}^0(x)$ l'image de de $x$ dans $H^{ptes}$ par l'application composée

$$\mathbb{Z}^{(\tilde{K} \backslash GL_2(\mathbb{Z}/N\mathbb{Z}))} \to \mathbb{Z}^{(\Gamma \backslash SL_2(\mathbb{Z}))} \to H^{ptes}.$$

**Théorème 4** *Soit $x \in \tilde{K} \backslash GL_2(\mathbb{Z}/N\mathbb{Z})$. On a*

$$T_{m,K}(\xi_{\tilde{K}}^0(x)) = \sum_M u_M \xi_{\tilde{K}}^0(xM).$$

*Démonstration.-* Soit $\lambda_{\tilde{K}} : \tilde{K} \backslash GL_2(\mathbb{Z}/N\mathbb{Z}) \to SL_2(\mathbb{Z})$ une section de la surjection canonique $SL_2(\mathbb{Z}) \to \tilde{K} \backslash GL_2(\mathbb{Z}/N\mathbb{Z})$. Soit $R_m$ un système de représentants de $A_m/SL_2(\mathbb{Z})$. Il existe $g \in SL_2(\mathbb{Z})$ tel que $x = \tilde{K}g$. On a

$$\sum_M u_M \xi_{\tilde{K}}^0(xM) = \sum_{r \in R_m} \sum_{M \in rSL_2(\mathbb{Z})} u_M \xi_{\tilde{K}}^0(xM)$$

$$= \sum_{r \in R_m} \sum_{M \in rSL_2(\mathbb{Z})} u_M \{\lambda_{\tilde{K}}(xM)0, \lambda_{\tilde{K}}(xM)\infty\}$$

$$= \sum_{r \in R_m} \sum_{M \in rSL_2(\mathbb{Z})} u_M \{\lambda_{\tilde{K}}(xr)r^{-1}M0, \lambda_{\tilde{K}}(xr)r^{-1}M\infty\}.$$

La dernière égalité provient du fait que $\lambda_{\tilde{K}}(xM)$ et $\lambda_{\tilde{K}}(xr)r^{-1}M$ diffèrent par multiplication à gauche par un élément de $\Gamma$. Utilisons maintenant la condition $(C)$ et les propriétés des symboles modulaires. On a

$$\sum_M u_M \xi_{\tilde{K}}^0(xM) = \sum_{r \in R_m} \{\lambda_{\tilde{K}}(xr)r^{-1}0, \lambda_{\tilde{K}}(xr)r^{-1}\infty\}$$

$$= \sum_{r \in R_m} \{\lambda_{\tilde{K}}(xr)r^{-1}g^{-1}\begin{pmatrix} m & 0 \\ 0 & m \end{pmatrix}g0,$$

$$\lambda_{\tilde{K}}(xr)r^{-1}g^{-1}\begin{pmatrix} m & 0 \\ 0 & m \end{pmatrix}g\infty\}.$$

Posons $s = gr$. Quand $r$ parcourt $R_m$, $s$ parcourt encore un système de représentants $S_m$ de $A_m/SL_2(\mathbb{Z})$. Il nous faut donc vérifier que

$$\lambda_{\tilde{K}}(\tilde{K}s)s^{-1}\begin{pmatrix} m & 0 \\ 0 & m \end{pmatrix}$$

parcourt un système de représentants de $\Gamma \backslash A_{m,K}$. On a

$$\lambda_{\tilde{K}}(\tilde{K}s)s^{-1}\begin{pmatrix} m & 0 \\ 0 & m \end{pmatrix} \in A_{m,K}$$

car $\phi_N(\lambda_{\tilde{K}}(\tilde{K}s)s^{-1}\begin{pmatrix} m & 0 \\ 0 & m \end{pmatrix})$ s'écrit $\tilde{k}s\tilde{s}$ avec $k \in K$ de déterminant $m^{-1}$ modulo $N$ ; on a donc $\tilde{k}s\tilde{s} = \tilde{k}\begin{pmatrix} m & 0 \\ 0 & m \end{pmatrix} = k^{-1} \in K$.

Deux éléments $s$ et $s'$ de $S_m$ définissent deux classes $\Gamma\lambda_{\tilde{K}}(\tilde{K}s)s^{-1}\begin{pmatrix} m & 0 \\ 0 & m \end{pmatrix}$ et $\Gamma\lambda_{\tilde{K}}(\tilde{K}s')s'^{-1}\begin{pmatrix} m & 0 \\ 0 & m \end{pmatrix}$ distinctes. Sinon il existerait $\gamma \in \Gamma$ tel que

$$s\lambda_{\tilde{K}}(\tilde{K}s)^{-1} = s'\lambda_{\tilde{K}}(\tilde{K}s')^{-1}\gamma$$

et les classes $sSL_2(\mathbb{Z})$ et $s'SL_2(\mathbb{Z})$ seraient égales.

Prouvons maintenant que $\lambda_{\tilde{K}}(\tilde{K}s)s^{-1}\begin{pmatrix} m & 0 \\ 0 & m \end{pmatrix}$ parcourt toutes les classes de $\Gamma \backslash A_{m,K}$ quand $s$ parcourt $S_m$. Il suffit de prouver que $\Gamma \backslash A_{m,K}$ et $S_m$ ont même cardinal. Cela résulte de la bijectivité de l'application canonique

$$\Gamma \backslash A_{m,K} \to SL_2(\mathbb{Z}) \backslash A_m$$

et de l'égalité des cardinaux de $SL_2(\mathbb{Z}) \backslash A_m$ et $A_m/SL_2(\mathbb{Z})$. Cela prouve le théorème.

*Remarque.-* Dans [**12**] et dans [**10**] nous avons prouvé des théorèmes très voisins relatifs dans le premier cas à $\Gamma = \Gamma_0(N)$ et dans le deuxième cas aux sous-groupes de congruence de $\Gamma(2)$.

*Exemple-.* Notons $\mathcal{S}_m$ l'ensemble des matrices $\begin{pmatrix} a & b \\ c & d \end{pmatrix} \in A_m$ vérifiant $a > b \geq 0$ et $d > c \geq 0$. Alors (voir [**11**]) $\sum_{M \in \mathcal{S}_m} M$ vérifie la condition $(C)$. Lorsque $m = p$ est un nombre premier, nous utilisons ces formules dans les conditions $iii^+)$ et $iii^-)$ du théorème 1. A titre d'exemple, lorsque $m = 2$, l'élément

$$\begin{pmatrix} 2 & 0 \\ 0 & 1 \end{pmatrix} + \begin{pmatrix} 1 & 0 \\ 0 & 2 \end{pmatrix} + \begin{pmatrix} 2 & 1 \\ 0 & 1 \end{pmatrix} + \begin{pmatrix} 1 & 0 \\ 1 & 2 \end{pmatrix}$$

vérifie la condition $(C)$.

**2.2 Action des opérateurs de Hecke sur** $H_{ptes}$. Rappelons que $m$ est un entier $> 0$ premier à $N$ et que $\sum_M u_M M$ est un élément de $\mathbb{Z}^{(A_m)}$ satisfaisant la condition $(C)$ introduite dans la section 2.1.

**Théorème 5** *Soit*

$$\sum_x \mu_x x \in \mathbb{Z}^{(\tilde{K} \backslash GL_2(\mathbb{Z}/N\mathbb{Z}))}$$

*tel que pour tout* $x \in \tilde{K} \backslash GL_2(\mathbb{Z}/N\mathbb{Z})$ *on ait* $\mu_x + \mu_{x\sigma} = \mu_x + \mu_{x\tau} + \mu_{x\tau^2} = 0$. *Alors* $\sum_M \sum_x u_M \mu_x x \tilde{M}$ *vérifie une condition analogue et on a*

$$T_{m,K}(\xi_0^{\tilde{K}}(\sum_x \mu_x x)) = \xi_0^{\tilde{K}}(\sum_M \sum_x u_M \mu_x x \tilde{M}).$$

*Démonstration.-* Reprenons la définition des opérateurs de Hecke donnée plus haut. L'ensemble $\tilde{K}$ déduit de $K$ est également un sous-groupe de $GL_2(\mathbb{Z}/N\mathbb{Z})$ avec déterminant surjectif. On a

$$\Gamma = \phi_N^{-1}(\tilde{K}) \cap SL_2(\mathbb{Z}).$$

Le groupe $\tilde{K}$ permet donc de définir une correspondance $T_{m,\tilde{K}}$ sur la courbe modulaire $X_\Gamma$ par un procédé analogue à celui qui définit $T_{m,K}$. On constate que les correspondances $T_{m,K}$ et $T_{m,\tilde{K}}$ sont adjointes. Elles définissent donc deux endomorphismes $T_{m,K}$ et $T_{m,\tilde{K}}$ sur $H_{ptes}$ et $H^{ptes}$ qui sont adjoints pour les accouplements définis par les produits d'intersection.

Soit $h \in SL_2(\mathbb{Z})$. Posons $y = Kh \in K \backslash GL_2(\mathbb{Z}/N\mathbb{Z})$. On a d'après les relations d'adjonction mentionnées ci-dessus les égalités

$$\xi_K^0(y) \bullet T_{m,K}(\xi_0^{\tilde{K}}(\sum_x \mu_x x)) = T_{m,\tilde{K}}(\xi_K^0(y)) \bullet \xi_0^{\tilde{K}}(\sum_x \mu_x x)$$
$$= (\sum_M u_M \xi_K^0(yM)) \bullet \xi_0^{\tilde{K}}(\sum_x \mu_x x)$$
$$= \sum_M u_M \xi_K^0(yM) \bullet \xi_0^{\tilde{K}}(\sum_x \mu_x x)$$
$$= \sum_M u_M \sum_x \mu_x [yM]_K^* \bullet [x]_*^{\tilde{K}}.$$

Démontrons que l'on a

$$[yM]_K^* \bullet [x]_*^{\tilde{K}} = [y)]_K^* \bullet [x\tilde{M}]_*^{\tilde{K}}.$$

Soit $g \in SL_2(\mathbb{Z})$ tel que $\tilde{K}g = x$. D'après la proposition 1, le membre de gauche (resp. de droite) est nul, sauf lorsque $\phi_N(g) \in KhM$ (resp. $\phi_N(h) \in \tilde{K}g\tilde{M}$), et dans ce cas il est égal à 1. Or on a $\phi_N(g) \in KhM$ si et seulement si $\phi_N(hMg^{-1}) \in K$, i.e. si et seulement si $\phi_N(g\tilde{M}h^{-1}) \in \tilde{K}$, i.e. si et seulement si $\phi_N(h) \in \tilde{K}g\tilde{M}$.

On en déduit l'égalité

$$\xi_K^0(y) \bullet T_{m,K}(\xi_0^{\tilde{K}}(\sum_x \mu_x x)) = \sum_M u_M \sum_x \mu_x [y]_K^* \bullet [x\tilde{M}]_*^{\tilde{K}}$$
$$= [y]_K^* \bullet (\sum_M \sum_x u_M \mu_x [x\tilde{M}]_*^{\tilde{K}}).$$

Si on écrit l'élément $\sum_M \sum_x u_M \mu_x x \tilde{M}$ de $\mathbb{Z}^{(\tilde{K}\backslash GL_2(\mathbb{Z}/N\mathbb{Z}))}$ sous la forme $\sum \nu_t \tilde{K} t$, où $t$ parcourt $\Gamma\backslash SL_2(\mathbb{Z})$, on a d'après ce qui précède

$$\nu_t = [t]^* \bullet \sum_M u_M \sum_x \mu_x [x\tilde{M}]_*^{\tilde{K}}$$
$$= \xi^0(t) \bullet T_{m,K}(\xi_0^{\tilde{K}}(\sum_x \mu_x x)).$$

Etant données les relations $\xi^0(t) + \xi^0(t\sigma) = \xi^0(t) + \xi^0(t\tau) + \xi^0(t\tau^2) = 0$, on a

$$\nu_t + \nu_{t\sigma} = \nu_t + \nu_{t\tau} + \nu_{t\tau^2} = 0.$$

L'élément $\sum_M \sum_x u_M \mu_x [x\tilde{M}]_*^{\tilde{K}}$ de $\mathrm{H}_{ptes}^{R\cup I}$ est donc l'image canonique d'un élément de $\mathrm{H}_{ptes}$, noté $\xi_0^{\tilde{K}}(\sum_M u_M \mu_x x \tilde{M})$. Par conséquent on a, pour tout $y \in K\backslash GL_2(\mathbb{Z}/N\mathbb{Z})$, l'égalité

$$\xi_K^0(y) \bullet T_{m,K}(\xi_0^{\tilde{K}}(\sum_x \mu_x x)) = \xi_K^0(y) \bullet \xi_0^{\tilde{K}}(\sum_M \sum_x u_M \mu_x x \tilde{M}).$$

Comme les $\xi_K^0(y)$ engendrent $\mathrm{H}^{ptes}$ et comme l'accouplement $\mathrm{H}^{ptes} \times \mathrm{H}_{ptes} \to \mathbb{Z}$ est non dégénéré cette égalité prouve le théorème 5.

2.3 *Application* à $\Gamma = \Gamma_0(N)$. Soit $N$ un entier $> 0$. Posons

$$K = \{\begin{pmatrix} a & b \\ 0 & d \end{pmatrix} \in GL_2(\mathbb{Z}/N\mathbb{Z})\}.$$

Le groupe $\Gamma = SL_2(\mathbb{Z}) \cap \phi_N^{-1}(K)$ est alors

$$\Gamma_0(N) = \{\begin{pmatrix} a & b \\ c & d \end{pmatrix} \in SL_2(\mathbb{Z})/N|c\}.$$

L'application $\begin{pmatrix} a & b \\ c & d \end{pmatrix} \mapsto (c,d)$ définit par passage aux quotients une bijection canonique

$$K\backslash GL_2(\mathbb{Z}/N\mathbb{Z}) \simeq \mathbb{P}^1(\mathbb{Z}/N\mathbb{Z})$$

compatible à la multiplication à droite par les matrices de $M_2(\mathbb{Z})$ de déterminant premier à $N$.

Remarquons qu'on a $K = \tilde{K}$. Pout tout entier $m > 0$ premier à $N$, les correspondances $T_{m,K}$ et $T_{m,\tilde{K}}$ sont donc égales.

Dans ces conditions le théorème 3 peut être reformulé ainsi. Soit $\sum_x \mu_x x \in \mathbb{Z}^{(\mathbb{P}^1(\mathbb{Z}/N\mathbb{Z}))}$. Alors l'élément $\sum_x \mu_x [x]_*$ de $\mathrm{H}^{R \cup I}_{ptes}$ est l'image canonique d'un élément de $\mathrm{H}_{ptes}$ si et seulement si pour tout $x \in \mathbb{P}^1(\mathbb{Z}/N\mathbb{Z})$ on a
$$\mu_x + \mu_{x\sigma} = 0 \text{ et } \mu_x + \mu_{x\tau} + \mu_{x\tau^2} = 0.$$
Dans ce cas on note $\xi_0(\sum_x \mu_x x)$ l'élément de $\mathrm{H}_{ptes}$ dont il est l'image.

On note $\xi^0(x)$ l'image d'un élément $x$ de $\mathbb{Z}^{(\mathbb{P}^1(\mathbb{Z}/N\mathbb{Z}))}$ par la surjection canonique
$$\mathbb{Z}^{(\mathbb{P}^1(\mathbb{Z}/N\mathbb{Z}))} \to \mathbb{Z}^{(K \backslash GL_2(\mathbb{Z}/N\mathbb{Z}))} \to \mathrm{H}^{ptes}.$$
Ajoutons un mot sur la conjugaison complexe. L'involution $z \mapsto -\bar{z}$ de $\mathfrak{H}$ définit une involution, la conjugaison complexe, sur la courbe modulaire $X_0(N) = X_{\Gamma_0(N)}$. Cette involution laisse stable l'ensemble des pointes de $X_0(N)$ ; de plus elle laisse stable l'image de $SL_2(\mathbb{Z})\{\rho, i\}$ dans $X_0(N)$. Elle définit des involutions de $\mathrm{H}_{ptes}$, $\mathrm{H}^{ptes}$, $\mathrm{H}^{R \cup I}_{ptes}$ et $\mathrm{H}^{ptes}_{R \cup I}$. On note ces involutions par une barre horizontale. Rappelons qu'on a posé dans l'introduction $c = \begin{pmatrix} -1 & 0 \\ 0 & 1 \end{pmatrix}$.

**Proposition 4** *Soit $x \in \mathbb{P}^1(\mathbb{Z}/N\mathbb{Z})$. On a*
$$\overline{[x]^*} = -[xc\sigma]^* \quad et \quad \overline{[x]_*} = [xc\sigma]_*.$$
*De plus on a*
$$\overline{\xi^0(x)} = \xi^0(xc).$$
*Soit $\sum_y \mu_y y \in \mathbb{Z}^{(\mathbb{P}^1(\mathbb{Z}/N\mathbb{Z}))}$ tel que $\mu_y + \mu_{y\sigma} = \mu_y + \mu_{y\tau} + \mu_{y\tau^2} = 0$. Alors $-\sum_y \mu_y yc$ satisfait une condition analogue et on a*
$$\overline{\xi_0(\sum_y \mu_y y)} = -\xi_0(\sum_y \mu_y yc).$$

*Démonstration.-* Remarquons d'abord que les chemins $\bar{\delta}$ et $\bar{\gamma}$ sont égaux à $\sigma\delta$ et au chemin opposé à $\sigma\gamma$ respectivement. Soit $g \in SL_2(\mathbb{Z})$ tel que $Kg$ s'identifie à $x$. On a
$$\overline{g\delta} = cgc\bar{\delta} \quad \text{et} \quad \overline{g\gamma} = cgc\bar{\gamma}.$$
Comme $\phi_N(c) \in K$, $Kcgc$ s'identifie à $xc$. On en déduit les deux premières égalités de la proposition.

Comme $-\xi^0(xc\sigma) = \xi^0(xc)$, on a $\overline{\xi^0(x)} = \xi^0(xc)$.

L'élément $\sum_y \mu_y [yc\sigma]_*$ de $\mathrm{H}^{R \cup I}_{ptes}$ est l'image canonique de l'élément $\overline{\xi_0(\sum_y \mu_y y)}$ de $\mathrm{H}_{ptes}$. Comme on a $yc\sigma = y\sigma c$ et $\mu_{y\sigma} = -\mu_y$ pour tout $y \in \mathbb{P}^1(\mathbb{Z}/N\mathbb{Z})$, cela prouve la dernière assertion.

Décrivons plus complètement l'action des opérateurs de Hecke pour $\Gamma_0(N)$. Pour le reste de cette section considérons un entier $m \geq 1$ non nécessairement premier à $N$. Posons
$$A_{m,N} = \{\begin{pmatrix} a & b \\ c & d \end{pmatrix} \in A_m/N | c, (N,a) = 1\}.$$

Cet ensemble coïncide avec $A_{m,K}$ si $m$ est premier à $N$. Le groupe $\Gamma_0(N)$ opère à gauche et à droite sur $A_{m,N}$. Soit $R_{m,N}$ un système de représentants de $\Gamma_0(N)\backslash A_{m,N}$. Notons $T_m$ la correspondance de Hecke définie sur $X_0(N)$ par

$$\Gamma_0(N)z \mapsto \sum_{r \in R_{m,N}} \Gamma_0(N)rz.$$

Elle définit comme dans la section 2.1 un endomorphisme encore noté $T_m$ de $\mathrm{H}^{ptes}$. Notons $\tilde{T}_m$ l'endomorphisme de $\mathrm{H}_{ptes}$ adjoint de $T_m$ ; c'est l'opérateur déduit de la correspondance $T_m$ comme dans la section 2.1 si $m$ et $N$ sont premiers entre eux.

Soient $M = \begin{pmatrix} a & b \\ c & d \end{pmatrix} \in M_2(\mathbb{Z})$ et $x = (u,v) \in \mathbb{P}^1(\mathbb{Z}/N\mathbb{Z})$. Nous dirons que $xM$ est défini si $au + cv$ et $bu + dv$ engendrent $\mathbb{Z}/N\mathbb{Z}$ et nous noterons dans ce cas $xM$ le point de $\mathbb{P}^1(\mathbb{Z}/N\mathbb{Z})$ de coordonnées homogènes $(au + cv, bu + dv)$. Lorsque le déterminant de $M$ est premier à $N$, $xM$ est toujours défini et l'on retrouve les conventions considérées dans l'introduction.

L'action des opérateurs de Hecke est alors donnée par le théorème suivant, qui est un corollaire du théorème 4 et du théorème 5 si $m$ et $N$ sont premiers entre eux.

**Théorème 6** *Soit $x \in \mathbb{P}^1(\mathbb{Z}/N\mathbb{Z})$. Soit $\sum_M u_M M$ satisfaisant la condition $(C)$. On a*

$$T_m(\xi^0(x)) = \sum_M {}'u_M \xi^0(xM),$$

*où $\sum'$ signifie que l'on somme sur tous les $M$ tels que $xM$ soit défini. Soit $\sum_x \mu_x x \in \mathbb{Z}^{(\mathbb{P}^1(\mathbb{Z}/N\mathbb{Z}))}$ vérifiant pour tout $x \in \mathbb{P}^1(\mathbb{Z}/N\mathbb{Z})$ les égalités $\mu_x + \mu_{x\sigma} = \mu_x + \mu_{x\tau} + \mu_{x\tau^2} = 0$, alors $\sum_M \sum_x \mu_x u_M x\tilde{M}$ vérifie également cette condition et on a*

$$\tilde{T}_m(\xi_0(\sum_x \mu_x x)) = \xi_0(\sum_{M,x} {}'\mu_{xM} u_M x),$$

*avec les mêmes conventions pour $\sum'$ que ci-dessus.*

*Démonstration.-* Notons $\tilde{A}_{m,N}$ l'image de $A_{m.N}$ par $M \mapsto \tilde{M}$. L'application canonique

$$\Gamma_0(N)\backslash A_{m,N} \to SL_2(\mathbb{Z})\backslash SL_2(\mathbb{Z})A_{m,N}$$

est surjective ; prouvons qu'elle est injective. Soient $\gamma = \begin{pmatrix} a & b \\ c & d \end{pmatrix} \in A_{m,N}$ et $t = \begin{pmatrix} a' & b' \\ c' & d' \end{pmatrix} \in SL_2(\mathbb{Z})$ tels que $t\gamma \in A_{m,N}$. Le coefficient inférieur gauche de $t\gamma$ est égal à $ac' + cd'$. Comme il est divisible par $N$, comme $N|c$ et comme $a$ est premier à $N$ on en déduit $N|c'$, i.e. $t \in \Gamma_0(N)$.

Cela permet de définir une application

$$\eta : \tilde{A}_{m,N} SL_2(\mathbb{Z}) \to \Gamma_0(N)\backslash SL_2(\mathbb{Z})$$
$$\gamma g \mapsto \Gamma_0(N)g.$$

Notons que l'ensemble $\tilde{A}_{m,N}SL_2(\mathbb{Z})$ est l'ensemble des matrices $\begin{pmatrix} a & b \\ c & d \end{pmatrix} \in A_m$ telles que $c$ et $d$ engendrent $\mathbb{Z}/N\mathbb{Z}$. L'image de $\eta\begin{pmatrix} a & b \\ c & d \end{pmatrix}$ dans $\mathbb{P}^1(\mathbb{Z}/N\mathbb{Z})$ est égale à $(c,d)$. Soit $g \in SL_2(\mathbb{Z})$ d'image $x$ dans $\mathbb{P}^1(\mathbb{Z}/N\mathbb{Z})$ et soit $M \in A_m$. On a $gM \in \tilde{A}_{m,N}SL_2(\mathbb{Z})$ si et seulement si $xM$ est défini, et alors l'image de $\eta(gM)$ dans $\mathbb{P}^1(\mathbb{Z}/N\mathbb{Z})$ est égale à $xM$. Soit $\bar{\eta} : \tilde{A}_{m,N}SL_2(\mathbb{Z}) \to SL_2(\mathbb{Z})$ un relèvement de $\eta$.

Démontrons la première égalité du théorème. Soit $g \in SL_2(\mathbb{Z})$ d'image $x$ dans $\mathbb{P}^1(\mathbb{Z}/N\mathbb{Z})$. Fixons un système de représentants $R$ de $g^{-1}\tilde{A}_{m,N}SL_2(\mathbb{Z})/SL_2(\mathbb{Z})$. Partons du terme de droite de l'égalité figurant dans le théorème. On a (comme dans la démonstration du théorème 4

$$\sum_M {}'u_M \xi^0(xM) = \sum_{M \in g^{-1}\tilde{A}_{m,N}SL_2(\mathbb{Z})} u_M\{\bar{\eta}(gM)0, \bar{\eta}(gM)\infty\}$$
$$= \sum_{r \in R} \sum_{M \in rSL_2(\mathbb{Z})} u_M\{\bar{\eta}(gr)r^{-1}M0,$$
$$\bar{\eta}(gr)r^{-1}M\infty\}$$
$$= \sum_{r \in R} \{\bar{\eta}(gr)r^{-1}g^{-1}\begin{pmatrix} m & 0 \\ 0 & m \end{pmatrix}g0, \bar{\eta}(gr)r^{-1}g^{-1}\begin{pmatrix} m & 0 \\ 0 & m \end{pmatrix}g\infty\}.$$

Il reste à vérifier que $\bar{\eta}(gr)r^{-1}g^{-1}\begin{pmatrix} m & 0 \\ 0 & m \end{pmatrix}$ parcourt un système de représentants de $\Gamma_0(N)\backslash A_{m,N}$ lorsque $r$ parcourt $R$. Posons $s = gr$. Les matrices $s$ parcourent un système de représentants de $\tilde{A}_{m,N}SL_2(\mathbb{Z})/SL_2(\mathbb{Z})$.

Vérifions que $\bar{\eta}(s)s^{-1}\begin{pmatrix} m & 0 \\ 0 & m \end{pmatrix} \in A_{m,N}$. La matrice $s$ s'écrit $\gamma\bar{\eta}(s)$ avec $\gamma \in \tilde{A}_{m,N}$. On a donc $\bar{\eta}(s)s^{-1}\begin{pmatrix} m & 0 \\ 0 & m \end{pmatrix} = \gamma^{-1}\begin{pmatrix} m & 0 \\ 0 & m \end{pmatrix} = \tilde{\gamma}$. Cela prouve que $\bar{\eta}(s)s^{-1}\begin{pmatrix} m & 0 \\ 0 & m \end{pmatrix}$ appartient à $A_{m,N}$.

Deux éléments $s$ et $s'$ de $\tilde{A}_{m,N}SL_2(\mathbb{Z})$ non congrus à droite modulo $SL_2(\mathbb{Z})$ définissent deux classes $\Gamma_0(N)\bar{\eta}(s)s^{-1}\begin{pmatrix} m & 0 \\ 0 & m \end{pmatrix}$ et $\Gamma_0(N)\bar{\eta}(s')s'^{-1}\begin{pmatrix} m & 0 \\ 0 & m \end{pmatrix}$ distinctes. En effet, dans le cas contraire il existerait $\gamma_0 \in \Gamma_0(N)$ tel que $s\bar{\eta}(s)^{-1}\gamma_0 = s'\bar{\eta}s'^{-1}$. On aurait alors $sSL_2(\mathbb{Z}) = s'SL_2(\mathbb{Z})$, ce qui est exclu par hypothèse

On déduit le fait $\bar{\eta}(s)s^{-1}\begin{pmatrix} m & 0 \\ 0 & m \end{pmatrix}$ parcourt un sytème de représentants de $\Gamma_0(N)\backslash A_{m,N}$ de la bijectivité de l'application composée

$$\begin{array}{ccc} \Gamma_0(N)\backslash A_{m,N} \to & SL_2(\mathbb{Z})\backslash SL_2(\mathbb{Z})A_{m,N} \to & \tilde{A}_{m,N}SL_2(\mathbb{Z})/SL_2(\mathbb{Z}) \\ \Gamma_0(N)\gamma \mapsto & SL_2(\mathbb{Z})\gamma & \mapsto \tilde{\gamma}SL_2(\mathbb{Z}) \end{array}$$

Cela prouve la première égalité du théorème.

La seconde égalité est une conséquence du fait que $\tilde{T}_m$ est par définition adjoint de $T_m$, de la première égalité du théorème et des formules de produits d'intersection établies par le corollaire 3.

Indiquons maintenant comment se comprennent les conditions évoquées sur les pointes dans la proposition 3 dans le cadre du groupe $\Gamma_0(N)$. Voir l'introduction pour la définition de l'ensemble $P_N$.

**Proposition 5** *Soit $\sum_x \lambda_x x \in \mathbb{Z}^{(\mathbb{P}^1(\mathbb{Z}/N\mathbb{Z}))}$. Le bord de l'élément $\sum_x \lambda_x \xi^0(x)$ de $H_1(X_0(N), \text{ptes}; \mathbb{Z})$ est nul si et seulement si on a dans $\mathbb{Z}^{(P_N)}$ l'égalité suivante :*
$$\sum_{(u,v)\in\mathbb{P}^1(\mathbb{Z}/N\mathbb{Z})} \lambda_{(u,v)}([(u,N),v^{-1}] - [(v,N),-u^{-1}]) = 0.$$

*Démonstration.-* On a une bijection
$$\Gamma_0(N)\backslash\mathbb{P}^1(\mathbb{Q}) \to P_N$$
$$\Gamma_0(N)\frac{x}{y} \mapsto [(y,N),x],$$

où par abus de notation $[(y,N),x]$ désigne l'élément $x \pmod{(y,N,N/(y,N))}$ dans $(\mathbb{Z}/(y,N,N/(y,N))\mathbb{Z})^* \subset P_N$ (voir l'introduction). Par conséquent $\mathbb{Z}^{(\Gamma_0(N)\backslash\mathbb{P}^1(\mathbb{Q}))}$ s'identifie à $\mathbb{Z}^{(P_N)}$. Soit $(u,v) \in \mathbb{P}^1(\mathbb{Z}/N\mathbb{Z})$. Soit $\begin{pmatrix} a & b \\ c & d \end{pmatrix} \in SL_2(\mathbb{Z})$ tel que la classe de $(c,d)$ dans $\mathbb{P}^1(\mathbb{Z}/N\mathbb{Z})$ soit égale à $(u,v)$. Le bord de l'élément $\xi^0((u,v)) = \{\frac{b}{d}, \frac{a}{c}\}$ dans $\mathbb{Z}^{(\Gamma_0(N)\backslash\mathbb{P}^1(\mathbb{Q}))}$ est égal à $(\frac{a}{c}) - (\frac{b}{d})$, ou encore dans $\mathbb{Z}^{(P_N)}$ à $[(c,N),a] - [(d,N),b]$, ou encore en utilisant l'égalité $ad - bc = 1$ et le fait que la classe de $(c,d)$ dans $\mathbb{P}^1(\mathbb{Z}/N\mathbb{Z})$ est égale à $(u,v)$ :
$$[(u,N),v^{-1}] - [(v,N),-u^{-1}].$$

On en déduit la proposition.

## 3 Paramétrisations modulaires.

*3.1 Degré des paramétrisations modulaires.* Soit $N$ un entier $> 0$. Dans cette partie on aura toujours $\Gamma = \Gamma_0(N)$. Notons $S_2(N)$ l'espace des formes modulaires paraboliques holomorphes de poids 2 pour $\Gamma_0(N)$. Soit $E$ une courbe elliptique modulaire définie sur $\mathbb{Q}$ de conducteur $N$. Soit $\phi$ :
$$X_0(N) \to E$$

une paramétrisation modulaire de $E$. L'image réciproque d'une forme différentielle de $E$ par ce morphisme est une forme différentielle sur $X_0(N)$. Elle définit une forme modulaire nouvelle normalisée $f \in S_2(N)$, propre pour tous les opérateurs de Hecke. La valeur propre de $T_p$ correspondant au nombre premier $p$ est $1 + p - |E(\mathbb{F}_p)|$.

Les parties invariantes et antiinvariantes de $H_1(E;\mathbb{Z})$ par la conjugaison complexe sont des $\mathbb{Z}$-modules libres de rang 1. Soient $\omega^+$ et $\omega^-$ deux générateurs respectivement de ces modules. On a un homomorphisme de groupes $\phi^*$ :
$$H_1(E;\mathbb{Z}) \to H_1(X_0(N);\mathbb{Z})$$

déduit fonctoriellement de $\phi$. Posons
$$c_0^+ = \phi^*(\omega^+) \text{ et } c_0^- = \phi^*(\omega^-).$$

Notons $H^0(X_0(N), \Omega^1)$ l'espace vectoriel complexe des formes différentielles holomorphes sur $X_0(N)$. On a un isomorphisme de $\mathbb{R}$-espaces vectoriels :

$$H_1(X_0(N); \mathbb{Z}) \otimes \mathbb{R} \to \mathrm{Hom}_{\mathbb{C}}(H^0(X_0(N), \Omega^1), \mathbb{C})$$

$$c \otimes t \mapsto (\omega \mapsto t \int_c \omega).$$

L'intégration de formes différentielles le long de cycles relatifs permet de définir un homomorphisme canonique de groupes (compatible aux actions des opérateurs de Hecke):

$$\mathrm{H}_1(X_0(N), ptes; \mathbb{Z}) \to \mathrm{Hom}_{\mathbb{C}}(\mathrm{H}^0(X_0(N), \Omega^1), \mathbb{C})$$
$$\simeq \mathrm{H}_1(X_0(N); \mathbb{Z}) \otimes \mathbb{R}.$$

Le théorème de Manin-Drinfeld nous assure que cet homomorphisme est à valeurs dans $\mathrm{H}_1(X_0(N); \mathbb{Z}) \otimes \mathbb{Q}$. Cet homomorphisme a été étudié pour $N$ premier dans [13] et fera l'objet d'une autre étude dans le cas général dans un article à paraître. Grâce aux produits d'intersection on a un homomorphisme canonique de groupes dual du précédent

$$\mathrm{H}_1(X_0(N); \mathbb{Z}) \to \mathrm{H}_1(Y_0(N); \mathbb{Z}) \otimes \mathbb{Q}.$$

Notons $c^-$ l'image de $c_0^-$ par ce dernier homomorphisme. L'image de $c^-$ par l'application canonique $\mathrm{H}_1(Y_0(N); \mathbb{Z}) \otimes \mathbb{Q} \to \mathrm{H}_1(X_0(N); \mathbb{Z}) \otimes \mathbb{Q}$ est $c_0^- \otimes 1$. Notons $c^+$ l'image de $c_0^+$ par l'injection canonique

$$\mathrm{H}_1(X_0(N); \mathbb{Z}) \to \mathrm{H}_1(X_0(N), ptes; \mathbb{Z}).$$

Les éléments $c^+$ et $c^-$ sont propres pour l'action des opérateurs de Hecke d'indice premier à $N$ avec les mêmes valeurs propres que $f$. Ils sont respectivement invariant et antiinvariant par la conjugaison complexe. D'après ce qui précède, il existe $\sum_x \lambda_x x \in \mathbb{Z}^{(\mathbb{P}^1(\mathbb{Z}/N\mathbb{Z}))}$ (bien défini modulo $R_N$, voir l'introduction) et $\sum_x \mu_x x \in \mathbb{Q}^{(\mathbb{P}^1(\mathbb{Z}/N\mathbb{Z}))}$ tels qu'on ait

$$\mu_x + \mu_{x\sigma} = \mu_x + \mu_{x\tau} + \mu_{x\tau^2} = 0$$

pout tout $x \in \mathbb{P}^1(\mathbb{Z}/N\mathbb{Z})$ et

$$\xi^0(\sum_x \lambda_x x) = c^+ \quad \text{et} \quad \xi_0(\sum_x \mu_x x) = c^-.$$

Notons $deg \phi$ le degré de $\phi$ et $\epsilon$ le nombre de composantes connexes de l'ensemble des points réels de $E$.

**Proposition 6** *On a*
$$deg\, \phi = \frac{\epsilon}{2} |\sum_x \lambda_x \mu_x|.$$

*Démonstration.-* Les produits d'intersection sur $X_0(N)$ et $E$ sont liés par la formule suivante

$$\phi^*(\omega^+) \bullet \phi^*(\omega^-) = \deg\phi\, \omega^+ \bullet \omega^-.$$

En examinant les cas où un parallélogramme fondamental d'un réseau complexe déterminant $E$ est un losange ou un rectangle on constate que l'on a
$$2 = \epsilon |\omega^+ \bullet \omega^-|.$$

On en déduit les égalités
$$\begin{aligned}\frac{2}{\epsilon}\deg\phi &= |c_0^+ \bullet c_0^-| \\ &= |c^+ \bullet c^-| \\ &= |\xi^0(\sum_x \lambda_x x) \bullet \xi_0(\sum_x \mu_x x)| \\ &= |\sum_x \lambda_x \mu_x|\end{aligned}$$

en utilisant la première partie. Cela prouve la proposition.

Signalons qu'une majoration polynômiale ne dépendant que de $N$ de $\deg\phi$ aurait d'intéressantes conséquences arithmétiques (voir [15], [3] et [17]).

Supposons maintenant que $E$ soit une courbe elliptique modulaire forte et que $\phi$ soit une paramétrisation modulaire forte. L'homomorphisme de groupes $\phi^*$ a pour image un facteur direct de $H_1(X_0(N); \mathbb{Z})$. Cela implique que $c_0^+$ et $c_0^-$ ne sont multiples entiers d'aucun élément de $H_1(X_0(N); \mathbb{Z})$ par un entier $> 1$.

Prouvons le théorème 1. Démontrons que les éléments $\sum_x \lambda_x x$ de $\mathbb{Z}^{(\mathbb{P}^1(\mathbb{Z}/N\mathbb{Z}))}$ et $\sum_x \mu_x x$ de $\mathbb{Q}^{(\mathbb{P}^1(\mathbb{Z}/N\mathbb{Z}))}$ dont $c^+$ et $c^-$ sont les images par $\xi^0$ et $\xi_0$ respectivement satisfont les conditions du théorème 1. La condition $i^-$) et l'expression de $R_N$ proviennent des présentations de $H_1(X_0(N), ptes; \mathbb{Z})$ et $H_1(Y_0(N); \mathbb{Z})$ respectivement (voir partie 1). Les conditions $ii^+$) et $ii^-$) expriment l'invariance et l'antiinvariance respectivement de $c^+$ et $c^-$ par la conjugaison complexe (voir section 2.3). Les conditions $iii^+$) et $iii^-$) traduisent que $c^+$ et $c^-$ sont vecteurs propres pour les opérateurs de Hecke d'indice premier à $N$ avec les valeurs propres associées à $f$. Comme $\xi^0(\sum_x \lambda_x x)$ provient de $H_1(X_0(N); \mathbb{Z})$, son bord est nul. On a donc la première partie de la condition $iv^+$). La première partie de la condition $iv^-$) exprime que l'image $c_0^-$ de $c^-$ dans $H_1(X_0(N); \mathbb{Z}) \otimes \mathbb{Q}$ est en réalité dans $H_1(X_0(N); \mathbb{Z})$, i.e. son intersection avec tout élément de $H_1(X_0(N); \mathbb{Z})$ est un entier (voir section 1.4). Les secondes parties des conditions $iv^+$) et $iv^-$) traduisent le fait que $c_0^+$ et $c_0^-$ ne sont multiples d'aucun élément de $H_1(X_0(N); \mathbb{Z})$ par un entier $> 1$.

Il reste à prouver que $\sum_x \lambda_x x + R_N$ et $\sum_x \mu_x x$ sont uniques au signe près. Soient $\sum_x \lambda'_x x \in \mathbb{Z}^{(\mathbb{P}^1(\mathbb{Z}/N\mathbb{Z}))}$ et $\sum_x \mu'_x x \in \mathbb{Q}^{(\mathbb{P}^1(\mathbb{Z}/N\mathbb{Z}))}$ vérifiant les conditions $i^+$), $ii^+$), $iii^+$), $iv^+$) et $i^-$), $ii^-$), $iii^-$), $iv^-$) respectivement. Considérons leurs images respectives $c'^+$ et $c'^-$ par $\xi^0$ et $\xi_0$ respectivement. D'après la condition $iv^-$), l'image $c'^-_0$ de $c'^-$ dans $H_1(X_0(N); \mathbb{Z}) \otimes \mathbb{Q}$ est un élément de $H_1(X_0(N); \mathbb{Z})$. D'après la condition $iv^+$), $c'^+$ est l'image $c'^+_0$ d'un élément de $H_1(X_0(N); \mathbb{Z})$. D'après les conditions $ii^+$), $iii^+$), $ii^-$), $iii^-$), $c'^+_0$ et $c'^-_0$ sont vecteurs propres pour les opérateurs de Hecke avec les valeurs propres correspondant à $f$ et sont respectivement invariant et antiinvariant par la conjugaison complexe. Les parties invariantes et antiinvariantes par la conjugaison complexe de $H_1(X_0(N); \mathbb{Z})$

sont isomorphes, après extension des scalaires à $\mathbb{C}$, à $S_2(N)$. Cet isomorphisme est compatible aux actions des opérateurs de Hecke d'indice premier à $N$ ([**16**]). Les systèmes de valeurs propres pour les opérateurs de Hecke associés à $c_0'^+$ et $c_0'^-$ correspondent à celui de $f$. Or $f$ est une forme modulaire nouvelle pour $\Gamma_0(N)$. On a donc un théorème de multiplicité 1 pour ce système de valeurs propres ([**1**]). On en déduit que $c_0'^+$ et $c_0'^-$ sont des multiples de $c_0^+$ et $c_0^-$. Les conditions de minimalité et d'intégralité $iv^+)$ et $iv^-)$ nous assurent que ces multiples sont égaux à $+1$ ou $-1$. Cela achève la preuve du théorème 1.

*Remarque.-* Manin a introduit dans [**6**] (de façon transcendante) les coefficients $\mu_x$. Si $g \in SL_2(\mathbb{Z})$ a pour image $x$ dans $\Gamma_0(N)\backslash SL_2(\mathbb{Z}) \simeq \mathbb{P}^1(\mathbb{Z}/N\mathbb{Z})$, $\mu_x$ est égal à la demi-somme des intégrales de la forme différentielle $2\pi i f(z)\,dz$ sur les chemins reliant $g0$ à $g\infty$ et $cgc0$ à $cgc\infty$ ($g \in SL_2(\mathbb{Z})$) dans $\bar{\mathfrak{H}}$ divisée par la période réelle de la courbe elliptique et multipliée par la constante de Manin. Grâce à ce procédé, Manin dresse des tables de valeurs des coefficients $\mu_x$ dans le cas des courbes modulaires $X_0(N)$ de genre 1.

*3.2 Nombre d'enroulement.* Reprenons toutes les notations de la section précédente. Notons $\omega_e$ l'image dans $H_1(E;\mathbb{Z}) \otimes \mathbb{Q}$ du symbole modulaire $\{0,\infty\}$ à travers les applications canoniques

$$H_1(X_0(N), ptes; \mathbb{Z}) \to H_1(X_0(N); \mathbb{Z}) \otimes \mathbb{Q} \to H_1(E;\mathbb{Z}) \otimes \mathbb{Q}.$$

Comme $\{0,\infty\}$ est invariant par la conjugaison complexe il en est de même de $\omega_e$. On a donc

$$\omega_e = n\omega^+ \quad (n \in \mathbb{Q}).$$

Le nombre rationnel $n_\phi = |n|$ est par définition le *nombre d'enroulement* de $\phi$. Ce nombre d'enroulement est égal à celui de [**9**] multiplié par $\epsilon$.

**Proposition 7** *On a*

$$n_\phi = \frac{\epsilon}{2}|\mu_{(0,1)}|.$$

*Démonstration.-* Notons que le symbole modulaire $\{0,\infty\}$ est égal à $\xi^0((0,1))$. Notons $e$ l'image de $\xi^0((0,1))$ dans $H_1(X_0(N);\mathbb{Z}) \otimes \mathbb{Q}$ ($e$ est le *winding element* de [**8**]). On a les égalités

$$\begin{aligned}
\frac{2}{\epsilon}n_\phi &= |n\omega^+ \bullet \omega^-| \\
&= |\omega_e \bullet \omega^-| \\
&= |e \bullet c_0^-| \\
&= |\xi^0((0,1)) \bullet c^-| \\
&= |\xi^0((0,1)) \bullet \xi_0(\sum_x \mu_x x)| \\
&= |\mu_{(0,1)}|.
\end{aligned}$$

Cela prouve la proposition.

Supposons maintenant que $E$ soit une courbe elliptique modulaire forte. On dispose alors d'une formule conjecturale pour $n_\phi$ provenant des conjectures de

Birch et Swinnerton-Dyer. Lorsque le groupe de Mordell-Weil $E(\mathbb{Q})$ de $E$ est infini on a
$$n_\phi = 0.$$
Lorsque $E(\mathbb{Q})$ est fini, notons $|E(\mathbb{Q})|$ son ordre, $|\text{III}|$ l'ordre (fini dans ce cas, voir [4]) du groupe de Shafarevich-Tate de $E$, $n_p$ le nombre de composantes rationnelles de la fibre d'un modèle de Néron en $p$ de $E$ ($p$ premier divisant $N$) et $c_\phi$ la constante de Manin de la paramétrisation (conjecturalement c'est toujours 1, voir [5]). On a alors ([9])
$$n_\phi = c_\phi \frac{\epsilon |\text{III}| \prod_p n_p}{|E(\mathbb{Q})|^2}.$$
Si on oublie la constante de Manin, tous les facteurs figurant dans le membre de droite de l'égalité ne font intervenir que les invariants intrinsèques de la courbe elliptique, indépendamment de toute paramétrisation modulaire. Essentiellement on a donc une formule donnant le coefficient $\mu_{(0,1)}$ donnée directement par la courbe elliptique.

*3.3 Deux exemples.* Considérons les deux courbes elliptiques modulaires fortes de conducteur 37. Donnons une liste de coefficients $\mu_x$ et $\lambda_x$ associés à ces courbes. Des coefficients $\mu_x$ associés à des courbes modulaires $X_0(N)$ de genre 1, qui sont nécessairement des courbes elliptiques modulaires fortes, sont donnés dans [6].

*La courbe (37C).* La courbe notée (37C) dans les tables d'Anvers [2] a pour équation minimale
$$y^2 + y = x^3 + x^2 - 23x - 50.$$
Son conducteur est 37 et son discriminant est $> 0$. On a donc $\epsilon = 2$. Des coefficients $\mu_x$ et $\lambda_x$ associés à la courbe elliptique par le théorème 1 sont donnés par la table suivante.

| $x$ | $(0,1)$ | $(1,1)$ | $(2,1)$ | $(3,1)$ | $(4,1)$ | $(5,1)$ | $(6,1)$ | $(7,1)$ |
|---|---|---|---|---|---|---|---|---|
| $\mu_x$ | $\frac{2}{3}$ | 0 | 2 | 1 | 1 | 1 | 0 | 1 |
| $\lambda_x$ | 0 | 0 | 1 | 0 | 0 | 0 | 0 | 0 |

| $x$ | $(8,1)$ | $(9,1)$ | $(10,1)$ | $(11,1)$ | $(12,1)$ | $(13,1)$ | $(14,1)$ | $(15,1)$ |
|---|---|---|---|---|---|---|---|---|
| $\mu_x$ | 0 | $-1$ | 0 | 0 | $-1$ | 1 | 0 | $-1$ |
| $\lambda_x$ | 0 | 0 | 0 | 0 | 0 | 0 | 0 | 0 |

| $x$ | $(16,1)$ | $(17,1)$ | $(18,1)$ | $(19,1)$ | $(20,1)$ | $(21,1)$ | $(22,1)$ | $(23,1)$ |
|---|---|---|---|---|---|---|---|---|
| $\mu_x$ | $-1$ | $-1$ | $-2$ | $-2$ | $-1$ | $-1$ | $-1$ | 0 |
| $\lambda_x$ | 0 | 0 | 0 | 0 | 0 | 0 | 0 | 0 |

| $x$ | $(24,1)$ | $(25,1)$ | $(26,1)$ | $(27,1)$ | $(28,1)$ | $(29,1)$ | $(30,1)$ | $(31,1)$ |
|---|---|---|---|---|---|---|---|---|
| $\mu_x$ | 1 | $-1$ | 0 | 0 | $-1$ | 0 | 1 | 0 |
| $\lambda_x$ | 0 | 0 | 0 | 0 | 0 | 0 | 0 | 0 |

| $x$ | $(32,1)$ | $(33,1)$ | $(34,1)$ | $(35,1)$ | $(36,1)$ | $(1,0)$ |
|---|---|---|---|---|---|---|
| $\mu_x$ | 1 | 1 | 1 | 1 | 0 | $-\frac{2}{3}$ |
| $\lambda_x$ | 0 | 0 | 0 | 0 | 0 | |

On en déduit la valeur de $n_\phi$ et de $d_\phi$

$$n_\phi = \frac{2}{3} \quad \text{et} \quad d_\phi = 2.$$

*La courbe (37A).* La courbe notée (37A) dans les tables de [**2**] a pour équation minimale

$$y^2 + y = x^3 - x.$$

Son conducteur est 37 et son discriminant est $> 0$. On a donc $\epsilon = 2$. Le rang du groupe de Mordell-Weil de la courbe (37A) est égal à 1. Dressons une liste de coefficients $\mu_x$ et $\lambda_x$.

| $x$ | $(0,1)$ | $(1,1)$ | $(2,1)$ | $(3,1)$ | $(4,1)$ | $(5,1)$ | $(6,1)$ | $(7,1)$ |
|---|---|---|---|---|---|---|---|---|
| $\mu_x$ | 0 | 0 | 0 | 0 | 0 | 1 | 0 | 1 |
| $\lambda_x$ | 0 | 0 | 0 | 1 | 0 | $-1$ | 0 | 0 |

| $x$ | $(8,1)$ | $(9,1)$ | $(10,1)$ | $(11,1)$ | $(12,1)$ | $(13,1)$ | $(14,1)$ | $(15,1)$ |
|---|---|---|---|---|---|---|---|---|
| $\mu_x$ | 1 | 0 | 0 | 0 | 0 | 0 | $-1$ | $-1$ |
| $\lambda_x$ | 0 | 0 | 0 | 0 | 0 | 0 | 1 | 0 |

| $x$ | $(16,1)$ | $(17,1)$ | $(18,1)$ | $(19,1)$ | $(20,1)$ | $(21,1)$ | $(22,1)$ | $(23,1)$ |
|---|---|---|---|---|---|---|---|---|
| $\mu_x$ | $-1$ | 0 | 0 | 0 | 0 | $-1$ | $-1$ | $-1$ |
| $\lambda_x$ | 0 | 0 | 0 | 0 | 0 | 0 | 0 | 0 |

| $x$ | $(24,1)$ | $(25,1)$ | $(26,1)$ | $(27,1)$ | $(28,1)$ | $(29,1)$ | $(30,1)$ | $(31,1)$ |
|---|---|---|---|---|---|---|---|---|
| $\mu_x$ | 0 | 0 | 0 | 0 | 0 | 1 | 1 | 0 |
| $\lambda_x$ | 0 | 0 | 0 | 0 | 0 | 0 | 0 | 0 |

| $x$ | $(32,1)$ | $(33,1)$ | $(34,1)$ | $(35,1)$ | $(36,1)$ | $(1,0)$ | | |
|---|---|---|---|---|---|---|---|---|
| $\mu_x$ | 1 | 0 | 0 | 0 | 0 | 0 | | |
| $\lambda_x$ | 0 | 0 | 0 | 0 | 0 | 0 | | |

Signalons que Manin a déjà donné cette liste de coefficients $\mu_x$ dans [**7**]. On déduit de cette table

$$n_\phi = 0 \quad \text{et} \quad d_\phi = 2.$$

*References*

[1] Atkin A.O.L., Lehner J., *Hecke operators on $\Gamma_0(m)$*, Math. Ann., 185:134–160, 1970.

[2] Atkin A.O.L. et al. *Numerical tables for elliptic curves. Modular functions of one variable IV*, number 476 in Lecture notes in mathematics. Springer-Verlag, 1972.

[3] Frey G. *Links between solutions of $a - b = c$ and elliptic curves*, In E. Wirsing H.P. Schlickewei, editor, *Number theory, Ulm 1987*, number 1380 in Lecture notes in mathematics, pages 31–62. Springer-Verlag, 1989.

[4] Kolyvagin V. A., *Finiteness of $E(\mathbb{Q})$ and $\mathrm{III}(E,\mathbb{Q})$ for a subclass of Weil curves*, Izvestija AN SSSR, Ser. Math., 52(3):522–540, 1988.

[5] Manin Y., *Cyclotomic fields and modular curves*, Russian Math. Survey, 26(6):7–71, 1971.

[6] Manin Y., *Parabolic points and zeta function of modular curves*, Math. USSR Izvestija, 6(1):19–64, 1972.

[7] Manin Y., *Modular forms and number theory*, In Ohli Lehto, editor, *Proceedings of the international congress of mathematicians 1978*, pages 177–186. 1980.

[8] Mazur B., *Modular curves and the Eisenstein ideal*, Pub. math. de l'IHES, 47:33–186, 1977.

[9] Mazur B., Swinnerton-Dyer H.P.F., *The arithmetic of Weil curves*, Inv. math., 25:1–61, 1974.

[10] Merel L., *Opérateurs de Hecke et sous-groupes de $\Gamma(2)$*, Journal of Number theory, A paraître. (= Thèse, chapitre 5).

[11] Merel L., *Universal fourier expansions of modular forms*, In Gerhard Frey, editor, *On Artin's conjecture for 2-dimensional, odd Galois representations*, Lecture Notes in Mathematics, 1585, Springer Verlag.

[12] Merel L., *Opérateurs de Hecke pour $\Gamma_0(N)$ et fractions continues*, Ann. Inst. Fourier, 41(3), 1991. (= Thèse, chapitre 2).

[13] Merel L., *L'accouplement de Weil entre le sous-groupe cuspidal et le sous-groupe de Shimura de $J_0(p)$*, 1992. A paraître. (= Thèse, chapitre 4).

[14] Merel L., *Intersections sur des courbes modulaires*, Manuscripta mathematica, 80:283–289, 1993.(= Thèse, chapitre 1).

[15] Oesterlé J., *Nouvelles approches du "théorème" de Fermat, séminaire Bourbaki $n^0$ 694*, In *Séminaire Bourbaki 1987-88*, volume 161-162, pages 165–186. Astérisque, 1988.

[16] Stevens G., *Arithmetic on modular curves*, Number 20 in Progress in mathematics. Birkhaüser, 1982.

[17] Zagier D., *Modular parametrization of elliptic curves*, Canad. Math. Bull., 28(3):377–389, 1985.

Conference on Elliptic Curves and Modular Forms
Hong Kong, December 18-21, 1993
Copyright ©1995 International Press

# Irreducible Galois representations arising from component groups of Jacobians

KENNETH A. RIBET
UC MATHEMATICS DEPARTMENT
BERKELEY, CA 94720-3840 USA
*E-mail address*: RIBET@MATH.BERKELEY.EDU

**1 Introduction.** Much has been written about component groups of Néron models of Jacobians of modular curves. In a variety of contexts, these groups have been shown to be "Eisenstein," which implies that they can be neglected in the study of irreducible two-dimensional representations of $\text{Gal}(\overline{\mathbf{Q}}/\mathbf{Q})$.

The first theorem to this effect may be extracted from Mazur's landmark article [11], which concerns the Jacobian $J_0(N)$ when $N$ is a prime. In this article, Mazur studies the action of Hecke operators on the "cuspidal subgroup" of the Jacobian, and obtains information about the relevant component group as an application [11], p. 98. More precisely, consider the fiber in characteristic $N$ of the Néron model of $J_0(N)$. This commutative group scheme is an extension of a finite étale group scheme $\Phi$ by an algebraic torus; one is interested in the functorial action of Hecke operators on $\Phi$. Mazur proves for all prime numbers $p \neq N$ that the $p$th Hecke operator $T_p$ acts on $\Phi$ by multiplication by $1+p$, and also that $T_N$ is the identity on $\Phi$. The "Eisenstein" terminology arsies from the fact that $1 + p$ is the $p$th coefficient of the standard Eisenstein series of weight two on $\Gamma_0(N)$.

In his first article on Serre's conjectures [13], the author generalized Mazur's result to the Jacobian $J_0(N)$ where $N$ is the product of a positive integer $M$ and a prime number $q$ prime to $M$. According to Theorem 3.12 of [13], one again has the identity $T_p = 1 + p$ for all prime numbers $p$ prime to $N$ on the group of connected components of the Néron reduction $J_0(N)_{/\mathbf{F}_q}$. (A slightly more refined statement appears as Theorem 3.22 of [13]; for the proof, see [14].) Subsequently, B. Edixhoven showed in [6] that an analogous result holds for every reduction $J_0(N)_{/\mathbf{F}_q}$ with $q > 3$; here, the new element is that $N$ is allowed to be divisible by an arbitrary power of $q$.

In another article on Serre's conjectures [15], the author discusses the component group $\Phi$ attached to the mod $q$ reduction of the Jacobian of the modular curve derived from $\Gamma_1(N) \cap \Gamma_0(q)$; here $N$ is a positive integer and $q$ is a prime number not dividing $N$. The analysis of [15], pp. 672–673 proves the identity

---

This work was partially supported by NSF Grant #DMS 93-06898. The author wishes to thank B. Edixhoven, H. W. Lenstra, Jr., D. Lorenzini, N. Skoruppa, D. Ulmer, and A. Wilkinson for helpful correspondence and suggestions. We also thank the International Scientific Committee and the local organizers of the Hong Kong conference on Fermat's Last Theorem for their generous invitation.

$T_p = 1 + p$ on $\Phi$ for all prime numbers $p \nmid qN$ and establishes at the same time that the "diamond bracket" operators $\langle d \rangle$ act as the identity on $\Phi$.

In this article, we consider the situation in which $\Gamma_1(N)$ is replaced by a subgroup $\Gamma$ intermediate between $\Gamma_0(N)$ and $\Gamma_1(N)$. In other words, we study the component group $\Phi_\Gamma$ associated to the mod $q$ reduction coming from $\Gamma \cap \Gamma_0(q)$. (We continue to assume that $q$ is prime to $N$.) Since $\Phi_\Gamma$ is Eisenstein in the two extreme situations $\Gamma = \Gamma_0(N)$ and $\Gamma = \Gamma_1(N)$, it is natural to ask whether $\Phi_\Gamma$ is Eisenstein for intermediate groups.

In order to rule out trivial counterexamples, we will generalize slightly our definition of "Eisenstein." Let $S$ be the complex vector space of weight-two cusp forms on $\Gamma \cap \Gamma_0(q)$, and let $\widetilde{\mathbb{T}}$ be the subring of $\text{End}\, S$ generated by the Hecke operators $T_n$ and $\langle d \rangle$ for $n$ a prime number prime to $qN$ and $d$ prime to $N$. As we shall recall below, there is a natural action of these operators on $\Phi_\Gamma$. We shall be interested in the set of maximal ideals $\mathfrak{m}$ of $\widetilde{\mathbb{T}}$ which lie in the support of $\Phi_\Gamma$ as a $\widetilde{\mathbb{T}}$-module, i.e., the set of maximal ideals in the image of $\widetilde{\mathbb{T}}$ in $\text{End}\, \Phi_\Gamma$.

To each $\mathfrak{m}$, one associates as usual a continuous semisimple representation

$$\rho_\mathfrak{m} : \text{Gal}(\overline{\mathbf{Q}}/\mathbf{Q}) \to \mathbf{GL}(2, \widetilde{\mathbb{T}}/\mathfrak{m}).$$

This representation is characterized up to isomorphism by the fact that the trace of $\rho_\mathfrak{m}(\text{Frob}_p)$ is $T_p$ mod $\mathfrak{m}$ and the determinant of this matrix is $p\langle p \rangle$ mod $\mathfrak{m}$, for all but finitely many prime numbers $p$. (Here, $\text{Frob}_p$ is an arithmetic Frobenius element for $p$ in $\text{Gal}(\overline{\mathbf{Q}}/\mathbf{Q})$. See, e.g., [15], Section 7 for some of the relevant background.) It is worth stressing that $\Phi_\Gamma$ gives rise through this construction to a collection of two-dimensional representations of $\text{Gal}(\overline{\mathbf{Q}}/\mathbf{Q})$, despite the fact there is no natural action of $\text{Gal}(\overline{\mathbf{Q}}/\mathbf{Q})$ on the group $\Phi_\Gamma$.

We shall say that $\Phi_\Gamma$ is *Eisenstein* if all $\mathfrak{m}$ in the support of $\Phi_\Gamma$ give rise to representations $\rho_\mathfrak{m}$ which are *reducible*. We say that $\Phi_\Gamma$ is *strongly Eisenstein* if the operators $\langle d \rangle$ are trivial on $\Phi_\Gamma$ and one has $T_p = 1 + p$ on $\Phi_\Gamma$ for all but finitely many prime numbers $p$. It is easy to show that "strongly Eisenstein" implies "Eisenstein" by using the Cebotarev Density Theorem and the Brauer-Nesbitt Theorem [13], 5.2c.

To be sure, one's guess that $\Phi_\Gamma$ is Eisenstein turns out to be not very far off the mark. Indeed, $\Phi_\Gamma$ is strongly Eisenstein in most cases, and the prime-to-6 part of $\Phi_\Gamma$ is strongly Eisenstein in *all* cases. Nonetheless, the blanket assertion that $\Phi_\Gamma$ is Eisenstein is definitely false. To convince the reader of this fact, we exhibit in the next section some non-Eisenstein component groups; our construction is a digression which is intended to motivate the more systematic study which follows.

This article's main contribution is an analysis of $\Phi_\Gamma$ in the case of general $\Gamma$ and $q$. As we indicated above, the $\ell$-primary parts of $\Phi_\Gamma$ are strongly Eisenstein (and cyclic as abelian groups) for all prime numbers $\ell \geq 5$. After recalling this simple result, we investigate the 2-primary and the 3-primary components of $\Phi_\Gamma$. In case $\Phi_\Gamma$ is non-Eisenstein, we identify the non-Eisenstein "pieces" and describe on these pieces the action of the Hecke operators $T_n$ and $\langle d \rangle$.

From the point of view of Galois representations, our main result is the identification of all irreducible representations with values in $\mathbf{GL}(2, \overline{\mathbf{F}}_2)$ and $\mathbf{GL}(2, \overline{\mathbf{F}}_3)$ that can be associated with component groups. These are induced representations, coming from certain characters $\text{Gal}(\overline{\mathbf{Q}}/\mathbf{Q}(\sqrt{-1})) \to \overline{\mathbf{F}}_2^*$ and $\text{Gal}(\overline{\mathbf{Q}}/\mathbf{Q}(\sqrt{-3})) \to \overline{\mathbf{F}}_3^*$, respectively. For a character $\theta$ to intervene in some component group, it is necessary and sufficient that all residue classes modulo its conductor be represented by rational integers.

This article is an outgrowth of the author's talks at the Hong Kong conference on Fermat's Last Theorem. These talks outlined portions of A. Wiles's manuscript [20]. This manuscript provides ample motivation for a detailed analysis of mod 3 representations of $\text{Gal}(\overline{\mathbf{Q}}/\mathbf{Q})$ which arise from modular forms. Indeed, Wiles's attack on the Taniyama-Shimura conjecture begins with a theorem of J. Tunnell [19] which implies that the mod 3 representation of $\text{Gal}(\overline{\mathbf{Q}}/\mathbf{Q})$ arising from an elliptic curve over $\mathbf{Q}$ is associated to some Hecke eigenform. Assuming that this mod 3 representation is irreducible, one shows that the eigenform may be chosen in accordance with the conjectures of [16]. For this, attention must be directed to the special problems posed by mod 3 representations; see F. Diamond's article in this volume [5], which provides an update on Serre's conjectures. In reflecting on mod 3 representations, the author was forced to abandon his "axiom" that component groups are Eisenstein; this led to the study which is presented below.

## 2 Some non-Eisenstein component groups.

Consider the irreducible continuous representation $\rho : \text{Gal}(\overline{\mathbf{Q}}/\mathbf{Q}) \to \text{GL}(2, \mathbf{F}_3)$ arising from the space of forms of weight two on $\Gamma_1(13)$. This space has dimension two, and the two Hecke eigenforms in the space are complex conjugates of each other. Their coefficients lie in the ring of integers of $\mathbf{Q}(\sqrt{-3})$; $\rho$ is the mod $(\sqrt{-3})$ representation of $\text{Gal}(\overline{\mathbf{Q}}/\mathbf{Q})$ associated with either of them. The determinant of $\rho$ is the product of the mod 3 cyclotomic character of $\text{Gal}(\overline{\mathbf{Q}}/\mathbf{Q})$ and the character of order two which corresponds to the quadratic extension $\mathbf{Q}(\sqrt{13})$ of $\mathbf{Q}$. Let $H$ be the group of squares in $(\mathbf{Z}/13\mathbf{Z})^*$. Then conjecture (3.2.4?) of Serre's article [16] predicts that $\rho$ should arise from a Hecke eigenform in the space of weight-two cusp forms on the group

$$\Gamma = \Gamma_H(13) := \left\{ \begin{pmatrix} a & b \\ c & d \end{pmatrix} \in \text{SL}(2, \mathbf{Z}) \,\Big|\, c \equiv 0 \bmod 13,\, d \in H \right\}.$$

However, as Serre pointed out in a letter to the author [17], this space of forms is zero—the conjecture is false as stated.

One way to deal with this apparent difficulty is to reformulate Serre's conjecture as a relation between mod $p$ Galois representations and mod $p$ modular forms in the sense of Katz [9]. Such a reformulation is presented as Conjecture 4.2 of Edixhoven's article [7], which attributes the reformulation to Serre. The characters which appear in the conjecture are then naturally $\overline{\mathbf{F}}_p^*$-valued, and the difficulty becomes invisible. On the other hand, this solution hides a problem that may be genuinely of interest, since one wants to characterize those spaces of forms which give rise to a given Galois representation.

One certainly knows that problems of the sort exemplified by $\rho$ occur only for mod 2 and mod 3 representations of $\text{Gal}(\overline{\mathbf{Q}}/\mathbf{Q})$. Further, among all mod 3 representations, only those which become abelian on $\text{Gal}(\overline{\mathbf{Q}}/\mathbf{Q}(\sqrt{-3}))$ can give trouble, cf. [2], p. 796. Finally, let $q > 2$ be a prime number which is congruent to 2 modulo 3. Then the difficulty in the case of $\rho$ "disappears" when the level is augmented by an auxiliary $\Gamma_0(q)$-structure: Lemme 1 of [**2**] implies that $\rho$ arises from the space of weight-two cusp forms on $\Gamma \cap \Gamma_0(q)$. (The author is grateful to N. Skoruppa for undertaking a numerical verification of this assertion when $q = 5$.)

Let $J$ be the Jacobian of the modular curve over $\mathbf{Q}$ associated with $\Gamma \cap \Gamma_0(q)$. (We again write $J$ for the Néron model of this abelian variety.) As above, we let $\Phi_\Gamma$ be the component group associated with the reduction of $J$ mod $q$.

**Theorem 1.** *The group $\Phi_\Gamma$ is non-Eisenstein.*

*Proof.* Let $V$ be a two-dimensional $\mathbf{F}_3$-vector space affording the representation $\rho$, and for each $r \nmid 13q$ let $a_r \in \mathbf{F}_3$ be the trace of $\rho(\text{Frob}_r)$, where $\text{Frob}_r$ is a Frobenius element for $r$ in $\text{Gal}(\overline{\mathbf{Q}}/\mathbf{Q})$. The fact that $\rho$ arises from $\Gamma \cap \Gamma_0(q)$ implies that there is a $\text{Gal}(\overline{\mathbf{Q}}/\mathbf{Q})$-equivariant embedding $V \hookrightarrow J(\overline{\mathbf{Q}})$ with the following property: For each prime number $r \nmid 13q$, the endomorphism $T_r$ of $J$ acts on $V$ as the homothety $a_r$. Since $q$ is prime to 39, $V$ is unramified at $q$. Thus (after choosing a place of $\overline{\mathbf{Q}}$ lying over $q$), we may view $V$ as embedded in $J_{/\mathbf{F}_q}(\overline{\mathbf{F}}_q)$.

We claim now that $V$ does not land entirely in the connected component $J^o_{/\mathbf{F}_q}$ of $J_{/\mathbf{F}_q}$. In other words, we assert that the image $W$ of $V$ in the group of connected components of $J_{/\mathbf{F}_q}$ is non-trivial. Since the group of connected components in question is none other than $\Phi_\Gamma$, our claim implies that $\Phi_\Gamma$ contains a non-zero subgroup of exponent 3 on which each Hecke operator $T_r$ acts as $a_r$. That the $a_r$ are traces of an *irreducible* representation indicates that this subgroup is non-Eisenstein, thereby proving the Theorem.

The claim concerning $V$ is proved by an appeal to results of Deligne-Rapoport [**4**] and general theorems of Grothendieck and Raynaud. This body of work is summarized in [**12**], [**13**], Section 2, [**15**], Section 8, and the discussion which occurs in Section 3 below. The main point is that by using [**4**], Th. 6.9, p. 286, one sees that $J^o_{/\mathbf{F}_q}$ is a certain extension of an abelian variety by a torus. The abelian variety in question appears as the product of two copies of the Jacobian of the modular curve associated with the group $\Gamma$. However, this modular curve has genus zero, so its Jacobian is zero. Thus $J^o_{/\mathbf{F}_q}$ coincides with its "toric part" $T$. Therefore, the statement that $V$ falls entirely in $J^o_{/\mathbf{F}_q}(\overline{\mathbf{F}}_q)$ is the statement that $V$ is a subgroup of $T(\overline{\mathbf{F}}_q)$. The argument given at the conclusion of [**15**], Section 8 shows that the inclusion $V \hookrightarrow T(\overline{\mathbf{F}}_q)$ is possible only when $q \equiv 1$ mod 3; the choice of $q$ thus guarantees that $W$ is non-zero. This proves the claim. □

**3 A concrete description of $\Phi_\Gamma$.** In this section, we present a nuts-and-bolts description of $\Phi_\Gamma$ as an abelian group furnished with a family of Hecke operators $T_r$ and $\langle d \rangle$. This material is now quite well known, at least in the case where $\Gamma = \Gamma_0(N)$ [**1, 12**].

Let $N$ be a positive integer. Subgroups $\Gamma$ between $\Gamma_1(N)$ and $\Gamma_0(N)$ are in 1-1 correspondence with subgroups of $(\mathbf{Z}/N\mathbf{Z})^*$. Let $H$ be such a subgroup, and set

$$\Gamma = \Gamma_H(N) = \left\{ \begin{pmatrix} a & b \\ c & d \end{pmatrix} \in \mathbf{SL}(2,\mathbf{Z}) \;\Big|\; c \equiv 0 \bmod N, \; d \in H \right\}.$$

Thus $\Gamma = \Gamma_1(N)$ in case $H$ is the trivial subgroup of $(\mathbf{Z}/N\mathbf{Z})^*$, while $\Gamma = \Gamma_0(N)$ if $H = (\mathbf{Z}/N\mathbf{Z})^*$. The modular curve corresponding to $\Gamma$ will be called $X_H(N)$; thus, $X_H(N)$ is the quotient of $X_1(N)$ by the image of $H$ in the Galois group $(\mathbf{Z}/N\mathbf{Z})^*/\{\pm 1\}$ of the covering $X_1(N) \to X_0(N)$. Postponing the assumption that $q$ is prime, we let $q$ be a positive integer prime to $N$, and put

$$\Gamma(H,q) = \Gamma_H(N) \cap \Gamma_0(q).$$

(A more precise, but less compact, name for this group would have been $\Gamma_H(N,q)$.)

The modular curve $X(H,q)$ corresponding to $\Gamma(H,q)$ is the quotient of $X_1(Nq)$ by the image of $H \times (\mathbf{Z}/q\mathbf{Z})^*$ in $(\mathbf{Z}/Nq\mathbf{Z})^*/\{\pm 1\}$. This image is unchanged if we augment $H$ by $\{\pm 1\}$; thus we can, and will, assume that $H$ contains $-1$. (If $N \leq 2$, we have $-1 = +1$ in $(\mathbf{Z}/N\mathbf{Z})^*$, and we are imposing no condition on $H$.) We view $X(H,q)$ as classifying elliptic curves $E$ which are furnished with a subgroup $C$ of order $q$ and a point $P$ of order $N$. In the classification, the point $P$ is considered "mod $H$" in the sense that the triples $(E,C,P)$ and $(E,C,hP)$ are identified for all $h \in H$. (Compare [4], IV, Section 3.) The orbit of $P$ mod $H$ will be called $\alpha_P$; we will say that $X(H,q)$ classifies triples $(E,C,\alpha)$ where $\alpha$ is a point of order $N$ on $E$ which is taken "mod $H$."

Define $J_H(N)$ and $J(H,q)$ to be the Jacobians of the modular curves $X_H(N)$ and $X(H,q)$, respectively.

From now on, we assume that $q$ is a *prime* number. For convenience, we impose the assumption $q \geq 5$ at this point. (The cases $q = 2$ and $q = 3$ presumably could be included with little modification in what follows.) Consider the reduction of $J(H,q)$ modulo $q$. The group of components associated with this reduction is the group $\Phi_\Gamma$ in the discussion above; of course, this group depends on $q$ as well as on $\Gamma$.

A description of $\Phi_\Gamma$ can be deduced in a standard way from the theorem of Deligne and Rapoport which we cited above [4], Th. 6.9, p. 286. The theorem provides a model $\mathcal{C}$ of $X(H,q)$ over $\mathbf{Z}_q$ which is an "admissible curve" in the sense of [8], Section 3 and [13], Section 2. The special fiber of $\mathcal{C}$ has two irreducible components, each isomorphic to the modular curve $X_H(N)$.

The set of singular points of $\mathcal{C}_{\overline{\mathbf{F}}_q}$ is in bijection with the set $S$ of supersingular $\overline{\mathbf{F}}_q$-valued points of $X_H(N)$, i.e., those points which arise from pairs $(E,P)$ where $E$ is a supersingular elliptic curve over $\overline{\mathbf{F}}_q$ and $P$ is a point of order $N$ on $E$. To each $s \in S$ we associate an integer $e(s) \in \{1,2,3\}$ in the following way: If $s$ is represented by the pair $(E,P)$, then $2e(s)$ is the number of automorphisms of $E$ which map $P$ to some point $hP$ with $h \in H$; in other words, $2e(s)$ is the number of automorphisms of $(E,\alpha_P)$. (Because $-1$ belongs to $H$, the automorphism "$-1$" of $E$ induces an automorphism of $(E,\alpha_P)$. The number of such automorphisms is then 2, 4, or 6 because we have assumed $q \geq 5$.)

The singular point of $\mathcal{C}$ corresponding to $s$ is then described in $\mathcal{C}$ by the local equation $XY = q^{e(s)}$.

Consider now the diagonal pairing on $\mathbf{Z}^S$ for which $\langle s, s \rangle$ is the positive integer $e(s)$. Let $L$ be the group of degree 0 elements in $\mathbf{Z}^S$, and let $\iota : L \hookrightarrow \text{Hom}(L, \mathbf{Z})$ be the linear map which describes the restriction of this pairing to $L$. The following result is a consequence of the theorem of Deligne and Rapoport and work of Grothendieck and Raynaud. (See [12], and perhaps the discussion in [13], p. 438.)

**Theorem 2.** *The component group $\Phi_\Gamma$ may be identified with the cokernel of $\iota$.*

One understands that the identification between $\Phi_\Gamma$ and coker $\iota$ is compatible with the functorial actions of Hecke operators and of the Galois group $\text{Gal}(\overline{\mathbf{F}}_q/\mathbf{F}_q)$. We shall discuss the Hecke operators.

As usual, Hecke operators act on $J(H, q)$ in two different ways, because the Jacobian construction is both covariant and contravariant. In what follows we adopt the contravariant point of view in which $J(H, q)$ is regarded as $\text{Pic}^0(X(H, q))$. We describe the action of Hecke operators on $\Phi_\Gamma$ without supplying any substantial justification. However, the reader may consult [12], which discusses in detail the special case case where $H = (\mathbf{Z}/N\mathbf{Z})^*$.

First, let $d$ be an integer prime to $N$. The association $(E, C, P) \mapsto (E, C, dP)$ defines a "diamond bracket" automorphism $\langle d \rangle$ on $X(H, q)$. Using the contravariant functoriality, we obtain automorphisms of $J(H, q)$ and then $\Phi_\Gamma$, both of which we shall call simply $\langle d \rangle$. In the description of $\Phi_\Gamma$ as the cokernel of an injection $L \hookrightarrow \text{Hom}(L, \mathbf{Z})$, the map $\langle d \rangle$ on $\Phi_\Gamma$ is induced from two automorphisms of $L$. Namely, consider the permutation of $S$ which takes the class of a pair $(E, P)$ to the class of $(E, dP)$. This permutation induces an automorphism $\langle d \rangle_L$ of $L$, which we regard as a subgroup of $\mathbf{Z}^S$. Let $\langle d \rangle'_L$ be the *inverse* of this permutation. Then we have a commutative diagram

$$0 \to L \xrightarrow{\iota} \text{Hom}(L, \mathbf{Z}) \to \Phi_\Gamma \to 0$$
$$\downarrow \qquad \downarrow \qquad \downarrow$$
$$0 \to L \xrightarrow{\iota} \text{Hom}(L, \mathbf{Z}) \to \Phi_\Gamma \to 0$$

in which the three vertical arrows are respectively $\langle d \rangle'_L$, $\text{Hom}(\langle d \rangle_L, \mathbf{Z})$ and $\langle d \rangle$.

Similarly, suppose that $p$ is a prime number which does not divide $qN$. The symbol $T_p$ then denotes a host of objects: the standard Hecke correspondence $T_p$ on $X(H, q)$, the endomorphism of $J(H, q)$ which this correspondence induces by contravariant functoriality, and finally the endomorphism $T_p$ of $\Phi_\Gamma$ which then arises from the functoriality of the Néron model. We seek to identify this latter endomorphism.

Fix $p$ for the moment and let $T$ be the correspondence on $S$ defined by the rule $(E, \alpha) \mapsto \sum (E/D, \alpha \mod D)$, where the sum is taken over the subgroups $D$ of $E$ having order $p$. In analogy with the situation above, we write $T_L$ for the induced endomorphism of $L \subset \mathbf{Z}^S$. The analogue of $\langle d \rangle'_L$ is then the endomorphism $T'_L := \langle p \rangle_L^{-1} \circ T_L$. We have a commutative diagram like the one above in which the three vertical arrows are respectively $T'_L$, $\text{Hom}(T_L, \mathbf{Z})$ and the endomorphism $T_p$ of $\Phi_\Gamma$. Since $p$ may vary in what follows, we restore $p$

to the notation, referring to the first of these operators as $T'_p$ and the second as $\text{Hom}(T_p, \mathbf{Z})$. In other words, we will permit $T_p$ to denote the endomorphism of $L$ induced from the correspondence $(E, \alpha) \mapsto \sum (E/D, \alpha \bmod D)$ on $S$ by $\mathbf{Z}$-linearity.

**4 Extra automorphisms.** We will now determine those points $s \in S$ with $e(s) > 1$. A point $s$ in $S$ is defined by a supersingular elliptic curve $E$ over $\overline{\mathbf{F}}_q$, together with a point $P$ on $E$ of order $N$. Write $\alpha$ for the orbit of $P$ under $H$ and $C$ for the subgroup of $E$ generated by $P$. Let $\overline{s}$ be the point on $X_0(N)$ defined by $(E, C)$.

Let $R$ be the subring $\mathbf{Z}[\text{Aut } E]$ of $\text{End } E$. If $e(s)$ is different from 1, then $\text{Aut } E$ is different from $\{\pm 1\}$. Since $\text{End}(E)$ is a definite quaternion algebra over $\mathbf{Q}$, $\text{Aut } E$ is different from $\{\pm 1\}$ if and only if $E$ has an automorphism of order 4 or 6. In fact, the condition $q > 3$ implies that the automorphism group of $E$ can only be cyclic of order 2, order 4, or order 6. Hence $R$ is either $\mathbf{Z}$, or else the integer ring in an imaginary quadratic field of discriminant $-3$ or $-4$. As is well known, $E$ has an automorphism of order 4 if and only if its $j$-invariant is 1728. Since $E$ is supersingular, the prime $q$ must be congruent to 3 mod 4. Similarly, $E$ has an automorphism of order 3 if and only if its $j$-invariant is zero; if the curve with this $j$-invariant is supersingular, we have $q \equiv 2 \bmod 3$. (Compare, for example, [**18**], page 103, pp. 143–144.)

Suppose that $\text{Aut } E \neq \{\pm 1\}$. Then $(E, C)$ has a non-trivial automorphism if and only if $\text{Aut } E$ coincides with its subgroup $\text{Aut}(E, C)$, i.e., if and only if $C$ is stable under $R$. If this condition is satisfied, then the group $C$ is free of rank 1 over $R/I$, where $I$ is the annihilator of $C$ in $R$. Since $C$ is cyclic, this puts a numerical constraint on $N$: No prime factor of $N$ can remain inert in the ring $R$, and the prime which ramifies in $R$ can occur in $N$ only to the first power if it occurs at all. Further, the inclusion of $\mathbf{Z}$ in $R$ induces an identification $R/I \overset{\sim}{\to} \mathbf{Z}/N\mathbf{Z}$. Thus the pair $(E, C)$ has a non-trivial automorphism if and only if $C$ is the kernel on $E$ of an ideal $I$ of $R$ such that $R/I$ is isomorphic to $\mathbf{Z}/N\mathbf{Z}$. Thus there is a 1-1 correspondence between cyclic subgroups $C$ of order $N$ which are stable by $R$ and ideals $I$ of $R$ such that the additive group of $R/I$ is cyclic of order $N$. One sees that distinct subgroups $C$ lead to distinct point of $X_0(N)$. Indeed, if $(E, C)$ and $(E, C')$ are isomorphic, then the isomorphism between them is induced by an automorphism of $E$, which preserves $C$ (and $C'$) by the definition of $R$.

The conjugation map $r \mapsto \overline{r}$ of $R$ induces an an involution on the set of $I$. This Atkin-Lehner style involution has no fixed points if $N > 3$. Indeed, suppose that $I = \overline{I}$. For each $r \in R$, there is an integer $n$ such that $r - n$ lies in $I$. Since $\overline{n} = n$, $r - \overline{r}$ belongs to $I + \overline{I} = I$. Hence $I$ contains $\sqrt{-3}$ or 2, according as the discriminant of $R$ is $-3$ or $-4$. Hence $N$ divides 3 in the former case and $N$ divides 2 in the latter.

Suppose now that $(E, C)$ has a non-trivial automorphism. Then $e(s) > 1$ if and only if $\text{Aut}(E, \alpha) = \text{Aut}(E, C)$. This equality translates into the statement that

(\*)  $$\mu \overset{?}{\subseteq} H \subseteq (\mathbf{Z}/N\mathbf{Z})^*,$$

where $\mu$ is the image in $(R/I)^*$ of the unit group of $R$. This condition involves only $(E, C)$ and $H$; hence it applies simultaneously to all points on $X_H(N)$ which lie above the point $\bar{s}$ on $X_0(N)$. One checks that when $(*)$ is satisfied, the number of points on $X_H(N)$ which lie above $\bar{s}$ is precisely $((\mathbf{Z}/N\mathbf{Z})^* : H)$.

In summary, to find those points $s \in S$ with $e(s) > 1$, we first look for the supersingular points $\bar{s}$ in $X_0(N)(\overline{\mathbf{F}}_q)$ which satisfy the analogous condition $e(\bar{s}) > 1$. Mark off those points (if any) which satisfy the supplementary condition $(*)$. Above each marked point, we find $((\mathbf{Z}/N\mathbf{Z})^* : H)$ points in $S$ with "extra automorphisms."

The condition $(*)$ may be true for some points $\bar{s}$ with $e(\bar{s}) > 1$ that arise from a given supersingular elliptic curve and false for others. For example, suppose that $E$ has six automorphisms and that $N = 7 \cdot 13 = 91$. There are four ideals $I$ of $R$ with $R/I \approx \mathbf{Z}/91\mathbf{Z}$. For two of these ideals, the image of $R^*$ in $(\mathbf{Z}/91\mathbf{Z})^*$ is the cyclic group generated by 10; for the other two, the image is generated by 17. If $H$ is one of these two groups, then $(*)$ will be satisfied for two of the $I$, but not for all. Notice, however, that the set of $I$ for which $(*)$ is satisfied is stable under the natural conjugation map on $R$. Indeed, $R^*$ is stable under this conjugation. Furthermore, we have observed that the inequality $N > 3$ implies that an ideal $I$ with $R/I \approx \mathbf{Z}/N\mathbf{Z}$ can never be its own conjugate. This gives

**Lemma 1.** *Suppose that $N \geq 4$. Then the set of points $\bar{s}$ for which $e(\bar{s}) > 1$ and for which $(*)$ is satisfied has an even number of elements. In particular, the number of such points is different from 1.*

The case $N \leq 3$ does not play an important role in the analysis which follows. In fact, one has

**Lemma 2.** *If $N \leq 12$ and there is at least one point $s$ with $e(s) > 1$, then $H = (\mathbf{Z}/N\mathbf{Z})^*$.*

*Proof.* Assume that $N < 13$. If $e(s) = 3$, then there is an ideal $I$ of $R = \mathbf{Z}[\frac{1+\sqrt{-3}}{2}]$ with $R/I \approx \mathbf{Z}/N\mathbf{Z}$. This implies that $N$ is 3 or 7. In both cases, $R^*$ maps onto $(\mathbf{Z}/N\mathbf{Z})^*$. If $e(s) = 2$, there is an ideal $I$ of $R = \mathbf{Z}[\sqrt{-1}]$ with $R/I \approx \mathbf{Z}/N\mathbf{Z}$. The possibilities for $N$ are 2, 5, and 10. Again, $R^*$ maps onto $(\mathbf{Z}/N\mathbf{Z})^*$ in all cases. □

The prime 13 splits both in $\mathbf{Z}[\frac{1+\sqrt{-3}}{2}]$ and in $\mathbf{Z}[\sqrt{-1}]$. If $R$ is one of these two rings, and if $I$ is a prime lying over 13 in $R$, then the image of $R^*$ in $(\mathbf{Z}/13\mathbf{Z})^*$ is strictly smaller than $(\mathbf{Z}/13\mathbf{Z})^*$. This circumstance (if not the discussion in Section 2) shows that 12 cannot be replaced by a larger integer in the statement of the Lemma.

## 5 A canonical subgroup of $\Phi_\Gamma$.

Let $\Phi = \Phi_\Gamma$. We shall discuss a cyclic subgroup $\Phi_0$ of $\Phi$ which has already been studied in case $\Gamma = \Gamma_0(N)$ (see [13] and [12]).

For each element $s$ of $S$, let $\varphi_s : L \to \mathbf{Z}$ be the linear form $\ell \mapsto \langle \ell, s \rangle$ and let $\omega_s$ be the linear form on $L$

$$\sum_t n_t t \mapsto n_s.$$

Clearly, $\mathrm{Hom}(L, \mathbf{Z})$ is generated by the $\omega_s$, so that $\Phi$ is generated by their images $\overline{\omega}_s$ in $\Phi = \mathrm{Hom}(L, \mathbf{Z})/L$. One has $e(s)\omega(s) = \varphi_s$.

The class $\overline{\varphi}$ of $\varphi_s$ in $\Phi = \mathrm{Hom}(L, \mathbf{Z})/L$ is independent of $s$, since $\varphi_s - \varphi_t$ is the image in $\mathrm{Hom}(L, \mathbf{Z})$ of $s - t \in L$, if $s, t \in S$. Let $\Phi_0$ be the subgroup of $\Phi$ generated by the canonically given element $\overline{\varphi}$. We set

$$Q = \Phi/\Phi_0.$$

In view of the formula $e(s)\omega(s) = \varphi_s$, we have $e(s)\overline{\omega}_s = \overline{\varphi}$ for each $s$. Hence, as noted in [13], Section 2, we have a surjection

$$\bigoplus_{s \in S} (\mathbf{Z}/e(s)\mathbf{Z}) \longrightarrow Q, \qquad (a_s) \mapsto \sum_s a_s \overline{\omega}_s.$$

One may view $\mathbf{Z}^S$ as embedded in $\mathrm{Hom}(\mathbf{Z}^S, \mathbf{Z})$ via the symmetric bilinear pairing $\langle\,,\,\rangle$ on $\mathbf{Z}^S$. Let $L^\perp \subset \mathrm{Hom}(\mathbf{Z}^S, \mathbf{Z})$ be the group of linear forms which vanish on $L$, and let $U = L^\perp \cap \mathbf{Z}^S$ be group of vectors in $\mathbf{Z}^S$ which are orthogonal to $L$ under the pairing. Since $\mathrm{Hom}(L, \mathbf{Z}) = \mathrm{Hom}(\mathbf{Z}^S, \mathbf{Z})/L^\perp$, we have $\Phi = \mathrm{Hom}(\mathbf{Z}^S, \mathbf{Z})/(L \oplus L^\perp)$. The image of $\mathbf{Z}^S/U$ in $\Phi$, i.e., the group $\mathbf{Z}^S/(L \oplus U)$, is clearly $\Phi_0$. (Indeed, the image of $s \in \mathbf{Z}^S$ in $\mathrm{Hom}(\mathbf{Z}^S, \mathbf{Z})$ is $\varphi_s$.)

Now $U$ is free of rank one over $\mathbf{Z}$; it consists of those multiples of $\sum \frac{1}{e(s)} s \in \mathbf{Q}^S$ which lie in $\mathbf{Z}^S$. Hence $U$ is generated by $\sum \frac{m}{e(s)} s$, where $m$ is the least common multiple of the $e(s)$. (Thus, $m$ divides 6.) It follows that the order of $\Phi_0 = \mathbf{Z}^S/(L \oplus U)$ is $m \cdot \sum e(s)^{-1}$, cf. [12], p. 16. Also, $L^\perp/U$ has order $m$, since $L^\perp$ is generated by the element $\sum \omega_s$ and since $m \sum \omega_s = \sum \frac{m}{e(s)} s$ in $\mathrm{Hom}(\mathbf{Z}^S, \mathbf{Z})$.

Consider the commutative diagram with exact rows

$$\begin{array}{ccccccccc} 0 & \to & L^\perp & \to & \mathrm{Hom}(\mathbf{Z}^S, \mathbf{Z}) & \to & \mathrm{Hom}(\mathbf{Z}^S, \mathbf{Z})/L^\perp & \to & 0 \\ & & \uparrow & & \uparrow & & \uparrow & & \\ 0 & \to & U & \to & \mathbf{Z}^S & \to & \mathbf{Z}^S/U & \to & 0 \end{array}$$

in which the central vertical map is the inclusion given by $\langle\,,\,\rangle$; the two other vertical maps are the evident, related inclusions. The cokernels of the vertical maps form the exact sequence $0 \to \mathbf{Z}/m\mathbf{Z} \to \bigoplus_{s \in S}(\mathbf{Z}/e(s)\mathbf{Z}) \to Q \to 0$, where the map from $m\mathbf{Z}$ to the direct sum takes $1 \in \mathbf{Z}/m\mathbf{Z}$ to the vector $(1, \ldots, 1)$. Note that $Q$ is now seen to have order $m^{-1} \prod e(s)$. Therefore, we recover the formula $\left(\prod_s e(s)\right) \cdot \left(\sum_s e(s)^{-1}\right)$ for the order of $\Phi$, cf. [1], Ch. 9, Prop. 10.

**Lemma 3.** *The group $\Phi_0$ is strongly Eisenstein. The identities $\langle d \rangle = 1$ and $T_p = 1 + p$ hold on $\Phi_0$ for $d \in (\mathbf{Z}/N\mathbf{Z})^*$ and $p$ prime to $qN$.*

*Proof.* We first consider the action of $T_p$ on $\overline{\varphi}$. Choose $s \in S$. From the perspective introduced at the end of Section 3, we see that the action of $T_p$ on this element of $\Phi$ is derived from the action of $\mathrm{Hom}(T_L, \mathbf{Z})$ on $\varphi_s \in \mathrm{Hom}(L, \mathbf{Z})$. The map $\varphi_s \circ T_L$ is the linear form mapping $\ell \in L$ to $\langle T_L \ell, s \rangle = \langle \ell, T'_L s \rangle = \langle \ell, \langle p \rangle^{-1} T_L s \rangle$. Since $\langle p \rangle^{-1} T_L s$ is an element of $\mathbf{Z}^S$ of degree $p+1$, the image in $\Phi$ of $\ell \mapsto \langle \ell, \langle p \rangle^{-1} T_L s \rangle$ is $(p+1)\overline{\varphi}$. This establishes the first identity. The second is proved by an analogous computation. $\square$

**Corollary.** *The group $\Phi$ is Eisenstein in each of the following situations:*
*(i)* $e(s) = 1$ for all $s \in S$;
*(ii)* $H = (\mathbf{Z}/N\mathbf{Z})^*$;
*(iii)* $N < 13$;
*(iv)* $H = \{\pm 1\}$.

*Proof.* In the first case, we have $\Phi_0 = \Phi$, so that the assertion to be proved is that given by the lemma. The assertion in the second case is given as Theorem 3.12 in [13]. In the third case, we are either in case (i) or case (ii) by Lemma 2. Suppose now that we are in the fourth case, which corresponds to the equality $X_H(N) = X_1(N)$. If $N$ is at least 4, we are in case (i) by [10], Corollary 2.7.4. If $N \leq 4$, then we are of course in case (iii). □

## 6 The structure of $Q$.

In view of the Corollary above, we will now impose the condition $N \geq 13$.

**Lemma 4.** *Suppose that $(E, C)$ represents a supersingular point of $X_0(N)_{/\overline{\mathbf{F}}_q}$. Let $D$ be a cyclic subgroup of $E(\overline{\mathbf{F}}_q)$ of order prime to $qN$. Assume that the automorphism group of the triple $(E, C, D)$ is larger than $\{\pm 1\}$. Then the pair $(E/D, C \bmod D)$ is isomorphic to $(E, C)$.*

*Proof.* This statement is proved as Proposition 2 in [14]; we recall the proof for the convenience of the reader. Let $\epsilon$ be an automorphism of $(E, C, D)$ which is different from $\pm 1$, and let $R$ be the subring of $\mathrm{End}(E, C, D)$ which is generated by $\epsilon$. Thus $R$ is isomorphic either to the ring of Gaussian integers or to the ring of integers of $\mathbf{Q}(\sqrt{-3})$. If $J = \mathrm{Ann}_R(D)$, then $J$ is a principal ideal $(r)$ of $R$, and $D$ is the kernel of $J$ on $E$. Thus the map "multiplication by $r$" on $E$ induces an isomorphism $(E/D, (C \oplus D)/D) \xrightarrow{\sim} (E, C)$ as required. □

Recall now that $Q$ is generated as an abelian group by the elements $\overline{\omega}_s$ with $s \in S$. Since $e(s)\overline{\omega}_s = 0$, it suffices to consider only those $\overline{\omega}_s$ with $e(s) > 1$. As we have seen, the inequality $e(s) > 1$ means that $s$ lies over a supersingular point $\overline{s}$ of $X_0(N)(\overline{\mathbf{F}}_q)$ which satisfies the numerical condition $e(\overline{s}) > 1$ and the supplementary condition $(*)$.

Let $\overline{s}$ be such a point. Define $Q_{\overline{s}}$ to be the subgroup of $Q$ generated by the $\overline{\omega}_s$ with $s$ lying over $\overline{s}$. Thus $Q_{\overline{s}}$ is a quotient of the elementary abelian group $\bigoplus (\mathbf{Z}/e(\overline{s})\mathbf{Z})$, where the sum is extended over the set of points $s$ lying over $\overline{s}$. As was mentioned above, the number of such points is the index $((\mathbf{Z}/N\mathbf{Z})^* : H)$. More precisely, the set of points $s$ is a principal homogeneous space over $(\mathbf{Z}/N\mathbf{Z})^*/H$, the action of this group being given by the "diamond bracket" operators.

**Proposition 1.** *The group $Q_{\bar{s}}$ is free of rank $((\mathbf{Z}/N\mathbf{Z})^* : H)$ over $\mathbf{Z}/e(\bar{s})\mathbf{Z}$.*

*Proof.* Let $S_0$ be the inverse image of $\bar{s}$ in $S$. We must show that the direct sum $\bigoplus_{s \in S_0}(\mathbf{Z}/e(s)\mathbf{Z})$ has trivial intersection with the kernel of the map

$$\bigoplus_{s \in S}(\mathbf{Z}/e(s)\mathbf{Z}) \to Q.$$

As we have seen, this kernel is the cyclic group generated by the vector $(1, \ldots, 1)$ in the full direct sum. Hence it suffices to check that there is a point $s \in S \setminus S_0$ with $e(s) > 1$. This follows from Lemma 1, since $N \geq 13$. $\square$

**Proposition 2.** *The subgroup $Q_{\bar{s}}$ of $Q$ is stable under the Hecke operators $T_p$ and $\langle d \rangle$.*

*Proof.* Let $p$ be a prime number not dividing $qN$. As at the end of Section 3, we write $T$ for the correspondence on $S$ defined by the formula $(E, \alpha) \mapsto \sum(E/D, \alpha \bmod D)$. This correspondence is summarized by the matrix of natural numbers $(a_{tu})$ which one constructs by writing $Tu = \sum_{t \in S} a_{tu} \cdot t$ for $u \in S$. Recall that the action of $T_p$ on $\Phi$ arises from the action of a map labeled $\mathrm{Hom}(T_L, \mathbf{Z})$ on $\mathrm{Hom}(L, \mathbf{Z})$. This group is generated by the elements $\omega_s$ for $s \in S$, and one finds that $\mathrm{Hom}(T_L, \mathbf{Z})$ maps $\omega_s$ to the sum $\sum_{t \in S} a_{st}\omega_t$.

To prove that $Q_{\bar{s}}$ is stable under $T_p$, it suffices to show that $a_{st}$ is divisible by $e(t)$ whenever $s$ maps to $\bar{s}$ and $t$ does *not*. In other words, we wish to show that $a_{tu}$ is divisible by $e(u)$ whenever $t$ and $u$ have distinct images on $X_0(N)$.

This divisibility can be established by the method used to prove [13], Th. 3.12. Indeed, suppose that $u$ is represented by $(E, \alpha)$. Let $C$ be the cyclic subgroup of order $N$ on $E$ which is associated with $\alpha$. The group $\mathrm{Aut}(E, \alpha)/\{\pm 1\}$ operates on the set of cyclic subgroups of $E$ of order $p$. Clearly, if $D$ and $D'$ are such subgroups which are equivalent under this action, then $(E/D, \alpha \bmod D)$ and $(E/D', \alpha \bmod D')$ are isomorphic. Moreover, suppose that the cyclic subgroup $D$ has a non-trivial stabilizer under this action. Then Lemma 4 shows that $(E, C)$ and $(E/D, (C \oplus D)/D)$ are isomorphic, i.e., that $(E, \alpha)$ and $(E/D, \alpha \bmod D)$ map down to the same point on $X_0(N)$. Now let $t$ be a point of $S$ whose image on $X_0(N)$ is distinct from that of $s$. Then the set of $D$ for which $(E/D, \alpha \bmod D)$ represents $t$ is a union of copies of the group $\mathrm{Aut}(E, \alpha)/\{\pm 1\}$, whose cardinality is $e(u)$. Hence $a_{tu}$ is divisible by $e(u)$, as was claimed.

Now let $d$ be an integer prime to $N$. The action of $\langle d \rangle$ on $\Phi$ is deduced from the automorphism $\mathrm{Hom}(\langle d \rangle_L, \mathbf{Z})$ introduced at the end of Section 3. This automorphism maps a given linear form $\omega_s$ to the linear form $\omega_{\langle d \rangle^{-1}s}$. Here, the automorphism labeled $\langle d \rangle^{-1}$ sends the class of $(E, P)$ to the class of $(E, d^{-1}P)$; the quantity $d^{-1}$ is computed mod $N$. Since $P$ and $d^{-1}P$ generated the same subgroup of $E$, it is clear that $\langle d \rangle$ permutes the $\bar{\omega}_s$ with $s$ having a given image on $X_0(N)$. Therefore, $\langle d \rangle$ preserves $Q_{\bar{s}}$. $\square$

Let $S_0$ again be the inverse image of $\bar{s}$ in $S$. We observed above that the diamond bracket operation of $(\mathbf{Z}/N\mathbf{Z})^*$ on $S_0$ makes $S_0$ a principal homogeneous space over the group $\Delta := (\mathbf{Z}/N\mathbf{Z})^*/H$. Let us identify $Q_{\bar{s}}$ with the direct sum $\bigoplus_{s \in S_0} (\mathbf{Z}/e(\bar{s})\mathbf{Z})$, i.e., with the space of functions from $S_0$ to $\mathbf{Z}/e(\bar{s})\mathbf{Z}$. Then the diamond bracket operation discussed in Proposition 2, which involves an inverse, may be viewed as the natural action coming from the diamond bracket action of $\Delta$ on $S_0$.

Let $\ell = e(\bar{s})$; thus $\ell = 2$ or $\ell = 3$. Let $\mathbb{T} = \mathbf{F}_\ell[\Delta]$ be the group algebra consisting of sums $\sum_{\delta \in \Delta} n_\delta[\delta]$ where the $n_\delta$ are integers mod $\ell$. Since $Q_{\bar{s}}$ is an $\mathbf{F}_\ell$-vector space with an action of $\Delta$, $Q_{\bar{s}}$ may be viewed as a $\mathbb{T}$-module in a natural way.

**Lemma 5.** *The $\mathbb{T}$-module $Q_{\bar{s}}$ is free of rank one.*

*Proof.* Proposition 1 shows that $Q_{\bar{s}}$ is an $\mathbf{F}_\ell$ vector space of dimension $\#(\Delta) = \dim_{\mathbf{F}_\ell} \mathbb{T}$. Further, it is clear that $Q_{\bar{s}}$ is a cyclic $\mathbb{T}$-module, since $\Delta$ permutes the generators $\omega_s$ of $Q_{\bar{s}}$. □

Now choose a supersingular elliptic curve $E$ over $\overline{\mathbf{F}}_q$ and a cyclic subgroup $C$ on $E$ so that the pair $(E, C)$ defines the point $\bar{s}$ of $X_0(N)$. As usual, let $R$ be the ring $\mathbf{Z}[\operatorname{Aut} E]$, and let $I$ be the annihilator of $C$ in $R$. The isomorphism $R/I \approx (\mathbf{Z}/N\mathbf{Z})^*$ gives a meaning to $[r]$ whenever $r \in R$ is prime to $I$. The element $[r]$ depends only on the ideal generated by $r$ in $R$, since $H$ contains the image of $R^*$ in $(\mathbf{Z}/N\mathbf{Z})^*$.

In fact, this might be a good time to relate $\Delta$ to the Galois group $\operatorname{Gal}(\overline{\mathbf{Q}}/\mathbf{Q})$. Let $K$ be the imaginary quadratic field $\mathbf{Q}[\operatorname{Aut} E] = R \otimes \mathbf{Q}$. Since all fractional ideals of $K$ are principal, the quotient of $(R/I)^*$ by $R^*$ is the ray class group of $K$ with conductor $I$. Via class field theory, this class group corresponds with an abelian extension $L$ of $K$. It will be useful to fix an embedding $K \hookrightarrow \overline{\mathbf{Q}}$ and to view $L$ as a subfield of $\overline{\mathbf{Q}}$. Since $\Delta$ is then a quotient of $\operatorname{Gal}(L/K)$, $\Delta$ becomes the Galois group of an abelian extension of $K$ in $\overline{\mathbf{Q}}$. The conductor of this extension divides $I$.

For each prime $p \nmid qN$, we define an element $\tau_p$ of $\mathbb{T}$ by the formula:

$$\tau_p = \begin{cases} 0 & \text{if } p \text{ is inert in } R, \\ [\pi] + [\pi'] & \text{if } p = (\pi)(\pi') \text{ is split in } R, \\ [\pi] & \text{if } p = (\pi)^2 \text{ is ramified in } R. \end{cases}$$

The third case occurs only when $p = \ell$.

**Theorem 3.** *For each $p \nmid qN$, the Hecke operator $T_p$ acts on $Q_{\bar{s}}$ by multiplication by $\tau_p$.*

*Proof.* Let $(a_{st})$ be the Brandt matrix which was introduced in the course of the proof of Proposition 2. Recall that if $t$ is the isomorphism class of the pair $(E, P)$, where $P$ is a point of order $N$ lying in $C$, then $a_{st}$ is the number of subgroups $D$ of order $p$ in $E$ such that $(E/D, P \bmod D)$ defines $s$. We have seen that $a_{st} \equiv 0 \bmod \ell$ if $s$ and $t$ do not have the same image on $X_0(N)$. We must now calculate $a_{st}$ mod $\ell$ in the case where $s$ and $t$ have the same image, namely $\bar s$.

Suppose first that $p$ is inert in $R$. Then there is no subgroup of order $p$ in $E$ which is stable under $\operatorname{Aut} E$. Accordingly, the proof of Proposition 2 shows that $a_{st}$ is divisible by $\ell$, since the group $(\operatorname{Aut} E)/\{\pm 1\}$ acts freely on the set of $D$ for which $(E/D, P \bmod D)$ defines $s$.

Next, let $p = (\pi)^2$ be ramified in $R$. Then $E[\pi]$ is the unique subgroup of order $p$ on $E$ which is stable under $R$. The pair $(E/D, P \bmod D)$ is isomorphic to $(E, \pi \cdot P)$ via $\pi$. With the obvious meaning of $\langle \pi \rangle$, we may write $a_{st} \equiv 0 \bmod \ell$ for $s \neq \langle \pi \rangle t$, while $a_{st} \equiv 1 \bmod \ell$ for $s = \langle \pi \rangle t$. Tracing through the definitions, one emerges with the desired assertion that $T_p$ operates on $Q_{\bar s}$ as $[\pi]$.

In the final case where $p = (\pi)(\pi')$, there are two distinct subgroups $D$ which are stable under $\operatorname{Aut} E$. The corresponding quotients of $(E, \alpha)$ are represented by $(E, \pi \cdot P)$ and $(E, \pi' \cdot P)$. Let $s_1 = \langle \pi \rangle t$ and $s_2 = \langle \pi' \rangle t$. If the two $s_i$ are distinct, the numbers $a_{st}$ mod $\ell$ are 1 for $s = s_1, s_2$ and zero otherwise. If $s_1 = s_2$, then $a_{st} = 2$ when $s = s_1 = s_2$ and $a_{st} = 0$ otherwise. This leads to the required formula $T_p = [\pi] + [\pi']$. □

Let $\widetilde{\mathbb{T}}$ be the subring of $\operatorname{End} J(H, q)$ generated by the endomorphisms $T_p$ and $\langle d \rangle$ of $J(H, q)$. (There is a functorial, faithful action of $\widetilde{\mathbb{T}}$ on the space of weight-two cusp forms for $\Gamma \cap \Gamma_0(q)$; thus $\widetilde{\mathbb{T}}$ may be defined alternatively as in the Introduction.) Theorem 3 states that the action of $\widetilde{\mathbb{T}}$ on $Q_{\bar s}$ factors through the action of $\mathbb{T}$ on $Q_{\bar s}$, in such a way that $\langle d \rangle \in \widetilde{\mathbb{T}}$ acts as $[d] \in \mathbb{T}$ for all $d \in (\mathbf{Z}/N\mathbf{Z})^*$ that $T_p \in \widetilde{\mathbb{T}}$ acts as $\tau_p$. Equivalently, there is a commutative triangle

in which the left-hand diagonal arrow describes the action of Hecke operators on $Q_{\bar s}$ and the right-hand diagonal arrow is the structural map making $Q_{\bar s}$ into a $\mathbb{T}$-module. The horizontal arrow is visibly surjective, since $\mathbb{T} = \mathbf{F}_\ell[\Delta]$ is generated by the various $[d]$. By Lemma 5, $Q_{\bar s}$ is a free rank-one $\mathbb{T}$-module. Therefore, the map $\mathbb{T} \to \operatorname{End} Q_{\bar s}$ is injective. Accordingly, the image of $\widetilde{\mathbb{T}}$ in $\operatorname{End} Q_{\bar s}$ may be identified with $\mathbb{T}$. Via this identification, the set of maximal ideals of $\widetilde{\mathbb{T}}$ in the support of $Q_{\bar s}$ becomes the set of maximal ideals of $\mathbb{T} = \mathbf{F}_\ell[\Delta]$.

This latter set may be viewed as a set of conjugacy classes of characters. Indeed, if $\mathfrak{m}$ is a maximal ideal of $\mathbb{T}$, then $\mathbb{T}/\mathfrak{m}$ is a finite field of characteristic $\ell$. Hence $\mathfrak{m}$ may be obtained as the kernel of some ring homomorphism $\mathbb{T} \to \overline{\mathbf{F}}_\ell$. On the other hand, restriction to $\Delta$ yields a 1-1 correspondence between ring homomorphisms $\mathbb{T} \to \overline{\mathbf{F}}_\ell$ and group homomorphisms $\theta : \Delta \to \overline{\mathbf{F}}_\ell^*$. By Galois theory, two ring homomorphisms $\mathbb{T} \rightrightarrows \overline{\mathbf{F}}_\ell$ have the same kernel if and only if they are conjugate under $\operatorname{Gal}(\overline{\mathbf{F}}_\ell/\mathbf{F}_\ell)$. It follows that the set of maximal ideals of $\mathbb{T}$ is in correspondence with the set of $\operatorname{Gal}(\overline{\mathbf{F}}_\ell/\mathbf{F}_\ell)$-conjugacy classes of characters $\theta$.

As was noted above, $\Delta$ may be viewed as the Galois group of an abelian extension of $K$ in $\overline{\mathbf{Q}}$. Accordingly, the characters $\theta$ become homomorphisms $\mathrm{Gal}(\overline{\mathbf{Q}}/K) \to \overline{\mathbf{F}}_\ell^*$ which factor through the Galois group we called $\mathrm{Gal}(L/K)$. For each such homomorphism, we let $\theta'$ be the map $\mathrm{Gal}(\overline{\mathbf{Q}}/K) \to \overline{\mathbf{F}}_\ell^*$ defined by

$$\theta'(x) = \theta(gxg^{-1}),$$

where $g$ is an element of $\mathrm{Gal}(\overline{\mathbf{Q}}/\mathbf{Q})$ which does not belong to $\mathrm{Gal}(\overline{\mathbf{Q}}/K)$.

**Lemma 6.** *We have $\theta = \theta'$ if and only if $\theta$ is identically 1.*

*Proof.* If $\theta$ is identically 1, it certainly coincides with $\theta'$. Conversely, suppose that $\theta = \theta'$. Let $\mathfrak{f}$ be the conductor of $\theta$. As is well known, the conductor of $\theta'$ is the image of $\mathfrak{f}$ under the non-trivial involution $K \to K$. Because $\theta = \theta'$, $\mathfrak{f}$ is invariant under this involution. It is in any case a divisor of $I$. It follows that $\mathfrak{f}$ divides $(\sqrt{-3})$ in the case where $K = \mathbf{Q}(\sqrt{-3})$ and that $\mathfrak{f}$ divides $(1+\sqrt{-1})$ in case $K = \mathbf{Q}(\sqrt{-1})$. In the former case, $\theta$ is a map $\mathbf{F}_3^* \to \overline{\mathbf{F}}_3^*$ which is trivial on $\{\pm 1\}$. In the latter case, $\theta$ is a map $\mathbf{F}_2^* \to \overline{\mathbf{F}}_2^*$. Hence $\theta$ is the trivial map in either case. $\square$

Now let $\mathrm{Ind}\,\theta$ be the induced representation

$$\mathrm{Ind}_{\mathrm{Gal}(\overline{\mathbf{Q}}/K)}^{\mathrm{Gal}(\overline{\mathbf{Q}}/\mathbf{Q})}\theta \; : \; \mathrm{Gal}(\overline{\mathbf{Q}}/\mathbf{Q}) \to \mathrm{GL}(2,\overline{\mathbf{F}}_\ell).$$

We can make an explicit model for $\mathrm{Ind}\,\theta$ by choosing $g \notin \mathrm{Gal}(\overline{\mathbf{Q}}/K)$ as above and defining

$$\mathrm{Ind}\,\theta(x) = \begin{cases} \begin{pmatrix} \theta(x) & 0 \\ 0 & \theta'(x) \end{pmatrix} & \text{for } x \in \mathrm{Gal}(\overline{\mathbf{Q}}/K), \\ \begin{pmatrix} 0 & \theta(xg) \\ \theta(g^{-1}x) & 0 \end{pmatrix} & \text{for } x \notin \mathrm{Gal}(\overline{\mathbf{Q}}/K). \end{cases}$$

In particular, one reads from this model the well known formula

$$\mathrm{tr}(\mathrm{Ind}\,\theta(x)) = \begin{cases} \theta(x) + \theta'(x) & \text{if } x \in \mathrm{Gal}(\overline{\mathbf{Q}}/K), \\ 0 & \text{if } x \notin \mathrm{Gal}(\overline{\mathbf{Q}}/K) \end{cases}$$

for the trace of $\mathrm{Ind}\,\theta$. The determinant of $\mathrm{Ind}\,\theta$ is the product of two characters, the first of which is the homomorphism $\mathrm{Gal}(\overline{\mathbf{Q}}/\mathbf{Q}) \to \{\pm 1\}$ which corresponds to the quadratic subfield $K$ of $\overline{\mathbf{Q}}$. This homomorphism may be described alternatively as the mod $\ell$ cyclotomic character $\chi_\ell$ which gives the action of $\mathrm{Gal}(\overline{\mathbf{Q}}/\mathbf{Q})$ on the group of $\ell$th roots of 1 in $\overline{\mathbf{Q}}$. It has order two if $\ell = 3$, while it is trivial if $\ell = 2$. The second of the two characters is $\theta \circ \mathrm{Ver}$, where $\mathrm{Ver}$ is the Verlagerung map from $\mathrm{Gal}(\overline{\mathbf{Q}}/\mathbf{Q})$ to the abelianization of $\mathrm{Gal}(\overline{\mathbf{Q}}/K)$. It is obtained by composing: the mod $N$ cyclotomic character $\mathrm{Gal}(\overline{\mathbf{Q}}/\mathbf{Q}) \to (\mathbf{Z}/N\mathbf{Z})^*$, the identification $(\mathbf{Z}/N\mathbf{Z})^* \xrightarrow{\sim} (R/I)^*$, and the character $\theta : (R/I)^* \to \overline{\mathbf{F}}_\ell^*$.

**Lemma 7.** *If $\theta$ is non-trivial, the representation $\operatorname{Ind} \theta$ is irreducible. Suppose instead that $\theta$ is trivial. Then the image of $\operatorname{Ind} \theta$ has order two; the kernel of $\operatorname{Ind} \theta$ is $\operatorname{Gal}(\overline{\mathbf{Q}}/K)$. When $\ell = 3$, $\operatorname{Ind} \theta$ is the direct sum of the trivial representation and the one-dimensional representation with character $\chi_3$. When $\ell = 2$, $\operatorname{Ind} \theta$ is indecomposable and its semisimplification is the direct sum of two copies of the trivial representation.*

*Proof.* Suppose that $\theta$ is non-trivial. Then by Lemma 6, $\theta$ and $\theta'$ are distinct. The restriction of $\operatorname{Ind} \theta$ to $\operatorname{Gal}(\overline{\mathbf{Q}}/K)$ is thus the direct sum of two *distinct* characters. It follows that there are precisely two lines in the representation space of $\operatorname{Ind} \theta$ which are invariant under $\operatorname{Gal}(\overline{\mathbf{Q}}/K)$. These lines are permuted by the element $g$; in particular, no line is stable under the full Galois group $\operatorname{Gal}(\overline{\mathbf{Q}}/\mathbf{Q})$. If $\theta$ is trivial, then the required assertions are clear from the model of $\operatorname{Ind} \theta$ presented above. $\square$

Note that, in the case where $\ell = 2$ and $\theta = 1$, the representation $\operatorname{Ind} \theta$ is that given by the unique non-trivial extension of $\mathbf{Z}/2\mathbf{Z}$ by $\mu_2$ in the category of commutative group schemes of type $(2, \ldots, 2)$ over $\operatorname{Spec} \mathbf{Z}$. (See [11], Ch. II, Section 12.)

Let $\mathfrak{m}$ be the maximal ideal of $\widetilde{\mathbb{T}}$ arising from $\theta$. The quotient $\widetilde{\mathbb{T}}/\mathfrak{m}$ becomes a subfield of $\overline{\mathbf{F}}_\ell$ via the embedding $\widetilde{\mathbb{T}}/\mathfrak{m} \hookrightarrow \overline{\mathbf{F}}_\ell$ induced by $\theta$. To $\theta$ we associate the representation

$$\rho_\theta = \rho_\mathfrak{m} \otimes_{\widetilde{\mathbb{T}}/\mathfrak{m}} \overline{\mathbf{F}}_\ell : \operatorname{Gal}(\overline{\mathbf{Q}}/\mathbf{Q}) \to \mathbf{GL}(2, \overline{\mathbf{F}}_\ell)$$

obtained by composing $\rho_\mathfrak{m}$ with the inclusion $\mathbf{GL}(2, \widetilde{\mathbb{T}}/\mathfrak{m}) \hookrightarrow \mathbf{GL}(2, \overline{\mathbf{F}}_\ell)$. By varying $\theta$ over the set of characters $\Delta \to \overline{\mathbf{F}}_\ell^*$, we obtain from $Q_{\overline{s}}$ a family of representations $\operatorname{Gal}(\overline{\mathbf{Q}}/\mathbf{Q}) \to \overline{\mathbf{F}}_\ell^*$. We seek to describe this family.

**Proposition 3.** *Let $p$ be a prime number which does not divide $qN\ell$. Then the trace of $\rho_\theta(\operatorname{Frob}_p)$ is*

$$\begin{cases} 0 & \text{if } p \text{ is inert in } K, \\ \theta(\pi) + \theta(\pi') & \text{if } p = (\pi)(\pi') \text{ is split in } K, \end{cases}$$

*while the determinant of $\rho_\theta(\operatorname{Frob}_p)$ is $p\theta(p) = \pm\theta(p)$.*

*Proof.* The matrix $\rho_\mathfrak{m}(\operatorname{Frob}_p)$ has trace $T_p$ mod $\mathfrak{m}$ and determinant $p\langle p \rangle$. Viewing $\widetilde{\mathbb{T}}/\mathfrak{m}$ as a quotient of $\mathbb{T}$, we can replace $\langle p \rangle$ by $[p]$ and $T_p$ by $\tau_p$. The map $\theta$ sends each $[\delta] \in \mathbb{T}$ to $\theta(\delta)$. The desired formulas now follow. $\square$

**Theorem 4.** *For each $\theta$, $\rho_\theta$ is the semisimplification of the induced representation $\operatorname{Ind} \theta$.*

*Proof.* This follows easily from Proposition 3, the Cebotarev Density Theorem, the Brauer Nesbitt Theorem and the formulas given above for the trace and determinant of $\operatorname{Ind} \theta$. $\square$

It is time now to recapitulate. The group $Q = \Phi/\Phi_0$ is the sum of subgroups $Q_{\bar{s}}$ belonging to the supersingular points $\bar{s}$ of $X_0(N)(\overline{\mathbf{F}}_q)$ which have extra automorphisms and which satisfy the condition $(*)$ of Section 4. To each such point is associated one of the quadratic fields $\mathbf{Q}(\sqrt{-3})$, $\mathbf{Q}(\sqrt{-1})$; call this field $K$. The point $\bar{s}$ and the group $H$ determine an abelian extension $L/K$. We induce those characters $\theta: \operatorname{Gal}(\overline{\mathbf{Q}}/K) \to \overline{\mathbf{F}}_\ell^*$ which factor through $\operatorname{Gal}(L/K)$ to obtain two-dimensional $\rho$ representations of $\operatorname{Gal}(\overline{\mathbf{Q}}/\mathbf{Q})$. According to Lemma 7, the representations $\rho$ are irreducible when $\theta$ is non-trivial. The semisimplifications of the induced representations $\rho$ are the $\overline{\mathbf{F}}_\ell$-linear two-dimensional representations of $\operatorname{Gal}(\overline{\mathbf{Q}}/\mathbf{Q})$ which arise from $Q_{\bar{s}}$.

From another point of view, let $K$ be one of the two fields $\mathbf{Q}(\sqrt{-3})$, $\mathbf{Q}(\sqrt{-1})$. To fix ideas, take the former field and embed it in $\overline{\mathbf{Q}}$. Let $R$ be the ring of integers of $K$, and let $\mathfrak{f}$ be a non-zero integral ideal of $R$ for which $R/\mathfrak{f}$ is cyclic as an abelian group. Consider the group $\mathcal{C}_{\mathfrak{f}}$ of ideal classes of $K$ modulo $\mathfrak{f}$, and let $\theta$ be a non-trivial homomorphism $\mathcal{C}_{\mathfrak{f}} \to \overline{\mathbf{F}}_3^*$. Using class field theory, regard $\theta$ as being defined on $\operatorname{Gal}(\overline{\mathbf{Q}}/K)$ and induce $\theta$ to obtain a two-dimensional representation $\rho$ of $\operatorname{Gal}(\overline{\mathbf{Q}}/\mathbf{Q})$. This representation is irreducible. Let $N$ be the norm of $\mathfrak{f}$, and let $H$ be the image of $R^*$ in $(R/\mathfrak{f})^* = (\mathbf{Z}/N\mathbf{Z})^*$.

**Theorem 5.** *The representation $\rho$ arises from the component group in the mod $q$ reduction of $J(H,q)$ whenever $q$ is a prime number which does not divide $2N$ and which is congruent to 2 modulo 3.*

*Proof.* Since $q$ is 2 modulo 3, we may choose a supersingular elliptic curve $E$ over $\overline{\mathbf{F}}_q$ whose automorphism group is cyclic of order six. Identify $\mathbf{Z}[\operatorname{Aut} E]$ with $R$ and let $C$ be the kernel of $\mathfrak{f}$ on $E$. The pair $(E,C)$ then represents a point $\bar{s}$ on $X_0(N)$ which has extra automorphisms and satisfies condition $(*)$. We have seen that $\rho$ arises from the subquotient $Q_{\bar{s}}$ of the component group associated with $J(H,q)$. □

As an application of the Theorem, we obtain the following statement, which could presumably be proved much more directly.

**Corollary.** *The representation $\rho$ arises from the space of weight-two cusp forms on $\Gamma_1(N)$.*

*Proof.* Let $q$ be a prime number as in the statement of Theorem 5. The conclusion of the Theorem implies that $\rho$ arises from the space of weight-two cusp forms on $\Gamma_H(N) \cap \Gamma_0(q)$, and hence from the space of cusp forms on $\Gamma_1(N) \cap \Gamma_0(q)$. Since $\rho$ is unramified at $q$, the desired conclusion now follows from "Mazur's Principle" [15], Theorem 8.1. □

In the situation of Theorem 5, one can ask whether or not $\rho$ arises from the space of weight-two cusp forms on $\Gamma_H(N)$. In the example treated in Section 2, this space is zero, so the response is negative. It might be interesting to answer this question in general.

*References*

[1] S. Bosch, W. Lütkebohmert and M. Raynaud, *Néron models*, Ergebnisse der Mathematik und ihrer Grenzgebiete, Springer-Verlag, Berlin and New York, 1990.

[2] H. Carayol, *Sur les représentations galoisiennes modulo $\ell$ attachées aux formes modulaires*, Duke Math. J., vol. 59 (1989), p. 785-801.

[3] H. Cohen, N. Skoruppa and D. Zagier, *Tables of coefficients of modular forms*, to appear.

[4] P. Deligne and M. Rapoport, *Les schémas de modules de courbes elliptiques*, Lecture Notes in Math., vol. 349, Springer-Verlag, Berlin and New York, 1973, p. 143-316.

[5] F. Diamond, *The refined conjecture of Serre*, This volume.

[6] B. Edixhoven, *L'action de l'algèbre de Hecke sur les groupes de composantes des jacobiennes des courbes modulaires est "Eisenstein"*, Astérisque, vol. 196-197 (1991), p. 159-170.

[7] B. Edixhoven, *The weight in Serre's conjectures on modular forms*, Invent. Math., vol. 109 (1992), p. 563-594.

[8] B. Jordan, R. Livné, *Local diophantine properties of Shimura curves*, Math. Ann., vol. 270 (1985), p. 235-248.

[9] N. M. Katz, *p-adic properties of modular schemes and modular forms*, Lecture Notes in Math., vol. 350, Springer-Verlag, Berlin and New York, 1973, p. 69-190.

[10] N. M. Katz, B. Mazur, *Arithmetic moduli of elliptic curves*, Ann. of Math. Studies **108**, Princeton Univ. Press, Princeton, 1985.

[11] B. Mazur, *Modular curves and the Eisenstein ideal*, Publ. Math. IHES, vol. 47 (1977), p. 33-186.

[12] M. Raynaud, *Jacobienne des courbes modulaires et opérateurs de Hecke*, Astérisque, vol. 196-197 (1991), p. 9-25.

[13] K. A. Ribet, *On modular representations of $\mathrm{Gal}(\overline{\mathbf{Q}}/\mathbf{Q})$ arising from modular forms*, Invent. Math., vol. 100 (1990), p. 431-476.

[14] K. A. Ribet, *On the Component Groups and the Shimura Subgroup of $J_o(N)$*, Sém. Th. des Nombres de l'Université de Bordeaux, 1987-88.

[15] K. A. Ribet, *Report on mod $\ell$ representations of $\mathrm{Gal}(\overline{\mathbf{Q}}/\mathbf{Q})$*, Proc. of Symposia in Pure Mathematics, vol. 55 (2) (1994), p. 639-676.

[16] J-P. Serre, *Sur les représentations modulaires de degré 2 de $\mathrm{Gal}(\overline{\mathbf{Q}}/\mathbf{Q})$*, Duke Math. J., vol. 54 (1987), p. 179-230.

[17] J-P. Serre, Letter to K. Ribet (15 April 1987), unpublished.

[18] J. H. Silverman, *The arithmetic of elliptic curves*, Graduate Texts in Math., vol. 106, Springer-Verlag, Berlin and New York, 1986.

[19] J. Tunnell, *Artin's conjecture for representations of octahedral type*, Bull. AMS, (new series), vol. 5 (1981), p. 173-175.

[20] A. Wiles, *Modular elliptic curves and Fermat's Last Theorem*, Annals of Math., to appear.

Conference on Elliptic Curves and Modular Forms
Hong Kong, December 18-21, 1993
Copyright ©1995 International Press

# Families of elliptic curves with constant mod $p$ representations

K. RUBIN
DEPARTMENT OF MATHEMATICS
OHIO STATE UNIVERSITY, COLUMBUS, OHIO 43210, USA
*E-mail address*: RUBIN@MATH.OHIO-STATE.EDU

A. SILVERBERG
DEPARTMENT OF MATHEMATICS
OHIO STATE UNIVERSITY, COLUMBUS, OHIO 43210, USA
*E-mail address*: SILVER@MATH.OHIO-STATE.EDU

**Introduction.** Suppose $E$ is an elliptic curve over $\mathbf{Q}$ and $p$ is 3 or 5. Then the collection of elliptic curves over $\mathbf{Q}$ having the same mod $p$ representation as $E$ forms an infinite family. In this paper we show how to construct these families explicitly. One reason for the interest in these families is the result announced by Wiles, that if $E$ and $E'$ are elliptic curves over $\mathbf{Q}$ with good reduction at an odd prime $p$ and with the same mod $p$ representation, and $E$ has complex multiplication, then $E'$ is modular.

The proofs of the results in Section 4 and Section 5 rely on symbolic computer computations, which were done using the programs Pari and Mathematica.

Note that all the results of this paper remain true with $\mathbf{Q}$ replaced by any number field.

The authors thank Tricia Pacelli and Glenn Stevens for asking a question which led us to simplify the polynomials in Theorems 4.1 and 5.1.

*Notation.* If $N$ is a positive integer and $E$ is an elliptic curve over a field $k$ with algebraic closure $\bar{k}$, let $E[N]$ denote the kernel of multiplication by $N$ on $E(\bar{k})$. Let $\mathfrak{H}$ denote the complex upper half plane,

$$\Gamma(N) = \{ \begin{pmatrix} a & b \\ c & d \end{pmatrix} \in SL_2(\mathbf{Z}) : \begin{pmatrix} a & b \\ c & d \end{pmatrix} \equiv \begin{pmatrix} 1 & 0 \\ 0 & 1 \end{pmatrix} \pmod{N} \},$$

and $G_\mathbf{Q} = \mathrm{Gal}(\bar{\mathbf{Q}}/\mathbf{Q})$.

If $E$ is an elliptic curve over $\mathbf{Q}$ and $p$ is a prime, write

$$\rho_{E,p} : G_\mathbf{Q} \to \mathrm{Aut}(E[p]) \cong GL_2(\mathbf{F}_p)$$

for the (isomorphism class of the) mod $p$ representation of $E$.

---

The authors thank the NSF for financial support.

**1 Modular curves.** Suppose $p$ is an odd rational prime. Denote by $Y_p$ the (non-compact) modular curve over $\mathbf{Q}$ which parametrizes isomorphism classes of pairs $(E, \phi)$ where $E$ is an elliptic curve and

$$\phi : \mathbf{Z}/p\mathbf{Z} \times \boldsymbol{\mu}_p \to E[p]$$

is an isomorphism with the property that

$$\langle \phi(a_1, \zeta_1), \phi(a_2, \zeta_2) \rangle = \zeta_2^{a_1}/\zeta_1^{a_2}$$

where $\langle , \rangle$ denotes the Weil pairing on $E[p]$. Equivalently, $Y_p$ parametrizes triples $(E, P, C)$ where $E$ is an elliptic curve, $P$ is a point of exact order $p$ on $E$, and $C$ is a subgroup of order $p$ on $E$, not containing $P$. (Given $(E, P, C)$, define $\phi$ by $\phi(a, \zeta) = aP + Q$ for the unique $Q \in C$ such that $\langle P, Q \rangle = \zeta$.) Let $Y(N)$ denote the modular curve which parametrizes elliptic curves with full level $N$ structure (see [5]). Fixing a $p$-th root of unity $\zeta$, the map

$$(E, \phi) \mapsto (E, \phi(1, 1), \phi(0, \zeta)),$$

induces an isomorphism (defined over $\mathbf{Q}(\zeta)$) from $Y_p$ onto one connected component of $Y(p)$. Thus $Y_p(\mathbf{C})$ is isomorphic to $\mathfrak{H}/\Gamma(p)$. Let $X_p$ and $X(N)$ denote the compactifications of $Y_p$ and $Y(N)$, respectively. Then $X_p$ has genus 0 when $p \leq 5$ and genus at least 3 when $p \geq 7$ (see p. 23 of [5]).

From now on we will identify $X(1)$ with $\mathbf{P}^1$ by the map which sends an elliptic curve $E$ to its $j$-invariant $j(E)$. From the description over $\mathbf{C}$ we see that the forgetful morphism $X_p \to X(1)$ induced by $(E, \phi) \mapsto E$ has degree

$$\frac{1}{2}[SL_2(\mathbf{Z}) : \Gamma(p)] = \frac{1}{2}\#(SL_2(\mathbf{F}_p)) = \frac{p^3 - p}{2}.$$

Let $W_p$ be a compactification of the elliptic surface associated to the universal elliptic curve over $Y_p$. Then $W_p$ is a variety over $\mathbf{Q}$, and $W_p$ is an elliptic surface over $X_p$ equipped with a morphism $W_p \to X_p$ and a zero section $\iota_p : X_p \to W_p$, both defined over $\mathbf{Q}$, and $p^2 - 1$ additional sections of order $p$ defined over $\bar{\mathbf{Q}}$. We will denote the $G_{\mathbf{Q}}$-module of these $p^2$ sections by $W_p[p]$. The definitions of $X_p$ and $W_p$ imply that $W_p[p]$ is isomorphic to $\mathbf{Z}/p\mathbf{Z} \times \boldsymbol{\mu}_p$ as a $G_{\mathbf{Q}}$-module. Note that $W_p$ is not uniquely determined, because there is a choice of compactification, but the choice is not important for our purposes. See [4] for the arithmetic of elliptic modular surfaces.

*1.1 Level three.* Let $A'_u$ denote the elliptic curve (the Hesse cubic)

$$X^3 + Y^3 + Z^3 = 3uXYZ$$

over $\mathbf{Q}(u)$, with origin $(0, 1, -1)$. This curve has a Weierstrass model

(1) $$A_u : y^2 = x^3 - 27u(u^3 + 8)x + 54(u^6 - 20u^3 - 8)$$

so in particular

(2) $$j(A'_u) = j(A_u) = 27\frac{u^3(u^3 + 8)^3}{(u^3 - 1)^3}.$$

Let $P_u = (-1, 1, 0) \in A'_u[3]$. If $\zeta$ is a primitive cube root of unity, then
$$C_u = \{(0, 1, -1), (0, \zeta, -1), (0, \zeta^2, -1)\}$$
is a $\text{Gal}(\overline{\mathbf{Q}(u)}/\mathbf{Q}(u))$-invariant subgroup of $A'_u[3]$. Thus the map
$$u \mapsto (A'_u, P_u, C_u)$$
induces a morphism $f : \mathbf{P}^1 \to X_3$ defined over $\mathbf{Q}$. To see that $f$ is an isomorphism it suffices to observe that by (2), the degree of the composition $\mathbf{P}^1 \xrightarrow{f} X_3 \to X(1)$ is 12 which is the same as the degree of $X_3$ over $X(1)$. From now on we will identify $X_3$ with $\mathbf{P}^1$ using the isomorphism $f$. With this identification we can take $W_3$ to be given by (1).

**1.2 Level five.** Following Klein (see [1] and p. 130 of [2]), let $B_u$ denote the elliptic curve

(3) $$y^2 = x^3 - \frac{u^{20} - 228u^{15} + 494u^{10} + 228u^5 + 1}{48}x$$
$$+ \frac{u^{30} + 522u^{25} - 10005u^{20} - 10005u^{10} - 522u^5 + 1}{864}$$

over $\mathbf{Q}(u)$. The $j$-invariant of $B_u$ is

(4) $$j(B_u) = \frac{-(u^{20} - 228u^{15} + 494u^{10} + 228u^5 + 1)^3}{u^5(u^{10} + 11u^5 - 1)^5}.$$

Let
$$x_0(u) = (u^{10} + 12u^8 - 12u^7 + 24u^6 + 30u^5 +$$
$$+ 60u^4 + 36u^3 + 24u^2 + 12u + 1)/12,$$
$$y_0(u) = (u^{13} + u^{12} + 4u^{11} + 5u^9 + 6u^8 + 21u^7 +$$
$$+ 29u^6 + 25u^5 + 15u^4 + 9u^3 + 4u^2 + u)/2,$$

and let $Q_u = (x_0(u), y_0(u))$. One can verify that $Q_u \in B_u[5]$. Let $\zeta$ denote a primitive fifth root of unity. Since the equation defining $B_u$ is invariant under $u \mapsto \zeta u$, we have $R_u = (x_0(\zeta u), y_0(\zeta u)) \in B_u[5]$ as well. Since $Q_u \in B_u(\mathbf{Q}(u))$ and $R_u \notin B_u(\mathbf{Q}(u))$, these are independent points of order 5. Further, one can verify that for $\sigma \in G_{\mathbf{Q}}$,
$$\sigma R_u = \epsilon(\sigma) R_u + (1 - \epsilon(\sigma)) Q_u$$
where $\epsilon : G_{\mathbf{Q}} \to (\mathbf{Z}/5\mathbf{Z})^\times$ is the cyclotomic character. Thus the subgroup $C_u$ of $B_u[5]$ generated by $Q_u - R_u$ is stable under $\text{Gal}(\overline{\mathbf{Q}(u)}/\mathbf{Q}(u))$. Therefore the map
$$u \mapsto (B_u, Q_u, C_u)$$
induces a morphism $g : \mathbf{P}^1 \to X_5$ defined over $\mathbf{Q}$. To see that $g$ is an isomorphism it is enough to observe that by (4), the degree of the composition $\mathbf{P}^1 \xrightarrow{g} X_5 \to X(1)$ is 60 which is the same as the degree of $X_5$ over $X(1)$. From now on we will identify $X_5$ with $\mathbf{P}^1$ using this isomorphism. With this identification we can take $W_5$ to be given by (3).

## 2  Twists of modular curves.

**Proposition 2.1.** *Suppose p is a prime and there are commutative diagrams*

(5)
$$\begin{array}{ccc} W_p \xleftarrow{\psi} W' & & W_p \xleftarrow{\psi} W' \\ \downarrow \quad \downarrow & & \iota_p \uparrow \quad \uparrow \iota' \\ X_p \xleftarrow{\psi_0} X' & & X_p \xleftarrow{\psi_0} X' \\ \searrow \swarrow & & \\ X(1) & & \end{array}$$

*where $X'$ is a curve and $W'$ is an elliptic surface over $X'$ with zero-section $\iota'$, all defined over $\mathbf{Q}$, $\psi$ and $\psi_0$ are isomorphisms defined over $\bar{\mathbf{Q}}$, the maps $W_p \to X_p \to X(1)$ are the natural ones (see §1), and the maps $W' \to X' \to X(1)$ are defined over $\mathbf{Q}$. Let $S \subset X'(\bar{\mathbf{Q}})$ denote the inverse image under $\psi_0$ of the (finite) set of cusps of $X_p$. Then there is a $G_\mathbf{Q}$-module $V$ such that for every point $t \in X'(\mathbf{C}) - S$, the fiber $E_t$ of $W'$ over $t$ is an elliptic curve defined over $\mathbf{Q}(t)$ and $E_t[p] \cong V$ as $\mathrm{Gal}(\overline{\mathbf{Q}(t)}/\mathbf{Q}(t))$-modules.*

*Proof.* Let $V$ be the group of $p^2$ sections from $X'$ to $W'$ of order dividing $p$,

$$V = \{\psi^{-1} \circ s \circ \psi_0 : s \in W_p[p]\}.$$

These sections are defined over $\bar{\mathbf{Q}}$, so $V$ is a $G_\mathbf{Q}$-module and the map from $V$ to $E_t[p]$ obtained by restricting the sections in $V$ to $t$ is a $\mathrm{Gal}(\overline{\mathbf{Q}(t)}/\mathbf{Q}(t))$-isomorphism. □

**Corollary 2.2.** *With notation as in Proposition 2.1, if $E_1, E_2$ are the fibers of $W'$ over $t_1, t_2 \in X'(\mathbf{Q}) - S$, then $E_1[p] \cong E_2[p]$ as $G_\mathbf{Q}$-modules.*

*Proof.* Immediate from Proposition 2.1. □

REMARK 2.3. Under the hypotheses of Proposition 2.1, the resulting module $V$ is endowed with a pairing coming from the Weil pairing on $W_p[p]$, and $X'$ is the moduli space for isomorphism classes of pairs $(E, \phi)$ where $E$ is an elliptic curve and $\phi : V \xrightarrow{\sim} E[p]$ is an isomorphism which takes the pairing on $V$ to the Weil pairing on $E[p]$.

REMARK 2.4. Fix an odd prime $p$, let $V_0 = \mathbf{Z}/p\mathbf{Z} \times \boldsymbol{\mu}_p$, and let $\eta_0$ be the $G_\mathbf{Q}$-equivariant pairing on $V_0$

$$\eta_0((a_1, \zeta_1), (a_2, \zeta_2)) = \zeta_2^{a_1}/\zeta_1^{a_2}.$$

There are natural $G_\mathbf{Q}$-equivariant maps

$$\mathrm{Aut}(V_0, \eta_0) \to \mathrm{Aut}(W_p) \to \mathrm{Aut}(X_p),$$

where $\mathrm{Aut}(V_0, \eta_0)$ is the group of automorphisms of $V_0$ which preserve $\eta_0$. Suppose $V$ is a two-dimensional $\mathbf{F}_p$-vector space with an action of $G_\mathbf{Q}$ and a non-degenerate skew-symmetric $\boldsymbol{\mu}_p$-valued $G_\mathbf{Q}$-equivariant pairing $\eta$. Let $\rho : G_\mathbf{Q} \to \mathrm{Aut}(V)$ and $\rho_0 : G_\mathbf{Q} \to \mathrm{Aut}(V_0)$ denote the representations induced by the $G_\mathbf{Q}$-actions, and fix a group isomorphism $\varphi : V_0 \to V$ which takes the pairing $\eta_0$ to the pairing $\eta$. Then $\sigma \mapsto \varphi^{-1}\rho(\sigma)\varphi\rho_0(\sigma)^{-1}$ defines a cocycle on $G_\mathbf{Q}$ with values in $\mathrm{Aut}(V_0, \eta_0)$, which induces cohomology classes in $H^1(G_\mathbf{Q}, \mathrm{Aut}(W_p))$ and $H^1(G_\mathbf{Q}, \mathrm{Aut}(X_p))$ that are independent of the choice of $\varphi$. The twists of $W_p$ and $X_p$ by these cohomology classes (see [3]) give the varieties $W'$ and $X'$ satisfying Proposition 2.1 for the given $G_\mathbf{Q}$-module $V$.

Suppose $E$ is an elliptic curve defined over $\mathbf{Q}$. Our aim is to explicitly construct $W'$, $X'$, and corresponding morphisms satisfying the hypotheses of Proposition 2.1, so that $E$ is a fiber of $W'$ over some rational point of $X'$. Then, by Corollary 2.2, the points of $X'(\mathbf{Q}) - S$ will correspond to elliptic curves $E'$ over $\mathbf{Q}$ with $\rho_{E',p} \cong \rho_{E,p}$.

From now on suppose that $p = 3$ or $5$, so that $X_p$ has genus 0. Recall that we have identified both $X_p$ and $X(1)$ with $\mathbf{P}^1$, and with these identifications the morphism from $X_p$ to $X(1)$ is given by the rational function $\mathcal{J}(u) \in \mathbf{Q}(u)$ defined by

$$\mathcal{J}(u) = \begin{cases} j(A_u) & \text{if } p = 3, \\ j(B_u) & \text{if } p = 5 \end{cases}$$

(see (2) and (4)). Let $X' = \mathbf{P}^1$. An isomorphism $\psi_0 : X' \to X_p$ defined over a number field $K$ is given by a linear fractional transformation

$$u = A(t) = \frac{at+b}{ct+d} \qquad \text{where} \qquad A = \begin{pmatrix} a & b \\ c & d \end{pmatrix} \in GL_2(K).$$

The map from $X'$ to $X(1)$ in (5) will be defined over $\mathbf{Q}$ if and only if

(6) $$\mathcal{J}(A(t)) \in \mathbf{Q}(t).$$

**3 Algorithm.** Fix $p = 3$ or $5$ and suppose $E$ is an elliptic curve over $\mathbf{Q}$. We will construct the family of all elliptic curves over $\mathbf{Q}$ with the same mod $p$ representation as $E$. When applying Wiles' result one is interested in curves $E$ with complex multiplication, but the method works in general.

The construction proceeds as follows. Define polynomials $a_4(u), a_6(u) \in \mathbf{Q}[u]$ so that

$$y^2 = x^3 + a_4(u)x + a_6(u)$$

is the elliptic curve $A_u$ of (1) if $p = 3$, or $B_u$ of (3) if $p = 5$. Also let $n = 1$ if $p = 3$ and $n = 5$ if $p = 5$, so that $a_4$ has degree $4n$ and $a_6$ has degree $6n$.

*Step 1.* Find $u_0 \in \bar{\mathbf{Q}}$ such that

$$\mathcal{J}(u_0) = j(E).$$

*Step* 2. Find $\mu \in \bar{\mathbf{Q}}^\times$ such that

$$\mu^{-2}a_4(u_0), \mu^{-3}a_6(u_0) \in \mathbf{Q}$$

and

$$y^2 = x^3 + \mu^{-2}a_4(u_0)x + \mu^{-3}a_6(u_0) \quad \text{is isomorphic over } \mathbf{Q} \text{ to } E.$$

*Step* 3. Find $\alpha, \gamma \in \bar{\mathbf{Q}}$ such that

(7) $$\mu^{-2}(1+\gamma t)^{4n}a_4(A(t)), \quad \mu^{-3}(1+\gamma t)^{6n}a_6(A(t)) \in \mathbf{Q}[t],$$

where

$$A = \begin{pmatrix} \alpha & u_0 \\ \gamma & 1 \end{pmatrix}.$$

REMARK 3.1. For Step 1, let $u_0$ be any root of the numerator of $\mathcal{J}(u) - j(E) \in \mathbf{Q}(t)$. For Step 2, after putting $E$ in Weierstrass form one can solve for $\mu$ directly. Step 2 ensures that the constant terms of the polynomials in (7) are rational numbers, and choosing rational values for the coefficients of $t$ gives values for $\alpha$ and $\gamma$.

REMARK 3.2. The condition in Step 3 implies (6). Conversely, if $\operatorname{Aut}(E) = \{\pm 1\}$, and we have $u_0, \mu, \alpha, \gamma \in \bar{\mathbf{Q}}$ and $A = \begin{pmatrix} \alpha & u_0 \\ \gamma & 1 \end{pmatrix}$ which satisfy (6) and the conditions in Steps 1 and 2, then it can be shown that the condition in Step 3 is satisfied.

**Theorem 3.3.** *Suppose $E$ is an elliptic curve defined over $\mathbf{Q}$, and we have elements $u_0, \mu, \alpha, \gamma \in \bar{\mathbf{Q}}$ and a matrix $A \in GL_2(\bar{\mathbf{Q}})$ satisfying the conditions in Steps 1, 2, and 3. Let $\mathcal{E}_t$ be the curve*

(8) $$y^2 = x^3 + \mu^{-2}(1+\gamma t)^{4n}a_4(A(t))x + \mu^{-3}(1+\gamma t)^{6n}a_6(A(t))$$

*over $\mathbf{Q}(t)$. Then for every rational number $t$ which is not a pole of $\mathcal{J} \circ A$, $\mathcal{E}_t$ is an elliptic curve over $\mathbf{Q}$ and $\mathcal{E}_t[p] \cong E[p]$ as $G_\mathbf{Q}$-modules.*

*Proof.* Let $X' = \mathbf{P}^1$ and let $W'$ be the elliptic surface over $X'$ defined by (8), with $W'$ mapping to $X'$ by $(x, y, t) \mapsto t$. Let $\psi_0 : X' \xrightarrow{\sim} X_p$ be given by $A$, and $\psi : W' \xrightarrow{\sim} W_p$ by

$$(x, y, t) \mapsto (\mu(1+\gamma t)^{-2n}x, \mu^{3/2}(1+\gamma t)^{-3n}y, A(t)).$$

Then all the hypotheses of Proposition 2.1 are satisfied. By Step 2, the fiber of $W'$ over $t = 0$ is isomorphic over $\mathbf{Q}$ to $E$. The theorem now follows from Corollary 2.2. □

## 4 Examples with $p = 3$.

### 4.1 Generic example.

**Theorem 4.1.** *Fix an elliptic curve $E$ over $\mathbf{Q}$,*
$$y^2 = x^3 + ax + b,$$
*and let $J = j(E)/1728 = 4a^3/(4a^3 + 27b^2)$. Define*
$$\begin{aligned}
a(t) &= (-3(J-1)^2 t^4 - 8(J-1)^2 t^3 + 6(J-1)t^2 + 1)a, \\
b(t) &= (-(J-1)^3(8J+1)t^6 + 6(J-1)^2(2J+1)t^5 + 15(J-1)^2 t^4 \\
&\quad + 20(J-1)^2 t^3 - 15(J-1)t^2 + 6t + 1)b
\end{aligned}$$
*and define $\mathcal{E}_t$ by*
$$y^2 = x^3 + a(t)x + b(t).$$
*Then for every rational number $t$ such that $\mathcal{E}_t$ is nonsingular, $\mathcal{E}_t[3]$ is isomorphic as a $G_\mathbf{Q}$-module to $E[3]$.*

*Proof.* If $a = 0$ then $\mathcal{E}_t$ is $y^2 = x^3 + (t+1)^6 b$ and if $b = 0$ then $\mathcal{E}_t$ is $y^2 = x^3 + ax$. In both cases the elliptic surface is isotrivial and the theorem holds. Now assume $a$ and $b$ are both nonzero and let $j = j(E)$. Using (2), a computation shows that
$$\mathcal{J}(u) - j = \frac{27u^{12} + (648-j)u^9 + 3(1728+j)u^6 + 3(4608-j)u^3 + j}{(u^3-1)^3}.$$
Let $u_0$ be a root of the numerator of $\mathcal{J}(u) - j$, so that Step 1 of §3 is satisfied. Let
$$\mu = \frac{a_6(u_0)a}{a_4(u_0)b} = \frac{-2(u_0^6 - 20u_0^3 - 8)a}{(u_0^3 + 8)u_0 b},$$
$$\alpha = \frac{\mu^3 b u_0}{144(u_0^3 - 1)^2},$$
$$\gamma = \frac{-\mu^3 b(u_0^3 + 2)}{864(u_0^3 - 1)^2}.$$
Since $a$ and $b$ are nonzero, we have $\mu \in \bar{\mathbf{Q}}^\times$. A symbolic computer calculation shows that with these choices, Steps 2 and 3 are satisfied, and the curve $\mathcal{E}_t$ is the same as the curve $\mathcal{E}_t$ of Theorem 3.3. Thus Theorem 3.3 completes the proof. □

REMARK 4.2. If $a$ and $b$ are both nonzero, then the elliptic surface $\mathcal{E}_t$ in Theorem 4.1 is not isotrivial and every elliptic curve $E'$ over $\mathbf{Q}$ such that $\rho_{E,3} \cong \rho_{E',3}$ is isomorphic over $\mathbf{Q}$ to $\mathcal{E}_t$ for some $t \in \mathbf{P}^1(\mathbf{Q})$ (see Remark 2.3).

REMARK 4.3. The values of $\alpha$ and $\gamma$ in the proof of Theorem 4.1 were obtained by choosing 0 and $6b$ for the linear coefficients of the polynomials in (7) and then solving for $\alpha$ and $\gamma$ (see Remark 3.1). This choice gives rise to relatively simple polynomials $a(t)$ and $b(t)$. Choosing other values (not both zero) for the linear coefficients leads to other $\alpha$, $\gamma$, $a(t)$, and $b(t)$, giving rise to an isomorphic elliptic surface.

## 4.2 $y^2 = x^3 - Dx$: supersingular at 3, CM by $\sqrt{-1}$.

**Theorem 4.4.** *Fix a nonzero integer $D$ and define $E_t$ by*

$$y^2 = x^3 + D(27D^2 t^4 - 18Dt^2 - 1)x + 4D^2 t(27D^2 t^4 + 1).$$

(i) *For every rational number $t$, $E_t$ is an elliptic curve over $\mathbf{Q}$ and $E_t[3]$ is isomorphic as a $G_{\mathbf{Q}}$-module to $E[3]$, where $E$ is the elliptic curve $y^2 = x^3 - Dx$.*

(ii) *If $D$ is prime to 3 and $t \in \mathbf{Q}$ is integral at 3 then $E_t$ has good reduction at 3.*

*Proof.* Using (2), a computation shows that

$$\mathcal{J}(u) - j(E) = \frac{27(u^2 - 2u - 2)^2 (u^4 + 2u^3 + 6u^2 - 4u + 4)^2}{(u^3 - 1)^3}.$$

Let $u_0$ be a root of $u^2 - 2u - 2$ and let $\nu$ be a fixed square root of $(1 + 2u_0)/D$. Let

$$\mu = 18\nu,$$
$$\alpha = (4 - u_0)\nu D,$$
$$\gamma = (1 - u_0)\nu D.$$

(We can take $u_0 = 1 + \sqrt{3}$, $\mu = \pm 18\sqrt{(3 + 2\sqrt{3})/D}$, $\alpha = \pm\sqrt{6D\sqrt{3}}$, and $\gamma = \pm\sqrt{(9 + 6\sqrt{3})D}$, where the signs of the square roots are taken appropriately and compatibly.) Then Steps 1, 2, and 3 are satisfied, and with these choices $E_t$ is the curve $\mathcal{E}_t$ of Theorem 3.3. The discriminant of $E_t$ is

$$\Delta(E_t) = -2^6 D^3 (27D^2 t^4 + 18Dt^2 - 1)^3,$$

which has no rational roots, so $E_t$ is an elliptic curve for every $t \in \mathbf{P}^1(\mathbf{Q})$. Theorem 3.3 implies (i). Statement (ii) follows immediately from the formula for the discriminant. □

REMARK 4.5. The $j$-invariant of the curve $E_t$ of Theorem 4.4 is

$$j(E_t) = \frac{1728(27D^2 t^4 - 18Dt^2 - 1)^3}{(27D^2 t^4 + 18Dt^2 - 1)^3}.$$

The result of Wiles mentioned in the introduction, together with Theorem 4.4, implies that if $E'$ is an elliptic curve over $\mathbf{Q}$ whose $j$-invariant belongs to the set

$$\left\{ \frac{1728(27D^2 t^4 - 18Dt^2 - 1)^3}{(27D^2 t^4 + 18Dt^2 - 1)^3} : D \in \mathbf{Z} \text{ is prime to 3 and } t \in \mathbf{Q} \text{ is integral at } 3 \right\}$$

then $E'$ is modular.

## 4.3 $y^2 = x^3 - 432D$: additive at 3, CM by $\sqrt{-3}$.

**Theorem 4.6.** *Fix a nonzero integer $D$ and define $E_t$ by*

$$y^2 = x^3 - 27Dt(Dt^3 + 8)x + 54D(D^2t^6 - 20Dt^3 - 8).$$

*If $t \in \mathbf{Q}$ and $Dt^3 \neq 1$, then $E_t$ is an elliptic curve over $\mathbf{Q}$ and $E_t[3]$ is isomorphic as a $G_{\mathbf{Q}}$-module to $E[3]$, where $E$ is the elliptic curve $y^2 = x^3 - 432D$.*

Note that when $D = 1$, this is the Weierstrass model (1) of the Hesse family. The proof of Theorem 4.6 is the same as that of Theorem 4.4, after letting

$$u_0 = 0, \quad \mu = D^{-1/3}, \quad \alpha = D^{1/3}, \quad \text{and} \quad \gamma = 0.$$

The discriminant and $j$-invariant of $E_t$ are given by

$$\Delta(E_t) = 2^{12}3^9 D^2(Dt^3 - 1)^3, \qquad j(E_t) = \frac{27Dt^3(Dt^3 + 8)^3}{(Dt^3 - 1)^3}.$$

## 4.4 $y^2 + y = x^3 - x^2 - 7x + 10$: ordinary at 3, CM by $\sqrt{-11}$.

**Theorem 4.7.** *Define polynomials*

$$a(t) = -5346t^4 + 2079t^3 - 99t^2 + 77t - 7,$$
$$b(t) = -154366t^6 + 73507t^5 - 26805t^4 - 694t^3 + 1091t^2 - 127t + 10$$

*and for each $t \in \mathbf{Q}$ let $E_t$ be the elliptic curve*

$$y^2 + (t^3 + t^2 + t + 1)y = x^3 - x^2 + a(t)x + b(t).$$

*For every rational number $t$, $E_t$ is an elliptic curve over $\mathbf{Q}$ and $E_t[3]$ is isomorphic as a $G_{\mathbf{Q}}$-module to $E[3]$, where $E$ is the elliptic curve $y^2 + y = x^3 - x^2 - 7x + 10$. If $t \in \mathbf{Q}$ is integral at 3 and $t \equiv 0$ or $1 \pmod{3}$ then $E_t$ has good reduction at 3.*

*Proof.* The elliptic curve $E$ has complex multiplication by $\mathbf{Q}(\sqrt{-11})$ and good reduction at 3. We apply Theorem 4.1 with the Weierstrass model

$$y^2 = x^3 - \frac{22}{3}x + \frac{847}{108}$$

of $E$. The change of variables

$$(x, y, t) \mapsto \left( \frac{56^2(3x - 1)}{3(56 - 147t)^2}, \frac{56^3(2y + t^3 + t^2 + t + 1)}{2(56 - 147t)^3}, \frac{27t}{56 - 147t} \right),$$

transforms the curve $\mathcal{E}_t$ of Theorem 4.1 (with $a = -22/3$ and $b = 847/108$) to the curve $E_t$. The assertion about good reduction follows by computing that the discriminant of $E_t$ is

$$-11^3(27t^2 - 8t + 1)^3(27t^2 + 36t + 1)^3. \qquad \square$$

REMARK 4.8. The curve $E_t$ of Theorem 4.7 has been chosen so that it is an integral model which is minimal when $t = 0$. The $j$-invariant of $E_t$ is

$$\left(\frac{-16(54t^2 - 27t + 2)(27t^2 + 3t + 1)}{(27t^2 - 8t + 1)(27t^2 + 36t + 1)}\right)^3.$$

## 5 Examples with $p = 5$.
### 5.1 Generic example.

**Theorem 5.1.** *Fix an elliptic curve $E$ over $\mathbf{Q}$,*

$$y^2 = x^3 + ax + b.$$

*Define*

$$a(t) = a \sum_{k=0}^{20} \alpha_k t^k, \quad b(t) = b \sum_{k=0}^{30} \beta_k t^k,$$

*where $\alpha_k, \beta_k \in \mathbf{Q}[J]$, with $J = j(E)/1728 = 4a^3/(4a^3 + 27b^2)$, are the polynomials given in the appendix. Define $\mathcal{E}_t$ by*

$$y^2 = x^3 + a(t)x + b(t).$$

*Then for every rational number $t$ such that $\mathcal{E}_t$ is nonsingular, $\mathcal{E}_t[5]$ is isomorphic as a $G_\mathbf{Q}$-module to $E[5]$.*

*Proof.* If $a = 0$ then $\mathcal{E}_t$ is $y^2 = x^3 + (t+1)^{30}b$ and if $b = 0$ then $\mathcal{E}_t$ is $y^2 = x^3 + ax$. In both cases the elliptic surface is isotrivial and the theorem holds. Now assume $a$ and $b$ are both nonzero and let $j = j(E)$. Using (4), a computation shows that the numerator of $\mathcal{J}(u) - j$ is

$$u^{60} + (j - 684)u^{55} + (55j + 157434)u^{50} + 5(241j - 2505492)u^{45}$$
$$+ 35(374j + 2213157)u^{40} + 3(23195j - 43563048)u^{35}$$
$$+ (134761j - 33211924)u^{30} - 3(23195j - 43563048)u^{25}$$
$$+ 35(374j + 2213157)u^{20} - 5(241j - 2505492)u^{15}$$
$$+ (55j + 157434)u^{10} - (j - 684)u^5 + 1.$$

Let $u_0$ be a root of this polynomial and let

$$\mu = \frac{a_6(u_0)a}{a_4(u_0)b} = \frac{-(u_0^{30} + 522u_0^{25} - 10005u_0^{20} - 10005u_0^{10} - 522u_0^5 + 1)a}{18(u_0^{20} - 228u_0^{15} + 494u_0^{10} + 228u_0^5 + 1)b},$$

$$\alpha = \frac{6\mu^3 b(57u_0^{15} - 247u_0^{10} - 171u_0^5 - 1)}{u_0^4(u_0^{10} + 11u_0^5 - 1)^4},$$

$$\gamma = \frac{6\mu^3 b(u_0^{15} - 171u_0^{10} + 247u_0^5 + 57)}{(u_0^{10} + 11u_0^5 - 1)^4}.$$

A symbolic computer calculation shows that with these choices, the polynomials of (7) are $a(t)$ and $b(t)$. Thus the theorem follows from Theorem 3.3. □

REMARK 5.2. If $ab \neq 0$, then the elliptic surface $\mathcal{E}_t$ in Theorem 4.1 is not isotrivial and every elliptic curve $E'$ over $\mathbf{Q}$ such that $\rho_{E,5} \cong \rho_{E',5}$ is isomorphic over $\mathbf{Q}$ to $\mathcal{E}_t$ for some $t \in \mathbf{P}^1(\mathbf{Q})$ (see Remark 2.3).

## 5.2 $y^2 = x^3 - Dx$: ordinary at 5, CM by $\sqrt{-1}$.

**Theorem 5.3.** *Fix a nonzero integer $D$ and define polynomials*

$$\begin{aligned}
a(t) &= D^5 t^9 - Dt, \\
b(t) &= -3125 D^{11} t^{20} - 39583 D^{10} t^{18} + 11875 D^9 t^{16} - 95000 D^8 t^{14} \\
&\quad + 61750 D^7 t^{12} + 41166 D^6 t^{10} + 12350 D^5 t^8 - 3800 D^4 t^6 \\
&\quad + 95 D^3 t^4 - 63 D^2 t^2 - D, \\
c(t) &= -521875 D^{16} t^{29} - 1355787 D^{15} t^{27} - 7366875 D^{14} t^{25} \\
&\quad + 9635000 D^{13} t^{23} - 8315875 D^{12} t^{21} - 3678639 D^{11} t^{19} \\
&\quad - 10560675 D^{10} t^{17} + 30400 D^9 t^{15} + 2091615 D^8 t^{13} \\
&\quad + 134479 D^7 t^{11} + 62583 D^6 t^9 - 14200 D^5 t^7 \\
&\quad + 2327 D^4 t^5 + 107 D^3 t^3 + 7 D^2 t,
\end{aligned}$$

*and for each $t \in \mathbf{Q}$ let $E_t$ be the elliptic curve*

$$y^2 = x^3 + a(t)x^2 + b(t)x + c(t).$$

*For every rational number $t$, $E_t$ is an elliptic curve over $\mathbf{Q}$ and $E_t[5]$ is isomorphic as a $G_{\mathbf{Q}}$-module to $E[5]$, where $E$ is the elliptic curve $y^2 = x^3 - Dx$. If $D$ is prime to 5, $t \in \mathbf{Q}$ is integral at 5, and $Dt^2 \not\equiv 3 \pmod{5}$ then $E_t$ has good reduction at 5.*

The elliptic curve $E$ has complex multiplication by $\mathbf{Q}(i)$ and good reduction at 5. Let $\nu$ be a fixed square root of $-(1+2i)/D$. The proof of Theorem 5.3 is the same as the proof of Theorem 4.4, using the values

$$u_0 = i, \quad \mu = (1+2i)^2 \nu / 2, \quad \alpha = -\nu D, \quad \gamma = -i\nu D,$$

and the change of variables $x \mapsto x + (D^5 t^9 - Dt)/3$. The discriminant of $E_t$ is

$$4^3 D^3 (5D^2 t^4 - 2Dt^2 + 1)^5 (25D^4 t^8 - 100D^3 t^6 - 210D^2 t^4 - 20Dt^2 + 1)^5$$

and the $j$-invariant is

$$\frac{4^3 (15D^2 t^4 + 10Dt^2 + 3)^3 (25D^4 t^8 + 70D^2 t^4 + 1)^3}{(5D^2 t^4 - 2Dt^2 + 1)^5} \times$$

$$\times \frac{(25D^4 t^8 + 300D^3 t^6 - 370D^2 t^4 + 60Dt^2 + 1)^3}{(25D^4 t^8 - 100D^3 t^6 - 210D^2 t^4 - 20Dt^2 + 1)^5}.$$

**5.3** $y^2 = x^3 + 16D$: supersingular at 5, CM by $\sqrt{-3}$.

**Theorem 5.4.** *Fix a nonzero integer $D$ and define polynomials*

$$\begin{aligned}
a(t) &= -15000D^7t^{19} - 106875D^6t^{16} + 85500D^5t^{13} - 185250D^4t^{10} \\
&\quad - 17100D^3t^7 - 4275D^2t^4 + 120Dt, \\
b(t) &= 50000D^{11}t^{30} + 3625000D^{10}t^{27} + 4893750D^9t^{24} \\
&\quad + 25012500D^8t^{21} - 47523750D^7t^{18} - 9504750D^5t^{12} \\
&\quad - 1000500D^4t^9 + 39150D^3t^6 - 5800D^2t^3 + 16D,
\end{aligned}$$

*and for each $t \in \mathbf{Q}$ let $E_t$ be the elliptic curve*

$$y^2 = x^3 + a(t)x + b(t).$$

*For every rational number $t$, $E_t$ is an elliptic curve over $\mathbf{Q}$ and $E_t[5]$ is isomorphic as a $G_\mathbf{Q}$-module to $E[5]$, where $E$ is the elliptic curve $y^2 = x^3 + 16D$. If $D$ is prime to 5 and $t \in \mathbf{Q}$ is integral at 5, then $E_t$ has good reduction at 5.*

The elliptic curve $E$ has $j$-invariant 0, complex multiplication by $\mathbf{Q}(\sqrt{-3})$, and good reduction at 5. The proof of Theorem 5.4 is the same as the proof of Theorem 4.4, after letting

$$\begin{aligned}
u_0 &\text{ be a root of the polynomial } u^4 - 3u^3 - u^2 + 3u + 1, \\
\mu &= 25(151711644u_0^3 - 6630528u_0^2 - 171313440u_0 \\
&\quad - 51318165)^{1/3}/(12D^{1/3}), \\
\alpha &= -((2u_0^3 - 6u_0^2 + 3)D)^{1/3}, \\
\gamma &= ((3u_0^3 - 6u_0 - 2)D)^{1/3},
\end{aligned}$$

using the real cube roots. The discriminant of $E_t$ is

$$-2^{12}3^3 D^2 (25D^4t^{12} - 275D^3t^9 - 165D^2t^6 + 55Dt^3 + 1)^5$$

and the $j$-invariant is

$$\frac{-15^3 Dt^3 (5D^2t^6 + 40Dt^3 - 1)^3 (5D^2t^6 - 5Dt^3 + 8)^3}{(25D^4t^{12} - 275D^3t^9 - 165D^2t^6 + 55Dt^3 + 1)^5} \times$$
$$\times (40D^2t^6 + 5Dt^3 + 1)^3.$$

**Appendix.** Since the polynomials $\alpha_k, \beta_k \in \mathbf{Q}[J]$ which occur in Theorem 5.1 are of marginal interest, we include them in this appendix.

$$\begin{aligned}
\alpha_0 &= 1, \\
\alpha_1 &= 0, \\
\alpha_2 &= 190(J-1), \\
\alpha_3 &= -2280(J-1)^2, \\
\alpha_4 &= 855(J-1)^2(-17+16J),
\end{aligned}$$

$$\alpha_5 = 3648(J-1)^3(17-9J),$$
$$\alpha_6 = 11400(J-1)^3(17-8J),$$
$$\alpha_7 = -27360(J-1)^4(17+26J),$$
$$\alpha_8 = 7410(J-1)^4(-119-448J+432J^2),$$
$$\alpha_9 = 79040(J-1)^5(17+145J-108J^2),$$
$$\alpha_{10} = 8892(J-1)^5(187+2640J-5104J^2+1152J^3),$$
$$\alpha_{11} = 98800(J-1)^6(-17-388J+864J^2),$$
$$\alpha_{12} = 7410(J-1)^6(-187-6160J+24464J^2-24192J^3),$$
$$\alpha_{13} = 54720(J-1)^7(17+795J-3944J^2+9072J^3),$$
$$\alpha_{14} = 2280(J-1)^7(221+13832J-103792J^2+554112J^3-373248J^4),$$
$$\alpha_{15} = 1824(J-1)^8(-119-9842J+92608J^2-911520J^3+373248J^4),$$
$$\alpha_{16} = 4275(J-1)^8(-17-1792J+23264J^2-378368J^3+338688J^4),$$
$$\alpha_{17} = 18240(J-1)^9(1+133J-2132J^2+54000J^3-15552J^4),$$
$$\alpha_{18} = 190(J-1)^9(17+2784J-58080J^2+2116864J^3-946944J^4+2985984J^5),$$
$$\alpha_{19} = 360(J-1)^{10}(-1+28J-1152J^2)(1+228J+176J^2+1728J^3),$$
$$\alpha_{20} = (J-1)^{10}(-19-4560J+144096J^2-9859328J^3-8798976J^4-226934784J^5+429981696J^6),$$
$$\beta_0 = 1,$$
$$\beta_1 = 30,$$
$$\beta_2 = -435(J-1),$$
$$\beta_3 = 580(J-1)(-7+9J),$$
$$\beta_4 = 3915(J-1)^2(7-8J),$$
$$\beta_5 = 1566(J-1)^2(91-78J+48J^2),$$
$$\beta_6 = -84825(J-1)^3(7+16J),$$
$$\beta_7 = 156600(J-1)^3(-13-91J+92J^2),$$
$$\beta_8 = 450225(J-1)^4(13+208J-144J^2),$$
$$\beta_9 = 100050(J-1)^4(143+4004J-5632J^2+1728J^3),$$
$$\beta_{10} = 30015(J-1)^5(-1001-45760J+44880J^2-6912J^3),$$
$$\beta_{11} = 600300(J-1)^5(-91-6175J+9272J^2-2736J^3),$$
$$\beta_{12} = 950475(J-1)^6(91+8840J-7824J^2),$$
$$\beta_{13} = 17108550(J-1)^6(7+926J-1072J^2+544J^3),$$
$$\beta_{14} = 145422675(J-1)^7(-1-176J+48J^2-384J^3),$$
$$\beta_{15} = 155117520(J-1)^8(1+228J+176J^2+1728J^3),$$
$$\beta_{16} = 145422675(J-1)^8(1+288J+288J^2+5120J^3-6912J^4),$$
$$\beta_{17} = 17108550(J-1)^8(7+2504J+3584J^2+93184J^3-283392J^4+165888J^5),$$

$$\beta_{18} = 950475(J-1)^9(-91-39936J-122976J^2-2960384J^3+11577600J^4-5971968J^5),$$

$$\beta_{19} = 600300(J-1)^9(-91-48243J-191568J^2-6310304J^3+40515072J^4-46455552J^5+11943936J^6),$$

$$\beta_{20} = 30015(J-1)^{10}(1001+634920J+3880800J^2+142879744J^3-1168475904J^4+1188919296J^5-143327232J^6),$$

$$\beta_{21} = 100050(J-1)^{10}(143+107250J+808368J^2+38518336J^3-451953408J^4+757651968J^5-367276032J^6),$$

$$\beta_{22} = 450225(J-1)^{11}(-13-11440J-117216J^2-6444800J^3+94192384J^4-142000128J^5+95551488J^6),$$

$$\beta_{23} = 156600(J-1)^{11}(-13-13299J-163284J^2-11171552J^3+217203840J^4-474406656J^5+747740160J^6-429981696J^7),$$

$$\beta_{24} = 6525(J-1)^{12}(91+107536J+1680624J^2+132912128J^3-3147511552J^4+6260502528J^5-21054173184J^6+10319560704J^7),$$

$$\beta_{25} = 1566(J-1)^{12}(91+123292J+2261248J^2+216211904J^3-6487793920J^4+17369596928J^5-97854234624J^6+96136740864J^7-20639121408J^8),$$

$$\beta_{26} = 3915(J-1)^{13}(-7-10816J-242352J^2-26620160J^3+953885440J^4-2350596096J^5+26796552192J^6-13329432576J^7),$$

$$\beta_{27} = 580(J-1)^{13}(-7-12259J-317176J^2-41205008J^3+1808220160J^4-5714806016J^5+93590857728J^6-70131806208J^7-36118462464J^8),$$

$$\beta_{28} = 435(J-1)^{14}(1+1976J+60720J^2+8987648J^3-463120640J^4+1359157248J^5-40644882432J^6-5016453120J^7+61917364224J^8),$$

$$\beta_{29} = 30(J-1)^{14}(1+2218J+77680J^2+13365152J^3-822366976J^4+2990693888J^5-118286217216J^6-24514928640J^7+509958291456J^8-743008370688J^9),$$

$$\beta_{30} = (J-1)^{15}(-1-2480J-101040J^2-19642496J^3+1399023872J^4-4759216128J^5+315623485440J^6+471904911360J^7-2600529297408J^8+8916100448256J^9).$$

*References*

[1] F. Klein, *Vorlesungen über das Ikosaeder und die Auflösung der Gleichungen fünften Grades*, Leipzig (1884) (= Lectures on the icosahedron and the solution of equations of the fifth degree, London (1913)).

[2] F. Klein, *Elementary mathematics from an advanced standpoint: Arithmetic, Algebra, Analysis*, Dover Publications, New York (1945).

[3] J-P. Serre, *Cohomologie galoisienne*, Lecture Notes in Mathematics **5**, Springer-Verlag, Berlin-New York (1965).

[4] G. Shimura, *Moduli and fibre systems of abelian varieties*, Ann. Math. **83** (1966) 294–338.

[5] G. Shimura, *Introduction to the arithmetic theory of automorphic functions*, Princeton Univ. Press, Princeton (1971).

# A review of non-archimedean elliptic functions

JOHN TATE

This expository article consists of two parts. The first is an old manuscript dating from 1959, entitled "Rational points on elliptic curves over complete fields" containing my first proof of the isomorphism $k^*/t^{\mathbb{Z}} \simeq E_t(k)$. The second part is a discussion of some further aspects of the theory. It begins with a sketch of some topics I had hoped to add to the old manuscript before publishing it, namely, the description of the rational functions on $E_t$ as "rigid analytic" meromorphic functions on $k^*$ with multiplicative period $t$, the construction of these via theta-functions, and the classification of isogenies between the $E_t$'s. Then, after a discussion of some consequences of the isogeny classification, there is a description of the kernel $E_t[m]$ of multiplication by $m$ on $E_t$ as finite flat group scheme, and an indication of its relevance to the main theme of this conference. Finally, curves $E_t$ over more general base rings than local fields are discussed, in particular, the "universal curve" $E_q$ over $\mathbb{Z}[[q]][q^{-1}]$, and the connection with moduli. First, here is the old manuscript.

---

**Rational Points on Elliptic Curves Over Complete Fields.**

Let $k$ be a field complete with respect to a non-trivial real valued valuation of the type satisfying

$$|x| \geq 0, \quad |xy| = |x|\,|y|, \quad |x+y| \leq |x| + |y|.$$

According to a theorem of Ostrowski, $k$ is either the real or complex field, with a valuation equivalent to the ordinary one, or else $k$ is 'non-archimedean' in the sense that the valuation satisfies the stronger condition

$$|x+y| \leq \max\{|x|, |y|\}.$$

In the 'classical' case when $k$ is the complex field, the group of points on an elliptic curve defined over $k$ can be represented as the quotient of the additive group of $k$ by a discrete subgroup generated by two independent 'periods' $\omega_1$ and $\omega_2$. Passing from the additive group to the multiplicative group by means of the exponential function one can absorb one of these periods and obtain a representation of the group of points as the quotient of the multiplicative group of $k$ by a discrete subgroup generated by one multiplicative period $t = e^{2\pi i \tau}$, where $\tau = \omega_2/\omega_1$. The explicit formulas giving this multiplicative representation are the well-known Fourier expansions of the Weierstrass functions $\wp, g_2, g_3$, etc. These Fourier expansions, suitably normalized, yield 'universal' identities among power series with rational integral coefficients. Our aim in this paper is to show that these identities can be used to obtain an exactly parallel 'multiplicative' representation (Theorem 1) for the group of rational points on certain elliptic curves over an arbitrary complete field $k$. Unfortunately when $k$ is non-archimedean this method works only for curves whose absolute invariant $j$ satisfies $|j| > 1$. In a second paper we will consider the case $|j| \leq 1$, which is of quite a different nature.[1]

Let $t$ be an element of $k$ such that $0 < |t| < 1$ and consider the series

$$(1) \qquad x(w) = \sum_{m=-\infty}^{\infty} \frac{t^m w}{(1-t^m w)^2} - 2\sum_{m=1}^{\infty} \frac{t^m}{(1-t^m)^2},$$

where $w$ is a non-zero variable in $k$. Using the identity

$$\frac{w}{(1-w)^2} = \frac{1}{w+w^{-1}-2} = \frac{w^{-1}}{(1-w^{-1})^2}$$

we can rewrite our series in the form

$$(2) \quad x(w) = \frac{w}{(1-w)^2} + \sum_{m=1}^{\infty}\left(\frac{t^m w}{(1-t^m w)^2} + \frac{t^m w^{-1}}{(1-t^m w^{-1})^2} - 2\frac{t^m}{(1-t^m)^2}\right)$$

which shows, by comparison with the geometric series $\sum_1^\infty t^m$, that the convergence is absolute for all $w \in k^*$ and is uniform for $w$ in a subset of the form $r_1 \leq |w| \leq r_2$ and $|w - t^m| \geq \varepsilon$ for all $m \in \mathbb{Z}$, with $0 < r_1 < r_2$ and $\varepsilon > 0$. The functional equations

$$(3) \qquad x(tw) = x(w) = x(w^{-1})$$

are now obvious from (1) and (2) respectively. In the restricted range $|t| < |w| < |t|^{-1}$ we have $|t^m w| < 1$ and $|t^m w^{-1}| < 1$ for all positive integers $m$ and can therefore expand the fractions under the summation sign in (2), obtaining

$$(4)$$
$$x(w) = \frac{w}{(1-w)^2} + \sum_{m=1}^{\infty}\sum_{n=1}^{\infty}(nt^{mn}w^n + nt^{mn}w^{-n} - 2nt^{mn})$$
$$= \frac{1}{w+w^{-1}-2} + \sum_{n=1}^{\infty} \frac{nt^n}{1-t^n}(w^n + w^{-n} - 2), \quad \text{for } |t| < |w| < |t|^{-1}.$$

In case $k$ is the complex field the classical Fourier expansions to which we alluded in the first paragraph can be written in the form

$$(5) \qquad \wp(u) = x(w) + \frac{1}{12} \qquad (w = e^u)$$

and

$$(6) \qquad \begin{cases} g_2 = \dfrac{1}{12} + 20\sum_{n=1}^{\infty} \dfrac{n^3 t^n}{1-t^n} \\ g_3 = -\dfrac{1}{216} + \dfrac{7}{3}\sum_{n=1}^{\infty} \dfrac{n^5 t^n}{1-t^n}, \end{cases}$$

---

[1] This "second paper," never written, was to have discussed kernel of the reduction map via the formal group, especially in the ordinary case when that group is of height 1, so twisted multiplicative.

where $\wp(u), g_2$ and $g_3$ are the Weierstrass functions belonging to the periods $\omega_1 = 2\pi i$ and $\omega_2 = \log t$ (any value). A good reference for this classical theory is [4], Abschnitt II; formulas (5) and (6) are proved there in 2, §12. However, we shall pause here to give a direct proof for the sake of completeness. The function $x(w)$ is obviously meromorphic in the domain $k^* = k - \{0\}$, with multiplicative period $t$, its only singularities being double poles at the points $w = t^m$, $m \in \mathbb{Z}$. Therefore the function $x(e^u)$ is meromorphic in the whole $u$ plane with additive periods $\omega_1 = 2\pi i$ and $\omega_2 = \log t$, its only singularities being double poles at the period points $m_1\omega_1 + m_2\omega_2$; $m_1, m_2 \in \mathbb{Z}$. The same is true of $\wp(u)$. The expansion of $\wp(u)$ in powers of $u$ is

$$(7) \quad \wp(u) = \frac{1}{u^2} + \frac{1}{20}g_2 u^2 + \frac{1}{28}g_3 u^4 + \cdots .$$

The corresponding expansion of $x(e^u)$ is readily obtained by substituting $w = e^u = 1 + u + \cdots$ in (4), namely

$$(8) \quad \begin{aligned} x(w) &= \frac{1}{u^2 + \frac{1}{12}u^4 + \frac{1}{360}u^6 + \frac{1}{20160}u^8 + \cdots} \\ &\quad + \sum_{n=1}^{\infty} \frac{nt^n}{(1-t^n)}\left(n^2 u^2 + \frac{n^4}{12}u^4 + \cdots\right) \\ &= \frac{1}{u^2} - \frac{1}{12} + \left(\frac{1}{240} + \sum_{n=1}^{\infty}\frac{n^3 t^n}{1-t^n}\right)u^2 \\ &\quad + \left(-\frac{1}{6048} + \frac{1}{12}\sum_{n=1}^{\infty}\frac{n^5 t^n}{1-t^n}\right)u^4 + \cdots . \end{aligned}$$

Since the pole terms cancel, the difference $\wp(u) - x(w)$ is an entire meromorphic function, hence constant. Formulas (5) and (6) now follow by comparison of coefficients in the two expansions (7) and (8).

Differentiating (5) we obtain

$$(9) \quad \begin{aligned} \wp'(u) = w\frac{d}{dw}x(w) &= \sum_{m=-\infty}^{\infty}\left(\frac{t^m w}{(1-t^m w)^2} + 2\frac{(t^m w)^2}{(1-t^m w)^3}\right) \\ &= x(w) + 2y(w) , \end{aligned}$$

where

$$(10) \quad y(w) = \sum_{m=-\infty}^{\infty}\frac{(t^m w)^2}{(1-t^m w)^3} + \sum_{m=1}^{\infty}\frac{t^m}{(1-t^m)^2}$$

$$(11) \quad = \frac{w^2}{(1-w)^3} + \sum_{m=1}^{\infty}\left(\frac{t^{2m}w^2}{(1-t^m w)^3} - \frac{t^m w^{-1}}{(1-t^m w^{-1})^3} + \frac{t^m}{(1-t^m)^2}\right) .$$

Substituting (5) and (9) in the identity

$$(12) \quad \wp'^2 = 4\wp^3 - g_2\wp - g_3$$

we find

$$(13) \quad y^2 + xy = x^3 - b_2 x - b_3 ,$$

where

(14)
$$\begin{cases} b_2 = \dfrac{1}{4}\left(g_2 - \dfrac{1}{12}\right) = 5\sum_{n=1}^{\infty} \dfrac{n^3 t^n}{1-t^n} = 5t + 45t^2 + 140t^3 + \cdots \\ b_3 = \dfrac{1}{4}\left(g_3 + \dfrac{g_2}{12} - \dfrac{1}{432}\right) = \sum_{n=1}^{\infty} \left(\dfrac{7n^5 + 5n^3}{12}\right) \dfrac{t^n}{1-t^n} \\ \phantom{b_3} = t + 23t^2 + 154t^3 + \cdots. \end{cases}$$

The coefficients of these power series for $b_2$ and $b_3$ are integers because

$$7n^5 + 5n^3 = 7n^3(n^2 - 1) + 12n^3 \equiv 12n^3 \pmod{24}.$$

We now abandon the assumption that $k$ is the complex field and see how much of the preceding theory carries over to an arbitrary complete field $k$, as described in the opening paragraph of this paper. Certainly the series (10) has the same convergence properties as (1) and can therefore be used to define a function $y(w)$ for non-zero $w \in k$. Trivial rearrangements of the defining series show that $y$ satisfies the functional equations

(15) $\qquad y(tw) = y(w) \quad \text{and} \quad y(w^{-1}) + y(w) = -x(w).$

Moreover we can use the power series (14) to define elements $b_2$ and $b_3$ in $k$, no matter what its characteristic, because the coefficients are rational integers; the convergence for $|t| < 1$ in the non-archimedean case is obvious because the coefficients have absolute value $\leq 1$. Thus it is natural to conjecture that for any complete field $k$ and any element $t \in k$ with $0 < |t| < 1$ the equation (13) defines an elliptic curve, and the map $w \to (x(w), y(w))$ gives a parametrization of the points on this curve by the transcendental variable $w \in k^*$.

Let us denote by $A$ the plane cubic curve which is defined over $k$ by equation (13). To show that $A$ is non-singular of genus 1 (rather than singular of genus 0) we must show that the discriminant $\Delta$ does not vanish. The expression for $\Delta$ as polynomial in $b_2$ and $b_3$, and hence as power series in $t$ can be obtained by way of the classical formula in terms of $g_2$ and $g_3$:

$$\Delta = g_2^3 - 27g_3^2 = \left(4b_2 + \dfrac{1}{12}\right)^3 - 27\left(4b_3 - \dfrac{1}{3}b_2 - \dfrac{1}{216}\right)^2$$

(16) $\qquad = b_3 + b_2^2 + 72b_2 b_3 - 432b_3^2 + 64b_2^3$

(17) $\qquad = t - 24t^2 + 252t^3 + \cdots.$

In the classical case, hence also in the real case, we know $\Delta \neq 0$. In the non-archimedean case we see that $\Delta \neq 0$ because $\Delta \equiv t \pmod{t^2}$. Although we shall not require it, we mention here that the classical expansion of $\Delta$ as infinite product,

(17') $\qquad \Delta = t \prod_{n=1}^{\infty} (1 - t^n)^{24}$

holds for all $t$ with $|t| < 1$ in any complete field, because the validity of this formula in the classical case ensures that it is a formal identity in the power series ring $\mathbb{Z}[\![t]\!]$, resulting from the substitution of (14) in (16). From (17') the non-vanishing of $\Delta$ follows without consideration of the different cases. At any rate, our curve $A$ is elliptic, with the absolute invariant

$$
\begin{aligned}
(18) \quad j &= \frac{(12g_2)^3}{\Delta} = \frac{(1+48b_2)^3}{\Delta} = \frac{1 + 240t + 2160t^2 + \cdots}{t - 24t^2 + 252t^3 + \cdots} \\
&= \frac{1}{t}(1 + 744t + 196884t^2 + \cdots),
\end{aligned}
$$

just as in the classical case.

In the projective plane our curve $A$ is complete and non-singular with just one point at infinity, where $x/y = 0$. We shall designate this infinite point by $0$ and shall from now on view $A$ as an abelian variety in the canonical way, with $0$ as origin. Thus the addition of points on $A$ is determined uniquely by the fact that the map which attaches to each point $P \in A$ the linear equivalence class of the divisor $(P) - (0)$ is an isomorphism between the group of points on $A$ and the group of divisor classes of degree $0$ on $A$; that this map is a bijection follows from the Riemann-Roch theorem which assures us that each divisor class of degree $1$ on $A$ contains one and only one point. For three points $P_i$ on $A$ we have $P_1 + P_2 + P_3 = 0$ if and only if the divisors $(P_1) + (P_2) + (P_3)$ and $3(0)$ are linearly equivalent. But $3(0)$ is the intersection divisor of $A$ with the infinite line; hence our condition is simply that $(P_1) + (P_2) + (P_3)$ be the intersection divisor of $A$ with some line in the projective plane.

Let $P_1 = (x_1, y_1)$ and $P_2 = (x_2, y_2)$ be points $\neq 0$ on $A$ and consider the line $L$ joining $P_1$ and $P_2$ (or tangent to $A$ at $P_1 = P_2$ if the points coincide). $L$ is parallel to the $y$-axis in the $(x, y)$ plane if and only if

$$
(19) \quad x_1 = x_2 \quad \text{and} \quad y_1 + y_2 = -x_1.
$$

Thus (19) is necessary and sufficient for $L$ to pass through $0$, or, equivalently, for $P_1 + P_2 = 0$. If $P_1 + P_2 \neq 0$, then $L$ has an equation of the form

$$
(20) \quad y = \lambda x + \nu,
$$

where

$$
(21) \quad \begin{aligned} \lambda &= \frac{y_1 - y_2}{x_1 - x_2} = \frac{x_1^2 + x_1 x_2 + x_2^2 - b_2 - y_2}{y_1 + y_2 + x_1} \\ \nu &= y_1 - \lambda x_1 = y_2 - \lambda x_2. \end{aligned}
$$

(Notice that the two expressions for $\lambda$ are identically equal when both are defined, and that at least one of them is defined since we are assuming (19) is false; in fact the second expression is needed only when $P_1 = P_2$, in which case it reduces to $dy/dx$ at $P = P_1$.) Let $(x_3, \lambda x_3 + \nu)$ be the third intersection of $L$ with $A$. Then $x_3$ is the third root of the cubic equation

$$
(22) \quad x^3 - (\lambda^2 + \lambda)x^2 - \big(\nu(2\lambda + 1) + b_2\big)x - (b_3 + \nu^2) = 0
$$

which results from substituting (20) in (13). Hence $x_1 + x_2 + x_3 = \lambda^2 + \lambda$. Taking the negative of the third intersection using (19) we find that the sum

$P_1 + P_2 = P_3$ has the coordinates

$$\begin{aligned}(23)\qquad x_3 &= \lambda^2 + \lambda - x_1 - x_2 \\ y_3 &= -x_3 - \lambda x_3 - \nu\ ,\end{aligned}$$

where $\lambda = \lambda(x_1, x_2, y_1, y_2)$ and $\nu = \nu(x_1, x_2, y_1, y_2)$ are given by (21).

Let $t^{\mathbb{Z}} = \{t^m \mid m \in \mathbb{Z}\}$ denote the infinite cyclic discrete subgroup of $k^*$ which is generated by our element $t$, and let $\varphi$ be the map of $k^*$ into the projective plane which is defined by

$$\begin{aligned}(24)\qquad \varphi(w) &= \bigl(x(w), y(w)\bigr)\ , &&\text{if } w \notin t^{\mathbb{Z}} \\ \varphi(w) &= 0\ , &&\text{if } w \in t^{\mathbb{Z}}\ .\end{aligned}$$

Let $A_k$ denote the group of points on $A$ which are rational over $k$.

**Theorem 1.** *The map $\varphi$ is a homomorphism of $k^*$ onto $A_k$ with kernel $t^{\mathbb{Z}}$.*

We prove first that $\varphi$ maps $k^*$ into $A$. Since $0 \in A$, this amounts to proving that equation (13) is an identity between the functions $x(w)$ and $y(w)$, holding for all values of the argument $w \in k^*$, $w \notin t^{\mathbb{Z}}$, for which the functions are defined. Since these functions have multiplicative period $t$, it is enough to consider values of $w$ such that $|t| < |w| \leq 1$ and $w \neq 1$. In this range we can use formula (4) which expresses $x$ as a power series in $t$ with coefficients which are rational functions of $w$. There is an analogous expression for $y$ which is obtained by expanding the fractions under the summation sign in (11) and which we don't bother to write out. Thus our contention will be proved if we can show that (13) is a *formal* identity when we interpret $x$ and $y$ (and $b_2$ and $b_3$) as *formal* power series in $t$ with coefficients which are rational functions of an *indeterminate* $w$. Now in fact the coefficients of the formal power series in question are expressed as elements of the ring $\mathbb{Z}[w, w^{-1}, (1-w)^{-1}]$, *i.e.*, the subring of $\mathbb{Q}(w)$ generated by $1, w, w^{-1}$ and $(1-w)^{-1}$. The canonical homomorphism $\mathbb{Z} \to k$ extends to a homomorphism $\mathbb{Z}[w, w^{-1}, (1-w)^{-1}] \to k(w)$. Hence the formal identity we are trying to establish is a 'universal' one, and will hold in any characteristic provided it holds in characteristic 0. From the classical theory we know that our equation (13) holds true *numerically* if we substitute any pair of complex numbers $w$ and $t$ in the domain of convergence $|t| < |w| < |t|^{-1}$, $w \neq 1$. Fixing first $w$ such that $|w| < 1$ and letting $t$ vary we conclude that the resulting power series in $t$ with complex coefficients are equal coefficient-wise; then letting $w$ vary, we conclude that the coefficients are formally equal as rational functions of an indeterminate, as was to be shown.

Next we prove that $\varphi$ is a homomorphism. Given $w_1, w_2 \in k^*$, we put $w_3 = w_1 w_2$ and must prove

$$(25)\qquad P_3 = P_1 + P_2\ , \qquad P_i = \varphi(w_i)\ ,\quad i = 1, 2, 3\ .$$

In view of the periodicity $\varphi(tw) = \varphi(w)$ we can restrict our consideration to values of $w_1$ and $w_2$ in the ranges $|t| < |w_1| \leq 1$ and $1 \leq |w_2| < |t|^{-1}$. Then $|t| < |w_3| < |t|^{-1}$, so that all three $w_i$ are within the domain of convergence of the power series expressions for $x$ and $y$ considered in the previous paragraph. Since $\varphi(1) = 0$ by definition, (25) holds trivially if $w_1 = 1$ or $w_2 = 1$. From (3),

(15), and (19) we see that it also holds if $w_1 w_2 = 1$. Thus we may assume all three points $P_i$ are different from 0 and write $P_i = (x_i, y_i)$, i.e., put $x_i = x(w_i)$, $y_i = y(w_i)$ for $i = 1, 2, 3$. Suppose $x_1 \neq x_2$. Then, using the first expression for $\lambda$ in (21), substituting (21) in (23), and clearing denominators we see that (25) is equivalent to the simultaneous identities

(25') $\begin{aligned} (x_1 - x_2)^2 x_3 &= (y_1 - y_2)^2 + (y_1 - y_2)(x_1 - x_2) - (x_1 - x_2)^2(x_1 + x_2) \\ (x_1 - x_2) y_3 &= -(x_1 - x_2)(y_1 + x_3) + (y_1 - y_2)(x_1 - x_3) . \end{aligned}$

Now we can argue just as in the preceding paragraph: (25') holds in the classical case (being in fact just the addition formulas for $\wp(u)$ and $\wp'(u)$). Hence (25') is an identity in the ring of formal power series in $t$ with coefficients in the ring

$$\mathbb{Z}\left[w_1, w_1^{-1}, w_2, w_2^{-1}, (1-w_1)^{-1}, (1-w_2)^{-1}, (1-w_1 w_2)^{-1}\right],$$

and is therefore a functional identity in any complete field $k$. We could take care of the remaining case $x_1 = x_2$ by other explicit formulas, or by a continuity argument, but perhaps the simplest way out is to observe that $x_1 = x_2$ if and only if $P_1 = \pm P_2$ and to appeal to

**Lemma 1.** *Let $\varphi$ be a map of a (multiplicative) group into an (additive) group which takes on an infinite number of distinct values and satisfies the identity $\varphi(w_1 w_2) = \varphi(w_1) + \varphi(w_2)$ whenever $\varphi(w_1) \neq \pm \varphi(w_2)$. Then $\varphi$ is a homomorphism.*

Indeed, given $w_1, w_2$ we can select $w$ so that

$$\varphi(w) \neq \pm \varphi(w_1) \quad , \quad \varphi(w) + \varphi(w_1) \neq \pm \varphi(w_2) \quad , \quad \varphi(w) \neq \pm \varphi(w_1 w_2)$$

and have then $\varphi(w w_1) = \varphi(w) + \varphi(w_1)$. Then

$$\varphi(w) + \varphi(w_1) + \varphi(w_2) = \varphi(w w_1) + \varphi(w_2) = \varphi(w w_1 w_2) = \varphi(w) + \varphi(w_1 w_2)$$

and the lemma follows upon cancelling $\varphi(w)$. It is obvious that our $\varphi$ takes an infinity of values; for example (2) shows that $|x(1+t^r)| = |t|^{-2r}$ in the non-archimedean case.

Thus $\varphi$ is a homomorphism. That its kernel is $t^{\mathbb{Z}}$ is apparent from its very definition (24), and to complete the proof of Theorem 1 we have only to show that $\varphi$ is surjective. This is true in the classical case, and the case in which $k$ is the real field can be obtained as a corollary by means of the following lemma, which will also be of assistance in the non-archimedean case.

**Lemma 2.** *If $w \neq 0$ is separably algebraic over $k$ and $\varphi(w) \in A_k$, then $w \in k$.*

Let $K$ be a finite Galois extension of $k$ containing $w$. Since $k$ is complete, there is a unique extension of its valuation to $K$, and $K$ is complete in the extended valuation. Thus we can consider the resulting map $\varphi : K^* \to A_K$. According to what we have already proved (applied to $K$ instead of $k$), $\varphi$ is a homomorphism with kernel $t^{\mathbb{Z}}$. Each automorphism $\sigma$ of $K$ over $k$ preserves the extended valuation and therefore commutes with $\varphi$. Thus if $\varphi(w) \in A_k$, then $\varphi(w^\sigma) = \varphi(w)$, and $w^\sigma/w$ is a power of $t$. But $w^\sigma$ and $w$ have the same absolute value so this power of $t$ must be 1. Thus $w^\sigma = w$ for all $\sigma$ and $w \in k$ as contended.

Before treating the case of non-archimedean $k$ we first recall some facts about power series in such a field. Any non-zero power series can be written in the canonical form

$$(26) \qquad f(z) = \alpha z^\nu (1 + a_1 z + a_2 z^2 + \cdots) \text{ with } \alpha \neq 0.$$

We define the *norm* of $f$ by

$$(27) \qquad \|f\| = \sup_{1 \leq n < \infty} \{|a_n|^{1/n}\}.$$

Notice that this is a sup rather than a lim sup; $\|f\|$ is the smallest real number such that

$$(28) \qquad |a_n| \leq \|f\|^n \text{ for all } n = 1, 2, 3, \ldots,$$

or is $+\infty$ if no such number exists. We call the open circle $|z| < \|f\|^{-1}$ the *inner* circle of convergence of $f$. Of course $f$ may converge in some larger circle, but if so it does so only hesitantly. In the inner circle the convergence is direct and business-like right from the start, for we have $|a_n z^n| < (\|f\| |z|)^n$ for $n = 1, 2, 3, \ldots$ there, and $\|f\| |z| < 1$. This shows also that we have $|f(z) - \alpha z^\nu| < |\alpha z^\nu|$ in the inner circle, because the valuation is non-archimedean.

**Proposition 1.** *The non-zero power series with norm less than or equal to a fixed real number $\rho \geq 0$ form a group under multiplication.*

Let $f(z)$ be as in (26) and $g(z) = \beta z^\mu (1 + b_1 z + b_2 z^2 + \cdots)$. Their quotient is

$$\frac{f(z)}{g(z)} = \frac{\alpha}{\beta} z^{\nu - \mu} (1 + c_1 z + c_2 z^2 + \cdots)$$

where the $c_n$ are determined recursively from the equation

$$a_n = b_n + b_{n-1} c_1 + b_{n-2} c_2 + \cdots + b_1 c_{n-1} + c_n.$$

From this equation we can prove inductively that $|c_n| \leq \rho^n$ if we assume $|a_n| \leq \rho^n$ and $|b_n| \leq \rho^n$ for $n = 1, 2, 3, \ldots$. Thus $\|f\| \leq \rho$ and $\|g\| \leq \rho$ implies $\|f/g\| \leq \rho$. This proves Proposition 1.

**Proposition 2.** *The power series of the form*

$$(29) \qquad f(z) = z + a_1 z^2 + a_2 z^3 + \cdots$$

*with norm less than or equal to a fixed real number $\rho \geq 0$ form a group with respect to the operation of composition $(f \circ g)(z) = f((g(z))$.*

Let $f(z)$ be given by (29), let $g(z) = z + b_1 z^2 + b_2 z^3 + \cdots$ be another series of the same form, and let $h = f \circ \overset{-1}{g}$ be the series such that $f(z) = h(g(z))$. The coefficients $c_i$ of $h(z) = z + c_1 z^2 + c_2 z^3 + \cdots$ are determined uniquely in terms of the $a$'s and $b$'s by comparing coefficients in the identity

$$z + a_1 z^2 + a_2 z^3 + \cdots = g(z) + c_1 \big(g(z)\big)^2 + c_2 \big(g(z)\big)^3 + \cdots .$$

Putting $g(z)^i = z^i + b_1^{(i)} z^{i+1} + b_2^{(i)} z^{i+2} + \cdots$ we can write the recursive equations for the $c$'s in the form

$$a_n = b_n + c_1 b_{n-1}^{(2)} + c_2 b_{n-2}^{(3)} + \cdots + c_{n-1} b^{(n)} + c_n$$

for $n = 1, 2, \ldots$. Assume now $\|f\| \leq \rho$ and $\|g\| \leq \rho$. By Proposition 1 we have $\|g^i\| \leq \rho$ for all $i$. Thus $|a_n| \leq \rho^n$, $|b_n| \leq \rho^n$ and $|b_n^{(i)}| \leq \rho^n$ for all $n \geq 1$, $i \geq 2$, and it follows by induction that $|c_n| \leq \rho^n$ for all $n$. Hence $\|h\| = \|f \circ \overset{-1}{g}\| \leq \rho$, and the proposition is proved.

In particular if our power series (29) has norm $\leq \rho$ then so does its inverse series

(30) $$\overset{-1}{f}(z) = z - a_1 z^2 - (a_2 - 2a_1^2) z^3 - (a_3 - 5a_2 a_1 + 5a_1^3) z^4 + \cdots ,$$

and since $f$ is the inverse of its inverse it follows that $f$ and $\overset{-1}{f}$ have the same norm and hence the same inner circle of convergence. Each maps that circle into itself, and the identities $z = \overset{-1}{f}(f(z)) = f(\overset{-1}{f}(z))$ are satisfied for all $z$ in the circle because the absolute convergence ensures the validity of the rearrangements of the series necessary to prove them. Hence:

**Corollary 1.** *A power series of the form (29) maps its inner circle of convergence $|z| < \|f\|^{-1}$ bijectively onto itself, the inverse mapping being given by the formal inverse series (30).*

Consider a power series of the form

(31) $$g(z) = \frac{1}{z} + a_1 + a_2 z + a_3 z^2 + \cdots .$$

Its reciprocal $1/g(z)$ is of the form (29) and has the same norm. Applying Corollary 1 to this reciprocal we obtain

**Corollary 2.** *A power series of the form (31) maps its inner circle of convergence $|z| < \|g\|^{-1}$ ($z = 0$ excluded) bijectively onto the domain $|z| > \|g\|$.*

These general facts about inversion of power series in a complete non-archimedean field will enable us to prove

**Lemma 3.** *If $k$ is non-archimedean then for each $x \in k$ there exists an element $w$ in a quadratic extension of $k$ such that $x = x(w)$. Moreover $w$ is separable over $k$ except possibly if $x = 0$ and the characteristic of $k$ is 2.*

The surjectivity of $\varphi : k^* \to A_k$ follows directly from Lemmas 2 and 3. Given a point $P = (x, y)$ in $A_k$ we choose $w$ as in Lemma 3 such that $x = x(w)$.

Then $\varphi(w)$ and $\varphi(w^{-1}) = -\varphi(w)$ are points on $A$ with the same $x$-coordinate as $P$. Since they are opposites, they are the only such points, so one of them is our given point $P$. By Lemma 2 we have $w \in k$ if $w$ is separable. Thus every point of $A_k$ is the image of some $w \in k^*$, except possibly the point $(0, \sqrt{b_3})$ in characteristic 2. But the image of $k^*$ under $\varphi$ is a subgroup of $A_k$ consisting of more than one element and its complement, which is a union of cosets, cannot possibly consist of one point. Thus the exception does not occur.

To prove Lemma 3 we treat the cases $|x| > |t|^{1/2}$ and $|x| < 1$ by different but analogous methods. In the first case we use the new variable

$$(32) \qquad r = w + w^{-1} - 2 .$$

Each fixed value $r \in k$ determines two reciprocal values of $w$, namely the roots of the quadratic equation

$$(33) \qquad w^2 - (r+2)w + 1 = 0$$

and these roots are separable over $k$ because $w = w^{-1}$ implies $w = \pm 1 \in k$. For all integers $n \geq 1$ we have

$$(34) \qquad w^n + w^{-n} - 2 = F_n(r) = r^n + c_{n,1} r^{n-1} + \cdots + c_{n,n-1} r$$

where $F_n(r)$ is a monic polynomial of degree $n$ with rational integral coefficients $c_{n,i}$ and without constant term. (For example, this follows from the recurrence relation $F_{n+1} = (z+2)F_n - F_{n-1} + 2r$.) Substituting (34) in (4) and rearranging the resulting series in powers of $r$ we obtain

$$(35) \qquad x(w) = f(r) = \frac{1}{r} + a_1 r + a_2 r^2 + \cdots ,$$

where the coefficients

$$a_n = \frac{n t^n}{1 - t^n} + c_{n+1,1} \frac{(n+1) t^{n+1}}{1 - t^{n+1}} + c_{n+2,2} \frac{(n+2) t^{n+2}}{1 - t^{n+2}} + \cdots$$

are elements of $k$ satisfying $|a_1| = |t|$, and $|a_n| \leq |t|^n$ for all $n \geq 1$. Notice that the numbering of the coefficients in (35) is shifted by one from that in the canonical form (31), so that the norm of $f$, being the supremum of the numbers $|a_n|^{1/(n+1)}$, is $|t|^{1/2}$ rather than $|t|$. Thus, although $f(r)$ converges for $|r| < |t|^{-1}$, the inner circle of convergence is only $|r| < |t|^{-1/2}$. Corollary 2 shows now that given $x \in k$, $|x| > |t|^{1/2}$, there exists $r \in k$, $|r| < |t|^{-1/2}$ such that $x = f(r)$ and consequently $x = x(w)$, where $w$ is a root of (33). For values of $x$ such that $|x| < 1$ we shall expand $x(w)$ in terms of another variable, namely

$$(36) \qquad s = w + tw^{-1} .$$

Each value of $s \in k$ determines two values of $w$, namely the roots of the quadratic equation

$$(37) \qquad w^2 - sw + t = 0 .$$

These roots are separable over $k$ except in the case $s = 0$, characteristic of $k = 2$, and $t \notin (k^*)^2$. For all integers $n \geq 1$ we have

$$(38) \qquad w^n + (tw^{-1})^n = G_n(s,t) = s^n + d_{n,2} t s^{n-2} + d_{n,4} t^2 s^{n-4} + \cdots ,$$

where $G_n(s,t)$ is a polynomial in $s$ and $t$ with rational integral coefficients $d_{n,i}$ of the form indicated. This follows for example from the recursion relation $G_{n+1} = sG_n - tG_{n-1}$ and the initial conditions $G_0 = 2$, $G_1 = s$. From (2) we have

$$
\begin{aligned}
(39) \quad x(w) + 2\sum_{m=1}^{\infty} \frac{t^m}{(1-t^m)^2} &= \sum_{m=0}^{\infty} \left( \frac{t^m}{(1-t^m w)^2} + \frac{t^m(tw^{-1})}{(1-t^m(tw^{-1}))^2} \right) \\
&= \sum_{m=0}^{\infty}\sum_{n=1}^{\infty} nt^{mn}w^n + nt^{mn}(tw^{-1})^n \\
&= \sum_{n=1}^{\infty} \frac{n}{1-t^n}(s^n + d_{n,2}ts^{n-2} + d_{n,4}t^2 s^{n-4} + \cdots) \\
&= a_0 + a_1 s + a_2 s^2 + \cdots,
\end{aligned}
$$

where

$$
(40) \quad a_m = \frac{m}{1-t^m} + d_{m+2,2}\frac{m+2}{1-t^{m+2}}t + d_{m+4,4}\frac{m+4}{1-t^{m+4}}t^2 + \cdots.
$$

In particular,

$$
a_0 = d_{2,2}\frac{2t}{1-t^2} + d_{4,4}\frac{4t^2}{1-t^4} + \cdots
$$

is divisible by $2t$, i.e., $|a_0| \leq |2t|$, and

$$
a_1 = \frac{1}{1-t} + d_{3,2}\frac{3t}{1-t^3} + d_{5,4}\frac{5t^2}{1-t^5} + \cdots
$$

is a unit, i.e., $|a_1| = 1$, and all $a_n$ are integers, i.e., $|a_n| \leq 1$ for all $n \geq 1$. Thus we have

$$
(41) \quad x(w) = 2t\alpha + \beta f(s),
$$

where $|\alpha| \leq 1$, $|\beta| = 1$, and where $f(s) = s + (a_2/a_1)s^2 + \cdots$ is a power series of the form (29) with norm 1. Since the map $z \to 2t\alpha + \beta z$ is a bijection of the circle $|z| < 1$ with itself we conclude from Corollary 1 that, given any $x \in k$ with $|x| < 1$, there exists $s \in k$ with $x = 2t\alpha + \beta f(s)$ and consequently $x = x(w)$ where $w$ is a root of (37). This concludes the proof of Lemma 3 and Theorem 1.

Let us have a closer look at our homomorphism $\varphi : k^* \to A_k$ in the non-archimedean case. Because of the periodicity $\varphi(tw) = \varphi(w)$ we can restrict our attention to values of $w$ such that $|t| < w \leq 1$. For these we have by formulas (2) and (8')

$$
(42) \quad \left| x(w) - \frac{w}{(1-w)^2} \right| < 1 \quad \text{and} \quad \left| y(w) - \frac{w^2}{(1-w)^3} \right| < 1.
$$

Hence

$$
(43) \quad \begin{aligned} &|t| < |w| < 1 \Longrightarrow |x(w)| < 1 \text{ and } |y(w)| < 1 \\ &|w| = 1 \Longrightarrow |x(w)| = \frac{1}{|1-w|^2} \geq 1 \text{ and } |y(w)| = \frac{1}{|1-w|^3} \geq 1. \end{aligned}
$$

Let us denote the ring of integers in $k$ by $R_k = \{z \in k; |z| \leq 1\}$, and the group of units of $R_k$ by $U_k = \{z \in k; |z| = 1\}$. Corresponding to each real number $\rho$, $0 < \rho \leq 1$, we have subgroups of $U_k$ defined and denoted by

(44)
$$U_k[\rho] = \{z \in U_k\,;\, |z-1| \leq \rho\}$$
$$U_k(\rho) = \{z \in U_k\,;\, |z-1| < \rho\}\,.$$

According to (43) the image of $U_k$ under $\varphi$ is the subset of $A_k$ consisting of 0 and the points $P = (x,y)$ such that $|x| \geq 1$ and $|y| \geq 1$. This subset, which we shall denote by $B_k$, is therefore a subgroup of $A_k$, isomorphic to $U_k$. It also follows from (43) that the images under $\varphi$ of the subgroups $U_k[\rho]$ and $U_k(\rho)$ are, respectively,

(45)
$$B_k[\rho] = \{P = (x,y) \in A_k\,;\, |x| \geq \rho^{-2}\}$$
$$B_k(\rho) = \{P = (x,y) \in A_k\,;\, |x| > \rho^{-2}\}\,.$$

The fact that these subsets of $A_k$ are subgroups can of course be proved directly from the algebraic addition formulas, without resort to our transcendental parameter $w$; the necessary argument becomes quite transparent if one uses the new variables $\xi = x/y$ and $\eta = 1/y$ which put 0 at the origin of the $(\xi,\eta)$-plane. Notice that $\xi$ is a uniformizing parameter at 0. $B_k$ is the subset of the $(\xi,\eta)$-plane where $|\xi| \leq 1$ and $|\eta| \leq 1$, and $B_k[\rho]$ is the subset of $B_k$ where $|\xi| \leq \rho$. We don't go into details here.

The quotient group $A_k/B_k$ is isomorphic to $k^*/\varphi^{-1}(B_k) = k^*/\langle t \rangle U_k$, and therefore to the quotient group of all absolute values of elements of $k^*$ (the "value group" of $k$) by the group of powers of $|t|$. Thus, $A_k/B_k$ is finite if and only if the valuation is discrete, in which case $A_k/B_k$ is a cyclic group of order $e$, where $e$ is the ordinal number of $t$ in $k$. In particular, we have $A_k = B_k$ if and only if $t$ is a prime element in $R_k$.

The quotient group $B_k/B_k(1)$ is isomorphic to $U_k/U_k(1)$ and therefore to the multiplicative group of the residue class field $\bar{k} = R_k/P_k$. (Here $P_k = \{z; |z| < 1\}$ is the maximal ideal of $R_k$.) Now the coefficients of the defining equation (9) of our curve $A$ lie in $R_k$; reducing them mod $y_k$ we obtain

(46)
$$\bar{y}^2 + \bar{x}\,\bar{y} = \bar{x}^3$$

as the equation for the "reduced curve" $\bar{A}$, defined over the residue class field. $\bar{A}$ is singular, having an ordinary double point at the origin $\bar{x} = 0$, $\bar{y} = 0$. There is one point at infinity on $\bar{A}$, which we shall denote by $\bar{0}$. The set of *non-singular* points on $\bar{A}$ forms a group with $\bar{0}$ as neutral element, the addition of non-singular points on $\bar{A}$ being defined geometrically in the same way as the addition of points on $A$, by the rule $\bar{P}_1 + \bar{P}_2 + \bar{P}_3 = \bar{0}$ if and only if $(\bar{P}_1) + (\bar{P}_2) + (\bar{P}_3)$ is the intersection of $\bar{A}$ with a line in the projective plane not passing through the singular point $(0,0)$. Being singular and cubic, $\bar{A}$ must be of genus 0. In fact, the map

(47) $\qquad \bar{\varphi}(\bar{w}) = (\bar{w}, \bar{y})\,,\ \text{where}\ \bar{x} = \dfrac{\bar{w}}{(1-\bar{w})^2}\,,\quad \bar{y} = \dfrac{\bar{w}^2}{(1-\bar{w})^3}\,,$

gives a birational transformation of the $\bar{w}$-line onto $A$, the inverse transform being $\bar{w} = \bar{y}^2/\bar{x}^3$. This transformation $\bar{\varphi}$ carries the points $\bar{w} = 0$ and $\bar{w} = \infty$

onto the singular point of $\bar{A}$; for the remaining points it is an isomorphism of the multiplicative group onto the group of non-singular points of $\bar{A}$. Applying the place $k \to \bar{k}$ to coordinates of points we obtain a *reduction map* $\theta : A \to \bar{A}$. Specifically, we have $\theta(P) = \bar{0}$ if and only if $P \in B(1)$, and for the remaining points $P = (x, y)$, $\theta(P) = (\bar{x}, \bar{y})$ where bar denotes the residue class of whatever is under it. Thus $B$ is just the set of points $P$ on $A$ such that $\theta(P)$ is a simple point of $\bar{A}$, and we shall denote the group formed by these simple points by $\bar{B}$. The diagram

(48)
$$\begin{array}{ccc} U_k & \xrightarrow{\varphi} & B_k \\ \downarrow & & \downarrow \theta \\ \bar{k}^* & \xrightarrow{\bar{\varphi}} & \bar{B}_{\bar{k}} \end{array}$$

is commutative, as one sees from (47) and (42). Thus our homomorphism $\varphi$ can be viewed as a transcendental "lifting" of the birational transformation $\bar{\varphi}$, and from this point of view the success of our method becomes more understandable.

---

In the following discussion we will denote the curve (13) of the above manuscript by $E$ or $E_t$ instead of $A$, and its group of $k$-rational points by $E(k)$ instead of $A_k$. We also assume that $k$ is non-archimedean, and let $C$ denote the completion of the algebraic closure of $k$.

### The field of rational functions on E via theta functions.

We will denote by $K = k(x, y)$ the field of rational functions on $E$ which are defined over $k$. In the classical case $k = \mathbb{C}$, $K$ can be identified with the field of meromorphic functions $f$ on $\mathbb{C}^*$ which have multiplicative period $t$, that is, satisfy $f(tw) = f(w)$ for all $w$ in $\mathbb{C}^*$. The same is true for our non-archimedean $k$ if we add "defined over $k$" and define "meromorphic function on $C^*$ defined over $k$" to be an element of the field of fractions of the ring of holomorphic ones, where by "holomorphic function on $C^*$ defined over $k$" we mean a function $f : C^* \to C^*$ which is representable by an everywhere convergent Laurent series with coefficients in $k$. For a non-zero such series, the number of zeros in an annulus $r \leq |w| \leq R$ is finite, and in fact, the exact number of such zeros (multiplicities counted) can be read off the Newton Polygon of the series in the usual way; see e.g., [2], Ch.II. The divisor of such a function is a collection of points $c_i$ in $C^*$ with multiplicities, such that each annulus contains only a finite number of $c_i$, and which is "defined over $k$" in the sense that the $c_i$ are algebraic over $k$, and points conjugate over $k$ have the same multiplicity, which is a multiple of the degree of inseparability of the point over $k$. Every such divisor is the divisor of a holomorphic function. In contrast to the classical case, the only holomorphic functions on $k^*$ with no zeros in $\bar{k}$ are the monomials $aw^n$, $a \in k^*$, and in the Weierstrass product

$$\prod_{i, |c_i| \geq 1} \left(1 - \frac{w}{c_i}\right) \prod_{i, |c_i| < 1} \left(1 - \frac{c_i}{w}\right)$$

showing the existence of a function with a given divisor $\{c_i\}$, no convergence factors are needed.

By an elliptic function (with period $t$) defined over $k$ we mean a function meromorphic on $C^*$ which is defined over $k$ and invariant under $w \mapsto tw$. The above considerations show that the divisor of such a function $f(w)$ is the difference of two disjoint divisors with multiplicities $> 0$, each of which is invariant by $w \mapsto tw$ and defined over $k$. Consequently, $f = g/h$, where $g$ and $h$ are holomorphic functions on $C^*$ defined over $k$, which satisfy a functional equation of the form $g(w) = aw^n g(tw)$. The monomial $aw^n$ is the same for $g$ and for $h$, since $f(w) = f(tw)$. Such functions are called theta functions of type $aw^n$, and $n$ is called their degree. The dimension of the space of theta functions of a given type $aw^n$ is $n$ for $n > 0$, and is 0 for $n < 0$ (for $n = 0$ it is 1 if $a = t^m$ for some $m$ in $\mathbb{Z}$, in which case the thetas are the monomials of degree $-m$, and is 0 otherwise, as is easily checked). Indeed, for $n \neq 0$, the space of *formal* Laurent series $f$ satisfying $f(w) = aw^n f(tw)$ is of dimension $|n|$, because $|n|$ successive coefficients of the series can be arbitrarily prescribed, and the remaining ones are then determined by the relation $f(w) = aw^n f(tw)$. The resulting series converge everywhere if $n > 0$, but diverge if $n < 0$. The Newton Polygon of such a series is invariant under the map $(x, y) \mapsto (x + n, y - \log|a| - x \log|t|)$. From this it follows easily that a theta function $g(w)$ of degree $n > 0$ has exactly $n$ zeros in an annulus of the form $r|t| < |w| \leq r$. Hence, an elliptic function $f$ has the same number of zeros as poles in such a "period annulus", and if that number is $n$, then $f$ is the quotient of two theta functions of degree $n$ of the same type. For $n = 0$ or $1$, the ratio of two such thetas is constant. Hence, an elliptic function with no pole is constant, and there is no elliptic function with just one simple pole in a period annulus.

Now one can proceed as in the classical Weierstrass theory. The role of the Weierstrass sigma function is played by a theta function of type $-w$, for example, the function

$$\theta(w) = \sum_{n \in \mathbb{Z}} (-1)^n t^{\frac{n^2-n}{2}} w^n .$$

The zeros of $\theta(w)$ are the points $w = t^n$, $n \in \mathbb{Z}$, each with multiplicity one. This follows from consideration of the Newton Polygon as above, together with the fact that $\theta(1) = 0$. It also follows from the classical identity

$$\theta(w) = (1-w) \prod_{m=1}^{\infty} [(1-t^m)(1-t^m w)(1-t^m w^{-1})]$$

Using that identity it is easy to show that $\theta(w)^2 x(w)$ and $\theta(w)^3 y(w)$ are holomorphic, and consequently, $x(w)$ and $y(w)$ are elliptic functions in the above sense. In fact, the differential operators $D := w d/dw$ and $D_2 := \frac{1}{2}(D^2 - D)$ act on the ring of holomorphic functions, and, putting

$$\zeta(w) = -\frac{D\theta(w)}{\theta(w)}$$

we have $\zeta(tw) = 1 + \zeta(w)$, and

$$x(w) = D\zeta(w) \quad , \quad y(w) = D_2 \zeta(w) .$$

Thus $Dx = 2y + x$, or, in terms of differentials,
$$dx/(2y+x) = dw/w \ .$$

Further details (e.g., the fact that elliptic functions are rational functions of $x$ and $y$, the expression of elliptic functions in terms of $\theta(w)$, etc.) are left to the reader; references are [9] and [13]. Everything is as in the classical case, except perhaps for some questions of rationality, like the surjectivity of the homomorphism $\varphi : k^* \to E(k)$. A shorter proof of that than the one in the manuscript above is as follows. Let $P = (x_0, y_0)$ be a point on $E(k)$. The function $x - x_0$ has zeros only at $P$ and $-P$. If $P$ is not in the image of $\varphi$, then the function $w \mapsto x(w) - x_0$ has no zero in $k^*$. That function has a pole at $w = 1$, hence it has a zero at some point $w_0$ in $C^*$. Then $\varphi(w_0) = P$ or $-P$, and, replacing $w_0$ by $1/w_0$ if necessary, we can assume $\varphi(w_0) = P$, i.e., that $w = w_0$ is a common zero of $x(w) - x_0$ and of $y(w) - y_0$. These two functions can't have two common zeros in a period annulus — otherwise the function $y(w)/x(w)$ would have exactly one pole there, which is impossible. Thus the divisor consisting of the points $w_0 t^n$, $n \in \mathbb{Z}$, with multiplicity one, is rational over $k$, i.e., $w_0$ is in $k^*$, as was to be shown.

At one point later on it will be convenient to consider the group of points of our curve $E_t$ not only with coordinates in $k$ or $C$, but also in a finite commutative $k$-algebra $A$. Any such algebra is a product of local ones, so it suffices to consider that case. We leave to the reader the exercise of showing that for a local $A$, our functions $x$ and $y$, or perhaps better, the theta functions of type $-w^3$, give a surjective homomorphism
$$\varphi : A^* \to E_t(A)$$
with kernel $t^{\mathbb{Z}}$, so induce an isomorphism
$$A^*/t^{\mathbb{Z}} \simeq E(A) \ ,$$
which is of course functorial in $A$. One can argue by induction on the length of $A$. If $I$ is an ideal of square zero and $B = A/I$, the induction step from $B$ to $A$ can be made with the diagram

$$\begin{array}{ccccccccc}
0 & \to & \mathrm{Lie}(\mathbb{G}_m) \otimes I & \to & A^*/t^{\mathbb{Z}} & \to & B^*/t^{\mathbb{Z}} & \to & 0 \\
& & \downarrow {\scriptstyle \mathrm{Lie}(\varphi) \otimes \mathrm{id}} & & \downarrow \varphi & & \downarrow \varphi & & \\
0 & \to & \mathrm{Lie}(E_t) \otimes I & \to & E_t(A) & \to & E_t(B) & \to & 0
\end{array}$$

For non local $A$ it follows that $E(A) = A^*/t^{\mathbb{Z}(A)}$, where $\mathbb{Z}(A)$ is the group of continuous maps $i : \mathrm{Spec}(A) \to \mathbb{Z}$.

### Isogenies.

Suppose $t_1, t_2 \in k$, with $0 < |t_1|, |t_2| < 1$. We will say $t_1$ and $t_2$ are "commensurable" if there exist non-zero integers $m, n$ such that $t_2^m = t_1^n$. When this is the case the left hand square of the following diagram is commutative

$$\begin{array}{ccccccccc}
0 & \to & \mathbb{Z} & \xrightarrow{1 \mapsto t_1} & C^* & \xrightarrow{\varphi} & E_{t_1}(C) & \to & 0 \\
& & \downarrow m & & \downarrow n & & \downarrow \alpha_{m,n} & & \\
0 & \to & \mathbb{Z} & \xrightarrow[1 \mapsto t_2]{} & C^* & \xrightarrow{\varphi} & E_{t_2}(C) & \to & 0
\end{array}$$

and consequently there is a unique homomorphism

$$\alpha_{m,n} : E_{t_1}(C) \to E_{t_2}(C)$$

making the right hand square commutative. This map $\alpha_{m,n}$ is in fact an isogeny defined over $k$, because composition with $w \mapsto w^n$ takes meromorphic functions on $C^*$ with period $t_2$ into those with period $t_1$ and preserves the field of definition of such functions, and therefore composition with $\alpha_{m,n}$ carries rational functions on $E_{t_2}$ defined over $k$ into those on $E_{t_1}$.

**Theorem.** *The degree of $\alpha_{m,n}$ is $mn$. Every isogeny between curves of type $E_t$ is of the form $\alpha_{m,n}$; the map $(m,n) \mapsto \alpha_{m,n}$ is an isomorphism*

$$\{(m,n) \in \mathbb{Z} \times \mathbb{Z} \mid t_2^m = t_1^n\} \simeq \operatorname{Hom}_k(E_{t_1}, E_{t_2}) = \operatorname{Hom}_C(E_{t_1}, E_{t_2}) \ .$$

*In particular, $E_{t_1}$ is isogenous to $E_{t_2}$ if and only if $t_1$ and $t_2$ are commensurable.*

*Proof.* If $t = t_2^m = t_1^n$, with $m > 0$, hence $n > 0$, then we have a commutative triangle.

$$\begin{array}{ccc} & E_t & \\ {}^{\alpha_{1,n}}\nearrow & & \searrow^{\alpha_{m,1}} \\ E_{t_1} & \xrightarrow{\alpha_{m,n}} & E_{t_2} \end{array}$$

The degree of $\alpha_{m,1}$ is $m$, because the corresponding function field extension is cyclic of degree $m$, with Galois group generated by the automorphism translation by $t_2$ on the field of meromorphic functions with period $t = t_2^m$. In case $m = n$ and $t_2 = t_1$ the map $\alpha_{n,n}$ is multiplication by $n$ on $E_{t_1}$, which shows that $\alpha_{n,1}$ is dual to $\alpha_{1,n}$ and has therefore degree $n$. Hence the degree of $\alpha_{m,n}$ is $mn$, as claimed. (Note that $\alpha_{-m,-n} = -\alpha_{m,n}$.)

To prove the rest we can assume the ground field is $C$, since we have already noted that $\alpha_{m,n}$ is defined over $k$ if $t_1$ and $t_2$ are in $k$. Since $C$ is algebraically closed, every isogeny over $C$ is a product of isogenies of prime degree. Since it is obvious that the composition of maps of the form $\alpha_{m,n}$ is of that same form, we are reduced to proving that an isogeny $\beta : E_t \to E_{t'}$ of prime degree $p$ is of the form $\alpha_{p,1}$, or $\alpha_{1,p}$. The $j$-invariants of our curves are not 0 or 1728 because they satisfy $|j| > 1$. Consequently the only automorphisms of the curves (preserving the group structure) are $\pm 1$, and it follows that an isogeny is determined up to sign by its kernel. Since $-\alpha_{m,n} = \alpha_{-m,-n}$, it suffices to show that the only finite subgroup-schemes of order $p$ of $E_t$ are the kernel of $\alpha_{1,p}$ and the kernels of the various $\alpha_{p,1}$'s. If $p$ is not the characteristic of $C$, then this amounts to the fact that a subgroup of order $p$ of $C^*/t^{\mathbb{Z}}$ is either the group of $p$th roots of 1, or is generated by one of the $p$th roots of $t$. Suppose $p = \operatorname{char}(C)$. Then $\alpha_{1,p} : E_t \to E_{t^p}$ is the Frobenius, and $\alpha_{p,1} : E_t \to E_{t^{1/p}}$ is the Verschiebung. The kernels of these two maps are the only subgroupschemes of order $p$, because their product is the kernel of multiplication by $p$ and they are not isomorphic, being isomorphic to $\mu_p$ and $\mathbb{Z}/p\mathbb{Z}$, respectively. This concludes the proof of the Theorem.

## Applications.

An immediate corollary is the fact that $\text{Hom}(E_t, E_t) = \mathbb{Z}$, i.e., our curves do not have nontrivial endomorphisms. When I first told Serre about the existence of the curves $E_t$, he noted that if the curve is defined over a finite extension of $\mathbb{Q}_p$ the non-existence of non-trivial endomorphisms can be seen from the action of Galois on points of finite order, and that this gives a nice proof that the $j$-invariant of an elliptic curve with complex multiplication is an algebraic integer. Indeed, if $j$ were not an integer at a place above a prime $p$, then there would be a finite extension of $\mathbb{Q}_p$ over which the curve is isomorphic to $E_t$ for $t = j^{-1} + 744j^{-2} + 750420j^{-3} + \cdots$, contradicting the fact that $E_t$ does not have non trivial endomorphisms. Incidentally, a generalization to abelian varieties of CM type is the fact that such a variety has potential good reduction at every place [**12**], Thm. 6(a).

Another corollary of the theorem is that if $\ell : k^* \to V$ is a homomorphism of $k^*$ into a $\mathbb{Q}$-vector space $V$, and if the valuation $v$ satisfies $v(k^*) \subset \mathbb{Q}$, then $\ell(t)/v(t)$ is an isogeny invariant of the curves of type $E_t$. The special case $k = \mathbb{Q}_p$, $V = k$, $\ell : k^* \to k$ is the $p$-adic logarithm $\log_p$, and $v = \text{ord}_p$ is of interest. In that case, if $E_t$ arises by base change from an elliptic curve $E$ over $\mathbb{Q}$, Greenberg and Stevens have proved [**3**], at least for $p > 3$, that

$$L_p'(1) = \frac{\log_p(t)}{\text{ord}_p(t)} \frac{L(1)}{\Omega}$$

where $L_p$ is the $p$-adic and $L$ the ordinary $L$-function of $E$, and $\Omega$ is the real period of $E$. This relation is predicted by the $p$-adic Birch and Swinnerton-Dyer conjecture proposed in [**7**]. The usual Birch and Swinnerton-Dyer conjecture involves the discriminant of a real-valued height pairing on the Mordell-Weil group $E(\mathbb{Q})$. The $p$-adic conjecture involves the discriminant of a $p$-adic height pairing on the so-called extended Mordell-Weil group which is the extension of $E(\mathbb{Q})$ by $\mathbb{Z}$ obtained by taking the inverse image of $E(\mathbb{Q})$ under the map $\varphi : \mathbb{Q}_p^* \to E(\mathbb{Q}_p)$. Call this inverse image $E'(\mathbb{Q})$, and let $E'(\mathbb{Q})_0$ denote its intersection with $\mathbb{Z}_p^*$. Then for $a, b\ E_0'(\mathbb{Q})$ and $c \in E'(\mathbb{Q})$, the $p$-adic height pairing satisfies

$$\langle a, b \rangle = \text{usual } p\text{-adic } \langle (a), (b) \rangle$$
$$\langle c, t \rangle = \log_p(c)/\text{ord}_p(t)$$

In particular, $\langle t, t \rangle$ is the isogeny invariant $\log_p(t)/\text{ord}_p(t)$ occurring in the result of Greenberg and Stevens mentioned above. In case the rank of $E(\mathbb{Q})$ is 0, their result is the $p$-adic Birch and Swinnerton-Dyer conjecture.

For each integer $n > 0$, let $K_n$ be the field of elliptic functions on $k^*$ with period $t^n$, isomorphic to the field of rational functions on $E_{t^n}$ defined over $k$. Let $K_\infty$ be the union of the $K_n$, and let $s$ denote the automorphism of $K_\infty$ given by translation by $t$. This automorphism $s$ is of infinite order and its fixed field is $K_1 = K$. The extension $K_\infty/K$ is Galois with group $\widehat{\mathbb{Z}}$ and the powers of $s$ form a canonically determined dense subgroup. ($K_n/k$, as abstract function field, determines the elliptic curve $E_{t^n}$ over $k$, its $j$-invariant, hence its period $t^n$, and then, up to sign, the parametrization $\varphi : k^* \to E_{t^n}$, and the automorphism $s$.) The finite subextensions of $K_\infty/K$ are the $K_n/K$, and they have the remarkable property to be not only unramified, but split completely at every place of $K$, even though $k$ may be far from algebraically closed.

**Points of finite order.**

For an integer $m > 0$ let $E_t[m]$ denote the kernel of multiplication by $m$ on $E_t$. This is a finite flat group scheme over $k$ whose structure is easy to describe. Let $R$ be an arbitrary ground ring, and let $u$ be an invertible element of $R$. Consider the functor from commutative $R$-algebras $A$ to abelian groups which associates to $A$ the quotient group of the group of pairs $(a,i) \in A^* \times \mathbb{Z}(A)$ such that $a^m = u^i$, modulo the subgroup of pairs of the form $(u^i, im)$. (Here $\mathbb{Z}(A)$ denotes the group of locally constant functions $i : \operatorname{Spec} A \to \mathbb{Z}$.) An element of that quotient is represented by a unique pair $(a,i)$ such that $0 \le i(x) < m$ for $x \in \operatorname{Spec} A$, and consequently the functor is represented by the scheme

$$\coprod_{i=0}^{m-1} \operatorname{Spec}(B_i) = \operatorname{Spec}\left(\prod_{i=0}^{m-1} B_i\right)$$

where $B_i = R[X]/(X^m - u^i)$. The group law on the functor makes this scheme a groupscheme which we denote by $G = G_{R,u,m}$. There is an exact sequence

$$0 \to \mu_m \to G_{R,u,m} \to \mathbb{Z}/m\mathbb{Z} \to 0,$$

the maps being $z \mapsto (z,0)$ and $(a,i) \mapsto i \pmod{m}$. Since $H^1(R, G_m) = 0$ in the flat topology ("Hilbert theorem 90"), every $(\mathbb{Z}/m\mathbb{Z})$-module scheme over $R$ which is an extension of $\mathbb{Z}/m\mathbb{Z}$ by $\mu_m$ is of the form $G_{R,u,m}$ for some $u \in R^*$. The class of the extension is determined by the image of $u$ in $R^*/(R^*)^m = H^1(R, \mu_m)$. It is easy to see that an isomorphism $G_{R,u,m} \to G_{R,v,m}$ which induces identity on $\mu_m$ and on $\mathbb{Z}/m\mathbb{Z}$ is of the form $(a,i) \to (ar^i, i)$, where $r \in R$ and $r^m = v/u$. More generally, for integers $e, f$, a homomorphism $G_{r,u,m} \to G_{r,v,m}$ which induces multiplication by $e$ on $\mu_m$ and by $f$ on $\mathbb{Z}/m\mathbb{Z}$ is of the form

$$(a,i) \mapsto (a^e r^i, fi), \quad \text{where } r \in R \text{ satisfies } u^e r^m = v^f.$$

If $u$ is of infinite order in $A^*$ in the sense that $i \mapsto u^i$ is an isomorphism $\mathbb{Z}(A) \simeq u^{\mathbb{Z}(A)}$, then a pair $(a,i)$ is determined by $a$ alone and

$$G_{R,u,m}(A) = \text{ points of order dividing } m \text{ in the group } A^*/u^{\mathbb{Z}(A)}.$$

Now take $R = k$, $u = t$. The functorial isomorphism $\varphi : A^*/t^{\mathbb{Z}(A)} \simeq E_t(A)$ for finite dimensional $k$-algebras $A$ induces an isomorphism of functors, and hence of finite flat group schemes,

$$G_{k,t,m} \simeq E_t[m].$$

Of course, if $m$ is prime to the characteristic of $k$, then we do not need the isomorphism $A^*/t^{\mathbb{Z}(A)} \simeq E_t(A)$ for all finite $A$, but only for $A$'s which are (products of) fields. Indeed in that case, the group scheme $E_t[m]$ is etale and is determined by the Galois module $E_t[m](\bar{k})$. In any case, there is a canonical exact sequence of finite flat $k$-group schemes, or Galois modules,

$$0 \to \mu_m \to E_t[m] \xrightarrow{a} \mathbb{Z}/m\mathbb{Z} \to 0 \ .$$

The Weil pairing

$$E_t[m] \times E_t[m] \to \mu_m$$

is given by $(P,Q) \mapsto a(P)Q - a(Q)P$, or by the negative of that expression, according to one's convention.

**The case $E_t[m]$ has good reduction.**

Let $R$ be the ring of integers in $k$. The elliptic curve $E_t$ over $k$ does not have good reduction, that is, there is no elliptic curve over $R$ whose general fiber is $E_t$. However, it may happen for some $m$ that $E_t[m]$ has good reduction, i.e., is the general fiber of a finite flat group scheme over $R$. From the above discussion it is clear that this happens if $|t| \in |k^*|^m$; then $t = ua^m$ with $u \in R^*$, $a \in k^*$, and $E_t[m]$ is isomorphic to the general fiber of $G_{R,u,m}$. This would be the situation at each place $\ell$ of bad reduction of the Frey curve corresponding to a counterexample to Fermat for a prime exponent $m$. That fact plays an essential role in the relation between the modularity of curves over $\mathbb{Q}$ and Fermat's last theorem suggested by Frey, made more precise by Serre [11], proved by Ribet [8] and recently exploited by Wiles.

**Serre's isogeny theorem.**

Suppose $p$ is a prime and $k$ is locally compact with residue characteristic $p$. Let $E_{t_1}$ and $E_{t_2}$ be two curves over $k$ of our type, and let $E_{t_i}[p^\infty]$ denote the $p$-divisible group of $E_{t_i}$ over $k$ for $i = 1, 2$. Then the natural map

$$\mathbb{Z}_p \otimes \operatorname{Hom}(E_{t_1}, E_{t_2}) \to \operatorname{Hom}(E_{t_1}[p^\infty], E_{t_2}[p^\infty]) \ ,$$

is an isomorphism. This was proved by Serre [10], A1.4 in case $k$ is of characteristic 0 and used by him to prove [10], 2.3 the global isogeny theorem for elliptic curves over a number field with non-integral $j$-invariant. The global theorem was later proved for all abelian varieties by Faltings, but Serre's result is all that is needed to show for semistable elliptic curves over $\mathbb{Q}$ that the modularity of their Galois representation implies that they are modular in the stronger sense of being covered by a modular curve.

We will prove the local isogeny theorem above by giving an explicit description of the two groups involved as indicated by the following diagram

$$\begin{array}{ccc}
\operatorname{Hom}(E_{t_1}, E_{t_2}) & \longrightarrow & \operatorname{Hom}(E_{t_1}[p^\infty], E_{t_2}[p^\infty]) \\
{\scriptstyle \alpha_{m,n}} \uparrow & \wr \uparrow & \downarrow \wr \qquad \qquad \downarrow h \\
(m,n) & & \qquad \qquad (e,f) \\
\{(m,n) \in \mathbb{Z} \times \mathbb{Z} \mid t_2^m = t_1^n\} & \longrightarrow & \{(e,f) \in \mathbb{Z}_p \times \mathbb{Z}_p \mid t_2^e = t_1^f \in \widehat{k^*}\} \\
& (m,n) \mapsto (m,n) &
\end{array}$$

Here,
$$\widehat{k^*} := \varprojlim_r (k^*/(k^*)^{p^r}) \ .$$

The right hand vertical map is explained by the diagram

$$\begin{array}{ccccccccc} 0 & \longrightarrow & \mu_{p^\infty} & \longrightarrow & E_{t_1}[p^\infty] & \longrightarrow & \mathbb{Q}_p/\mathbb{Z}_p & \longrightarrow & 0 \\ & & \downarrow f & & \downarrow h & & \downarrow e & & \\ 0 & \longrightarrow & \mu_{p^\infty} & \longrightarrow & E_{t_2}[p^\infty] & \longrightarrow & \mathbb{Q}_p/\mathbb{Z}_p & \longrightarrow & 0 \end{array}$$

Here $h$ is an arbitrary homomorphism of the $p$-divisible groups. It must respect the filtrations because there is no non-trivial homomorphism $\mu_{p^\infty} \to \mathbb{Q}_p/\mathbb{Z}_p$. (In characteristic $p$, $\mu_{p^\infty}$ is connected and the connected component of $\mathbb{Q}_p/\mathbb{Z}_p$ is trivial; in characteristic 0 it is clear from the fully faithful representation of the situation by Galois modules, since the Galois action on $\mu_{p^\infty}$ is non-trivial.) The endomorphisms induced by $h$ on the submodule and quotient module are multiplications by $p$-adic integers $f$ and $e$, respectively, because those groups have no other endomorphisms than such multiplications. Thus the restriction of $h$ to the finite group schemes $E_t[p^r] \simeq G_{k,t}, p^r$ is of the type considered above and consequently $t_2^f = t_1^e$ in $k^*/(k^*)^{p^r}$ for all $r$. In other words, the equality $t_2^f = t_1^e$ holds in the multiplicatively written $\mathbb{Z}_p$-module

$$\widehat{k^*} := \varprojlim_r (k^*/(k^*)^{p^r})$$

The map $h \mapsto (e, f)$ is injective because there is no non-trivial homomorphism $\mathbb{Q}_p/\mathbb{Z}_p \to \mu_{p^\infty}$. It is surjective because by the above discussion, for given $(e, f)$ satisfying $t_2^e = t_1^f$ in $\widehat{k^*}$, there is a homomorphism

$$h_j : E_{t_1}[p^j] \longrightarrow E_{t_2}[p^j]$$

for each $j$ inducing $e$ on the quotient module and $f$ on the submodule, and $h_j$ is unique up to the addition of homomorphisms $\mathbb{Z}/p^j\mathbb{Z} \to \mu_{p^i}$. Therefore if $p^c = \#(\mu_{p^\infty}(k))$ is the number of $p$-power roots of 1 in $k$, then all choices of $h_j$ (for a given pair $(e, f)$) induce the same homomorphism on $E_{t_1}[p^{j-c}]$. Hence these latter homomorphisms are coherent and give the desired $h$ which maps to $(e, f)$.

Now the key idea in Serre's proof is that the valuation gives a homomorphism $v : \widehat{k^*} \to \mathbb{Z}_p$ which applied to the condition on $e$ and $f$ gives

$$ev(t_2) = fv(t_1)$$

Since $v(t_1)$ and $v(t_2)$ are non-zero rational integers, it follows that we can multiply the pair $(e, f)$ by a $p$-adic unit so that it becomes a pair of ordinary integers. Then $t_2^e t_1^{-f}$ is in the kernel of the map $k^* \to \widehat{k^*}$ so is a root of unity of order prime to $p$ in $k$. Multiplying the pair $(e, f)$ by that order, we get a pair $(m, n)$ such that and our original pair is a $p$-adic unit times $(m, n)$, which is what we needed to show.

## More general ground rings.

Suppose $R$ is a commutative ring and $I$ an ideal in $R$ such that $R$ is $I$-adically complete and separated. Suppose $t \in I$ is a non-zero-divisor in $R$, and put $k = R[1/t]$. Formulas (14) define elements $b_2$ and $b_3$ in $I$, and equation (13) defines an elliptic curve over $k$ which we will denote by $E_t$. Putting

$$u = \frac{y+x}{y} = 1 + \frac{x}{y} \quad , \quad v = \frac{1}{y}$$

the equation for $E_t$ becomes

$$F(u,v) := uv - (u-1)^3 + b_2(u-1)v^2 + b_3 v^3 = 0 \ .$$

Let $E_t^0(k) = \{P = (u,v) \in E_t(k) \mid u \in R^* \text{ and } v \in R\}$. Then the map $(u,v) \mapsto u$ is a bijection from $E_t^0(k) \to R^*$. The inverse map can be defined by $v = v(u) := \lim(v_n(u))$, where the functions $v_n : R^* \to R$ are defined inductively by

$$v_1(u) = \frac{(u-1)^3}{u} \ , \quad \text{and } v_{n+1}(u) = v_n(u) - F(u,v_n(u))/u \ , \quad \text{for } n > 0 \ .$$

By induction, one proves that $F(u, v_n(u))$ is in $I^n$.

In fact, $E_t^0(k)$ is a subgroup of $E_t(k)$, "analytically" isomorphic to $R^*$ in the following sense. There is a Laurent series $f(u)$ with coefficients in $R$, convergent for all $u \in R^*$, such that $f(u) \equiv u \pmod{I}$, and such that the map $P = (u,v) \mapsto f(u)$ is an isomorphism $E_t^0(k) \to R^*$. In case $R$ is a discrete valuation ring this map is just the inverse of the negative of the map $\varphi : R^* \to E_t^0(k)$, and we can use the same formulas in the present more general setting. Formulas (2) and (11) yield

$$x + y = \frac{w}{(1-w)^3} + g(w)$$
$$y = \frac{w^2}{(1-w)^3} + h(w)$$

where $g(w)$ and $h(w)$ are Laurent series with coefficients in $I$ which are Laurent polynomials modulo $I^n$ for every $n$. Hence

$$u = (x+y)/y = (w^{-1} + g(w)(1-w)^3/w^2)/(1 + h(w)(1-w)^3/w^2)$$
$$= F(w^{-1}) \ , \text{ say,}$$

where $F(z)$ is a Laurent series in $z$ with coefficients in $R$, $\equiv z \pmod{I}$ and polynomial (mod $I^n$) for every $n$. Such an $F$ maps $R^*$ bijectively to $R^*$, and the inverse function is of the same type. To see this, note that all positive and negative powers of $F(z)$ are functions of a similar type, the monomial $z$ being replaced by $z^n$ in the $n$th power, so that we can define functions $f_n(Z)$ inductively by

$$f_1(z) = 1 \text{ and } f_{n+1}(z) = f_n(z) + H_n(z) \ , \quad \text{where } H_n(z) := z - f_n(F(z)) \ ,$$

for $n > 0$. Then $H_{n+1}(z) = H_n(z) - H_n(F(z))$, hence $H_n(z) \equiv 0 \pmod{I^n}$, and the $f_n$'s converge to an $f$ such that $f(F(z)) = z$.

We leave to the reader the task of showing that the map : $w \mapsto P = (u, v(u))$, with $u = f(w)$ is a group homomorphism. I do not know to what extent this isomorphism $R^* \simeq E_t^0(k)$ can be extended to $k^*/q^{\mathbb{Z}} \to E_t(k)$ in this more general situation, as it can in the case $k$ is a local field. Presumably it can at least be extended to points $w \in k^*$, some power of which is in $R^* q^{\mathbb{Z}}$, with the image being the points in $E_t(k)$, some multiple of which is in $E_t^0(k)$. That would enable us to show that the isomorphism $G_{k,t,m} \simeq E_t[m]$ still holds in this more general situation. But that is known anyway, as we indicate in the next section.

**The modular point of view.**

All of the elliptic curves which we have discussed so far are obtainable by base change from one single curve, $E_q$, defined over the ring $\mathbb{Z}[[q]][1/q]$ of finite-tailed Laurent series in an indeterminate $q$ with coefficients in $\mathbb{Z}$. Indeed, given any $R$, $t$ as above, there is a unique homomorphism from $\mathbb{Z}[[q]][1/q]$ to $k = R[1/t]$ taking $q \to t$, and continuous from the $q$-adic topology in $\mathbb{Z}[[q]]$ to the $I$-adic topology in $R$, and thus $E_t$ can be obtained as a base change of $E_q$. A more sophisticated and algebraic construction of $E_q$ than the one we give here, due to Raynaud, is explained in detail in [1], VII. That construction, which involves no specific power series at all, uses the notion of "generalized elliptic curve" and Grothendieck's existence theorem. It produces $E_q$ as the generalized elliptic curve over $\mathbb{Z}[[q]]$ which algebrizes a formal such curve which is obtained as the limit of schemes over $\mathbb{Z}[q]/(q^n)$ which are quotients by the action of $\mathbb{Z}$ on schemes whose geometric fibers are infinite strings of copies of $P^1$ indexed by $\mathbb{Z}$, the point $\infty$ of the $i$th copy being identified with the point 0 on the $(i+1)$th. This approach yields an isomorphism $G_{k,q,m} \to E_q[m]$ for each $m$, where here $k = \mathbb{Z}[[q]][1/q]$.

The curve $E_q$ is used to study the modular curves in the neighborhood of cusps. For example, it identifies $\mathbb{Z}[[q]]$ with the formal completion of $\mathcal{M}_1$ at $j = \infty$, and if a modular form $f$ of level 1 defined over a ring $R$ is viewed as a function which attaches to every pair $(E, \omega)$ consisting of an elliptic curve $E$ over an $R$-algebra $A$, together with a nowhere vanishing differential $\omega$ on $E$, an element $f(E, \omega) \in A$, then the Fourier expansion of $f$ is the series $f(E_q \otimes R, dq/q) \in \mathbb{Z}[[q]][1/q] \otimes R$. Thus the coefficients of the expansion are contained in a finitely generated $\mathbb{Z}$-submodule of $R$. Working over $\mathbb{Z}[[q^{1/n}]][1/q][z]/(z^n - 1)$ one can similarly define the expansions of forms of higher level. But the story of the arithmetic of modular curves and modular forms is very long and all we can do here is to refer the reader to such accounts as [1], [5] and [6], where it is extensively explained.

*References*

[1] Deligne, P. and Rapoport, M., *Les schémas de modules de courbes elliptiques*, in *Modular Functions of One Variable II*, Springer Lecture Notes in Math. **349** (1973), 143–316.

[2] Dwork, B., Gerotto, G. and Sullivan, F.J., *An Introduction to G-functions*, Annals of Math. Studies **133**, Princeton U. Press, 1994.

[3] Greenberg, R. and Stevens, G., *p-adic L-functions and p-adic periods of modular forms*, Invent. Math. **111** (1993), 407–447.

[4]  A. Hurwitz and R. Courant, *Funktionentheorie*, Springer, Berlin, 1929.

[5]  Katz, N., *p-adic properties of modular schemes and modular forms*, in Modular Functions of One Variable III, Springer Lecture Notes in Math. **350** (1973), 69–190.

[6]  Katz, N. and Mazur, B., *Arithmetic Moduli of Elliptic Curves*, Annals of Math. Studies **108**, Princeton U. Press, 1985.

[7]  Mazur, B., Tate, J., Teitelbaum, J., *On p-adic analogs of the conjectures of Birch and Swinnerton-Dyer*, Invent. Math. **84** (1986), 1–48.

[8]  Ribet, K. A., *On modular representations of $Gal(\overline{\mathbb{Q}}/\mathbb{Q})$ arising from modular forms*, Invent. Math. **100** (1989), 359–407.

[9]  Roquette, P., *Analytic theory of elliptic functions over local fields*, Hamburger Math. Einzelschriften, Neue Folge - Heft 1, Vandenhoeck & Ruprecht, Gottingen.

[10]  Serre, J-P., *Abelian l-adic representations and elliptic curves*, W. A. Benjamin, 1968.

[11]  Serre, J-P., *Sur les représentations modulaires de degré 2 de $Gal(\overline{\mathbb{Q}}/\mathbb{Q})$*, Duke Math J. **54** (1987), 179–230.

[12]  Serre, J-P. and Tate, J., *Good Reduction of Abelian Varieties*, Annals of Math. **88** (1968), 492–517.

[13]  Silverman, J., *Advanced Topics in the Arithmetic of Elliptic Curves*, Springer, 1994.

Conference on Elliptic Curves and Modular Forms
Hong Kong, December 18-21, 1993
Copyright ©1995 International Press

# On Galois representations associated to Hilbert modular forms II

RICHARD TAYLOR

D.P.M.M.S., CAMBRIDGE UNIVERSITY,
16 MILL LANE, CAMBRIDGE, CB2 1SB, U.K.

**Introduction.** Let $K$ be a totally real field. In our previous paper we [**12**] we attached a system of $\lambda$-adic representations of $\mathrm{Gal}\,(\overline{K}/K)$ to a cuspidal automorphic representation $\pi$ of $GL_2(\mathbb{A}_K)$ for which $\pi_\infty$ is regular algebraic. Based on the work of Carayol ([**2**]) one can determine the restriction of these representations to the decomposition group at all primes $\wp \nmid l$. One can ask what happens at primes $\wp | l$. One would expect these representations always to be de Rham and to be crystalline if $\wp$ does not divide the conductor of $\pi$. When there is a direct geometric construction of these $\lambda$-adic representations (see [**2**] and [**1**]) results of this sort follow from the work of Faltings ([**4**]). Tony Scholl raised the question of what can be said in the remaining cases where the only construction is via a congruence argument (see [**12**]). In this paper we give a partial answer to this question. We show that for all but finitely many $\lambda$ the $\lambda$-adic representations are crystalline. In fact the argument is rather easy given a recent, important, but entirely algebraic result of Carayol (proposition 1 of [**3**]). In fact we shall show that if $\pi_\infty$ is the lowest discrete series representation, $\wp$ does not divide the conductor of $\pi$, and the mod $\lambda$ representation associated to $\pi$ is irreducible then the $\lambda$-adic representation associated to $\pi$ is crystalline. This suffices as the mod $\lambda$ representation associated to $\pi$ can only be irreducible for finitely many $\lambda$. As this latter result is not explicitly in the literature we sketch a proof, but we stress that the result is well known to experts and the arguments are immediate generalisations of arguments of Ribet ([**9**]) and Serre ([**10**]).

**1 Notation and Results.** We shall let $K$ denote a totally real field; $\mathbb{A}_K$ the ring of adeles of $K$; $I$ the set of embeddings of $K \hookrightarrow \mathbb{R}$; and for $\tau \in I$ we shall let $v_\tau$ denote the corresponding place of $K$. We shall also let $\pi = \otimes_v \pi_v$ denote a cuspidal automorphic representation of $GL_2(\mathbb{A}_K)$ and $N(\pi)$ its conductor of $\pi$.

We shall let $c$ denote complex conjugation. If $k \in \mathbb{Z}_{\geq 2}$ we shall let $\sigma_k$ denote the irreducible admissible representation of $GL_2(\mathbb{R})$ with Langlands parameter

$$W_\mathbb{R} = \langle \mathbb{C}^\times, j | \ j^2 = -1; \ jzj^{-1} = {}^c z \rangle \longrightarrow GL_2(\mathbb{C})$$

$$z \longmapsto \begin{pmatrix} z^{1-k} & 0 \\ 0 & {}^c z^{1-k} \end{pmatrix} |z|$$

$$j \longmapsto \begin{pmatrix} 0 & 1 \\ (-1)^{1-k} & 0 \end{pmatrix}.$$

We shall say that $\pi$ has weight $\mathbf{k} \in \mathbb{Z}^I_{\geq 2}$ if for all $\tau \in I$ the component $\pi_{v_\tau} = \sigma_{k_\tau}$. For the rest of this paper we shall assume that $\pi$ has some such weight $\mathbf{k}$.

If $N$ is an ideal of $\mathcal{O}_K$ then we shall let $\mathbb{T}_N$ denote the polynomial algebra over $\mathbb{Z}$ in variables $T_\wp$ and $S_\wp$ for primes $\wp$ of $\mathcal{O}_K$ not dividing $N$. If $N(\pi)|N$ then we get a character $\theta_\pi : \mathbb{T}_N \to \mathbb{C}$ which sends $T_\wp$ (resp. $S_\wp$) onto the eigenvalue of the Hecke operator $[GL_2(\mathcal{O}_{K,\wp}) \begin{pmatrix} 1 & 0 \\ 0 & \varpi_\wp \end{pmatrix} GL_2(\mathcal{O}_{K,\wp})]$ (resp. $[GL_2(\mathcal{O}_{K,\wp})\varpi_\wp GL_2(\mathcal{O}_{K,\wp})]$) on $\pi_\wp^{GL_2(\mathcal{O}_{K,\wp})}$. Here $\varpi_\wp$ denotes a uniformiser in $\mathcal{O}_{K,\wp}$. It is well known that the field generated by the image of $\theta_\pi$ is a number field, which we denote $E_\pi$, and that in fact $\theta_\pi$ is valued in the ring of integers of $E_\pi$. In the rest of this paper $\lambda$ will denote a prime of $E_\pi$ and $l$ will denote its residue characteristic.

Then the main theorem of [12] is as follows.

**Theorem 1.1** *There is a continuous representation*

$$\rho_{\pi,\lambda} : \mathrm{Gal}\,(\overline{K}/K) \to GL_2(E_{\pi,\lambda})$$

*such that if $\wp \nmid N(\pi)l$ then $\rho_{\pi,\lambda}$ is unramified at $\wp$ and $\rho_{\pi,\lambda}(\mathrm{Frob}\,_\wp)$ has characteristic polynomial*

$$X^2 - \theta_\pi(T_\wp)X + (\mathbf{N}\wp)\theta_\pi(S_\wp).$$

*In fact if $\wp$ is any prime of $F$ not dividing $l$ then the representation $\sigma_{\pi,\wp,\lambda}$ of the Weil-Deligne group of $F_\wp$ corresponding to the restriction of $\rho_{\pi,\lambda}$ to the decomposition group at $\wp$ (see [11]) is rational over $E_\pi$ and has Frobenius semi-simplification $\sigma_{\pi,\wp}$, the representation of the Weil-Deligne group of $F_\wp$ associated to $\pi_\wp$ by the local Langlands correpondence (see [2]).*

The behaviour of $\rho_{\pi,\lambda}$ at primes dividing $l$ would be at least partly described by the following conjecture.

**Conjecture 1** *If $\wp$ is a prime of $\mathcal{O}_K$ above $l$ then the restriction of $\rho_{\pi,\lambda}$ to the decomposition group at $\wp$ is de Rham. If further $\wp \nmid N(\pi)$ then it is crystalline.*

Combining the work of Carayol [2] with that of Faltings [4] it is not hard to see that the conjecture is true if either $[K : \mathbb{Q}]$ is odd or if $\pi_v$ is discrete series at some finite place $v$. That is we have the following theorem.

**Theorem 1.2 (Carayol, Faltings)** *Suppose that either $[F : \mathbb{Q}]$ is odd or that $\pi_v$ is discrete series for some finite place $v$. Suppose also that $\wp$ is a prime of $\mathcal{O}_K$ above $l$. Then the restriction of $\rho_{\pi,\lambda}$ to the decomposition group at $\wp$ is de Rham and if $\wp \nmid N(\pi)$ then it is crystalline.*

In the case that $k_\tau > 2$ for some $\tau \in I$, Blasius and Rogawski [1] gave a second construction of the representations $\rho_{\pi,\lambda}$ and hence, using Faltings' results [4], can show the following result.

**Theorem 1.3 (Faltings, Blasius, Rogawski)** *Suppose that for some $\tau \in I$ $k_\tau > 2$.*
  1. *If $\wp|l$ then the restriction of $\rho_{\pi,\lambda}$ to the decomposition group at $\wp$ is de Rham.*

2. *For all but finitely many primes $\lambda$ of $E_\pi$ and for all primes $\wp$ of $\mathcal{O}_K$ dividing $l$ (which depends on $\lambda$) the restriction of $\rho_{\pi,\lambda}$ to the decomposition group at $\wp$ is crystalline.*

It is not clear to us whether their method gives the rest of conjecture 1 in the case that $k_\tau > 2$ for some $\tau \in I$. In this paper we shall prove the following result.

**Theorem 1.4** *For all but finitely many primes $\lambda$ of $E_\pi$ and for all primes $\wp$ of $\mathcal{O}_K$ dividing $l$ (which depends on $\lambda$) the restriction of $\rho_{\pi,\lambda}$ to the decomposition group at $\wp$ is crystalline.*

From theorems 1 and 2 we see that we can restrict our attention to the case $[F : \mathbb{Q}]$ even and $k_\tau = 2$ for all $\tau$. We shall assume this for the rest of this paper.

In fact, in this case, we shall prove a slightly more precise result. Recall that $\rho_{\pi,\lambda}$ is conjugate to a representation valued in $GL_2(\mathcal{O}_{E_\pi,\lambda})$ and hence we can form its reduction modulo $\lambda$, which we shall denote $\overline{\rho_{\pi,\lambda}}$. This may depend on the choice of $GL_2(E_{\pi,\lambda})$-conjugate we take, but its semi-simplification $\overline{\rho_{\pi,\lambda}}^{ss}$ depends only on $\pi$ and $\lambda$. In section 3 we shall prove the following result.

**Proposition 1.5** *$\overline{\rho_{\pi,\lambda}}^{ss}$ is irreducible for all but finitely many primes $\lambda$.*

Our main theorem will then be the following.

**Theorem 1.6** *Suppose that $\wp$ is a prime of $\mathcal{O}_K$ and that $\lambda$ is a prime of $E_\pi$, both lying above $l$. Suppose moreover that $\wp \nmid N(\pi)$ and that $\overline{\rho_{\pi,\lambda}}^{ss}$ is irreducible. Finally suppose that $k_\tau = 2$ for all $\tau \in I$. Then there is an $l$-divisible group $A/\mathcal{O}_{K,\wp}$ such that $\rho_{\pi,\lambda}$ is isomorphic to $VA$ as $\mathbb{Q}_l[\mathrm{Gal}\,(\overline{K_\wp}/K_\wp)]$ modules.*

Here $VA$ denotes the Tate module $(\varprojlim A[l^n](\overline{K})) \otimes_{\mathbb{Z}_l} \mathbb{Q}_l$. We note that the theorem and the proposition together imply theorem 1.4.

## 2 The Main Theorem.

We now turn to the proof of theorem 1.6. We first recall a result about finite flat group schemes.

**Lemma 2.1** *Suppose that $L/\mathbb{Q}_l$ is a finite extension.*
1. *Suppose that $A/L$ is an $l$-divisible group. Then $A$ extends to an $l$-divisible group $\widetilde{A}/\mathcal{O}_L$ if and only if for all $n$, $A[l^n]$ extends to a finite flat group scheme $\widetilde{A[l^n]}/\mathcal{O}_L$.*
2. *Suppose that $\widetilde{G}/\mathcal{O}_L$ is a finite flat group scheme. If $H$ is a subquotient of the generic fibre of $\widetilde{G}$ then $H$ extends to a finite flat group scheme $\widetilde{H}/\mathcal{O}_L$.*
3. *Suppose that $A$ and $B$ are isogenous $l$-divisible groups over $L$. Then $A$ extends to an $l$-divisible group $\widetilde{A}/\mathcal{O}_L$ if and only if $B$ extends to an $l$-divisible group $\widetilde{B}/\mathcal{O}_L$.*
4. *Suppose that $T/\mathbb{Z}_l$ is a finite algebra and that $\rho : \mathrm{Gal}\,(\overline{L}/L) \to GL_d(T)$ is a continuous homomorphism. Suppose that there is an $l$-divisible group $\widetilde{A}/\mathcal{O}_L$ such that $V\widetilde{A} \cong (\rho \otimes_{\mathbb{Z}_l} \mathbb{Q}_l)$ as $\mathbb{Q}_l[\mathrm{Gal}\,(\overline{L}/L)]$-modules. Suppose finally that $I$ is an ideal of $T$ of finite index. Then there is a finite flat group scheme $\widetilde{G}/\mathcal{O}_L$ such that $\widetilde{G}(\overline{L}) \cong (\rho \bmod I)$ as $\mathrm{Gal}\,(\overline{L}/L)$-modules.*

The first part of this proposition is a theorem of Raynaud (see [8]). We thank Fontaine for pointing this out to us. For the second part there are sub-finite flat group schemes $G_1 \subset G_2$ of the special fibre of $\widetilde{G}$. Let $\widetilde{G_i}$ denote the scheme theoretic closure of $G_i$ in $\widetilde{G}$. Then $\widetilde{G_1}/\widetilde{G_2}$ is a finite flat group scheme over $\mathcal{O}_L$ with generic fibre $H$. For the third part suppose that $A$ extends to an $l$-divisible group scheme over $\mathcal{O}_L$ and that there is a surjection $A \twoheadrightarrow B$. Then for all $n$, $B[l^n]$ is a sub-quotient of $A[l^{n+r}]$ for suitable $r$. Hence $B[l^n]$ extends to a finite flat group scheme over $\mathcal{O}_L$ for all $n$, and so by the first part $B$ extends to a $l$-divisible over $\mathcal{O}_L$.

For the final part note that $(T \otimes_{\mathbb{Z}_l} \mathbb{Q}_l)^d / I^d$ with the action of $\text{Gal}(\overline{L}/L)$ defines an $l$-divisible group over $L$ isogenous to the special fibre of $\widetilde{A}$. Hence it extends to an $l$-divisible group $\widetilde{B}/\mathcal{O}_L$. Then for sufficiently large $n$, the finite flat group scheme over $L$ defined by $\rho \bmod I$ is a sub-group scheme of the generic fibre of $\widetilde{B}[l^n]$, and hence extends to a finite flat group scheme over $\mathcal{O}_L$, as desired.

We now return to theorem 1.6. Because $\overline{\rho_{\pi,\lambda}}^{ss}$ is irreducible any two $GL_2(E_{\pi,\lambda})$-conjugates of $\rho_{\pi,\lambda}$ which are valued in $GL_2(\mathcal{O}_{E_\pi,\lambda})$ are $GL_2(\mathcal{O}_{E_\pi,\lambda})$-conjugate. Hence $\rho_{\pi,\lambda}$ is well defined as a representation into $GL_2(\mathcal{O}_{E_\pi,\lambda})$ as is its reduction modulo $\lambda^n$, which we shall denote $\rho_\pi/\lambda^n$. It suffices to prove that for infinitely many $n$ there is a finite flat group scheme $A_n/\mathcal{O}_{K,\wp}$ such that $A_n(\overline{K_\wp}) \cong \rho_\pi/\lambda^n$ as $\text{Gal}(\overline{K_\wp}/K_\wp)$-modules. We now fix $n$ for the rest of this section.

We must recall something of our construction of $\rho_{\pi,\lambda}$ (see [12]). We shall let $\widehat{\mathcal{O}_K}$ denote the product of all completions of $\mathcal{O}_K$ at finite places. If $N$ is an ideal of $\mathcal{O}_K$ we let $U_1(N)$ denote the subgroup of $GL_2(\widehat{\mathcal{O}_K})$ consisting of matrices $\begin{pmatrix} a & b \\ c & d \end{pmatrix}$ with $c$ and $d-1 \in N\widehat{\mathcal{O}_K}$. We let $\mathbb{T}(N)$ denote the image of $\mathbb{T}_N$ in the ring of endomorphisms of $\bigoplus_\pi (\pi^\infty)^{U_1(N)}$, where $\pi$ runs over cuspidal automorphic representations of $GL_2(\mathbb{A}_F)$ of weight $\mathbf{2}$ and $\pi^\infty$ denotes the product over all finite places $v$ of $\pi_v$. If a prime $\mathfrak{r}$ divides $N$ exactly once then we let $\mathbb{T}(N)^{\mathfrak{r}-new}$ denote the quotient of $\mathbb{T}(N)$ which acts faithfully on $\bigoplus_\pi (\pi^\infty)^{U_1(N)}$, where $\pi$ runs over cuspidal automorphic representations of $GL_2(\mathbb{A}_F)$ of weight $\mathbf{2}$ and with $\pi_\mathfrak{r}$ special.

If $N(\pi)|N$ then $\theta_\pi$ factors through $\mathbb{T}(N)$ and we get an embedding

$$\prod_\pi \theta_\pi : \mathbb{T}(N) \hookrightarrow \prod_\pi \mathcal{O}_{E_\pi},$$

where the product is over a set of representatives of $\text{Gal}(\overline{\mathbb{Q}}/\mathbb{Q})$ orbits of cuspidal automorphic representations $\pi$ of weight two and conductor dividing $N$. This map becomes an isomorphism after tensoring with $\mathbb{Q}$. Similarly if $\mathfrak{r}$ divides $N$ exactly once one get a similar map from $\mathbb{T}(N)^{\mathfrak{r}-new}$ to a similar product but with the added restriction that only those $\pi$ for which $\pi_\mathfrak{r}$ is special occur.

In [12] we prove that there is a prime $\mathfrak{r} \nmid N(\pi)l$ and a homomorphism $\theta : \mathbb{T}(N(\pi)\mathfrak{r})^{\mathfrak{r}-new} \longrightarrow \mathcal{O}_{E_\pi}/\lambda^n$ such that

$$\begin{array}{ccc} \mathbb{T}(N(\pi)\mathfrak{r}) & \longrightarrow & \mathbb{T}(N(\pi)\mathfrak{r})^{\mathfrak{r}-new} \\ \downarrow & & \downarrow \\ \mathcal{O}_{E_\pi} & \longrightarrow & \mathcal{O}_{E_\pi}/\lambda^n \end{array}$$

commutes. Here the horizontal arrows are the natural surjections and the vertical arrows are the maps $\theta_\pi$ and $\theta$ respectively.

Let $\mathfrak{m}$ denote the kernel of the composite map

$$\mathbb{T}(N(\pi)\mathfrak{r})^{\mathfrak{r}-new} \xrightarrow{\theta} \mathcal{O}_{E_\pi}/\lambda^n \longrightarrow \mathcal{O}_{E_\pi}/\lambda,$$

and let $\mathbb{T}$ denote the completion of $\mathbb{T}(N(\pi)\mathfrak{r})^{\mathfrak{r}-new}$ at $\mathfrak{m}$. Then $\theta$ factors through $\mathbb{T}$. Let $I$ denote the kernel of $\mathbb{T} \xrightarrow{\theta} \mathcal{O}_{E_\pi}/\lambda^n$. Now $\mathbb{T} \otimes_{\mathbb{Z}_l} \mathbb{Q}_l \cong \prod_j E_{\pi_j, \lambda_j}$, where $j$ runs over some finite set; $\pi_j$ is a cuspidal automorphic representation of $GL_2(\mathbb{A}_F)$ of weight **2**, of conductor dividing $N(\pi)\mathfrak{r}$ and with $\pi_{j,\mathfrak{r}}$ special; and $\lambda_j$ is a prime of $E_{\pi_j}$ above $l$. For each $j$ there is a continuous representation $\rho_j : \mathrm{Gal}\,(\overline{K}/K) \to GL_2(E_{\pi_j, \lambda_j})$ such that if $\mathfrak{q}$ is a prime of $K$ not dividing $N(\pi)\mathfrak{r}l$ then $\rho_j$ is unramified at $\mathfrak{q}$ and $\rho_j(\mathrm{Frob}_\mathfrak{q})$ has characteristic polynomial

$$X^2 - \theta_{\pi_j}(T_\mathfrak{q})X + (\mathbf{N}\mathfrak{q})\theta_{\pi_j}(S_\mathfrak{q}).$$

Moreover Carayol ([**2**]) has shown that $\rho_j$ occurs in the Tate module of a Jacobian of a Shimura curve defined over $K$ and which has good reduction at $\wp$. Hence there is an $l$-divisible group $A_j/\mathcal{O}_{K,\wp}$ such that $VA_j \cong \rho_j$ as $\mathbb{Q}_l[\mathrm{Gal}\,(\overline{K_\wp}/K_\wp)]$-modules. Thus we get a continuous representation $\rho : \mathrm{Gal}\,(\overline{K}/K) \to GL_2(\mathbb{T} \otimes_{\mathbb{Z}_l} \mathbb{Q}_l)$ and an $l$-divisible group $A/\mathcal{O}_{K,\wp}$ such that $\rho \cong VA$ as $\mathbb{Q}_l[\mathrm{Gal}\,(\overline{K_\wp}/K_\wp)]$-modules and such that if $\mathfrak{q}$ is a prime of $K$ not dividing $N(\pi)\mathfrak{r}l$ then $\rho$ is unramified at $\mathfrak{q}$ and $\rho(\mathrm{Frob}_\mathfrak{q})$ has characteristic polynomial

$$X^2 - T_\mathfrak{q}X + (\mathbf{N}\mathfrak{q})S_\mathfrak{q}.$$

The representations $\overline{\rho_j}^{ss}$ have the same trace as $\overline{\rho_{\pi,\lambda}}^{ss}$ on Frobenius elements and hence are isomorphic. Thus they are all irreducible by assumption. Crucially, we can now apply proposition 1 of [**3**] to conclude that the representation $\rho$ is conjugate to one valued in $GL_2(\mathbb{T})$. Now applying part 4 of our proposition we see that there is a finite flat group scheme $G/\mathcal{O}_{K,\wp}$ such that

$$G(\overline{K_\wp}) \cong (\theta \bmod I) \cong \theta \circ \rho \cong \rho_\pi/\lambda^n$$

as $\mathrm{Gal}\,(\overline{K_\wp}/K_\wp)$-modules. This completes the proof of theorem 1.6.

**3 Irreducibility Results.** In this section we remind the reader how to prove proposition 1. First we record the following result.

**Proposition 3.1** *If $\pi$ is a cuspidal automorphic representation of $GL_2(\mathbb{A}_K)$ of some weight $\mathbf{k}$ then $\rho_{\pi,\lambda}$ is irreducible.*

The proof is just Ribet's argument from [**9**], we repeat it as we know no reference in this context. Suppose that $\rho_{\pi,\lambda}$ were reducible. Then $\rho_{\pi,\lambda}^{ss}$ is a continuous, semi-simple abelian representation of $\mathrm{Gal}\,(\overline{K}/K)$ such that all but finitely many Frobenius elements have characteristic polynomials which are rational over the number field $E_\pi$. By the results of Serre and Henniart (see [**5**]) we see that there are two grossencharacters of type $A_0$, $\chi_1, \chi_2 : \mathbb{A}_K^\times \to \mathbb{C}^\times$ such that for $\wp \nmid N(\pi)l$ the characteristic polynomial of $\rho_{\pi,\lambda}(\mathrm{Frob}_\wp)$ is

$$(X - \chi_1(\varpi_\wp))(X - \chi_2(\varpi_\wp)) \in E_\pi[X].$$

Because $K$ has a real place and the $\chi_i$ are of type $A_0$ we know that there is an integer $w_i \in \mathbb{Z}$ such that for all $\tau \in I$,

$$\chi_i|_{(K_{v_\tau}^\times)_{>0}} : t \longmapsto t^{w_i};$$

and hence that $|\chi_i(\varpi_\wp)| = |\mathbf{N}\wp|^{w_i}$. By corollary 2.5 of [7] we know that $(k-2)/2 < w_i < k/2$ and hence that $w_i = (k-1)/2$ for $i = 1, 2$. In particular $\chi_2/\chi_1$ is a unitary (in fact finite order) character. Let $S$ denote the set of primes dividing $lN(\pi)$. Then we have an equality of partial L-functions

$$L_S(\pi \otimes \chi_1^{-1}, s) = \zeta_{K,S}(s) L_S(\chi_2/\chi_1, s).$$

As $\pi$ is a cusp form $L_S(\pi \otimes \chi_1^{-1}, s)$ is entire (see [6]), while $\zeta_{K,S}(s)$ has a pole at $s = 1$. Thus $L_S(\chi_2/\chi_1, s)$ must vanish at $s = 1$, but this is impossible as $\chi_2/\chi_1$ is unitary. This contradiction proves the proposition.

We now turn to proposition 1. First we have the following lemma.

**Lemma 3.2** *Suppose that $\pi$ is a cuspidal automorphic representation of $GL_2(\mathbb{A}_K)$ of weight 2. Suppose that $\lambda$ is a prime of $E_\pi$ above $l$. Suppose that $\wp$ is an unramified prime of $K$ above $l$ which does not divide $N(\pi)$. Let $I$ denote the inertia group at $\wp$ and let $\psi_i$ denote the natural character $I \twoheadrightarrow \mathbb{F}_{l^i}^\times$ for $i \in \mathbb{Z}_{\geq 0}$. Then over an algebraically closed field the semisimplification of $\overline{\rho_{\pi,\lambda}}$ is either $\psi_0 \oplus \psi_1$ or $\psi_2 \oplus \psi_2^l$.*

From the classification of the simple finite flat group schemes over the maximal unramified extension of $\mathbb{Q}_l$, which extend to finite flat group schemes over the ring of integers of the maximal unramified extension of $\mathbb{Q}_l$ (see [8]), we know that the simple factors of $\overline{\rho_{\pi,\lambda}}^{ss}$ must be of the form $\psi_i^{l^j}$. Also the product must be $\psi_1$ because the determinant of $\overline{\rho_{\pi,\lambda}}$ is the product of a character unramified at $\wp$ with the cyclotomic character. This leaves only the two possibilities mentioned in the lemma.

Proposition 1 follows immediately from this lemma, proposition 2 and the following result, which is essentially due to Serre.

**Proposition 3.3** *Let $K$ and $E$ be number fields and let $\Gamma$ denote the set of embeddings of $K$ into the algebraic closure of $E$. Let $N$ be an ideal of $\mathcal{O}_K$.*

*Let $\mathcal{L}$ be an infinite set of primes $\lambda$ of $\mathcal{O}_E$. Let $\lambda$ lie above the rational prime $l(\lambda)$ and have residue field $k_\lambda$. For $\lambda \in \mathcal{L}$ let $\rho_\lambda : \mathrm{Gal}\,(\overline{K}/K) \to GL_d(\mathcal{O}_{E,\lambda})$ be a continuous representation. If $\wp | l(\lambda)$ suppose that the conductor of $\rho_\lambda|_{\mathrm{Gal}\,(\overline{K_\wp}/K_\wp)}$ is less than or equal to the $\wp$-adic valuation of $N$. If $\wp \nmid N$ suppose that there exists a polynomial $P_\wp(T) \in \mathcal{O}_E[T]$ such that $P_\wp(T)$ is the characteristic polynomial of $\rho_\lambda(\mathrm{Frob}\,_\wp)$ for all $\lambda$ with residue characteristic distinct from that of $\wp$.*

*Also suppose that for each $\lambda$, $\overline{\rho_\lambda}^{ss}$ is the sum of $d$ characters $\theta_{\lambda,i}$. Restricting these characters to the inertia groups at primes above $l(\lambda)$ and composing with the local reciprocity map we get maps $\theta^*_{\lambda,i} : \mathcal{O}^\times_{K,l(\lambda)} \to k_\lambda^\times$. Suppose finally that there are integers $n_{\lambda,i}(\sigma)$ for $\sigma \in \Gamma$ which are bounded independently of $\lambda$ and such that $\theta^*_{\lambda,i}$ is the reduction of $\alpha \mapsto \prod_{\sigma \in \Gamma}(\sigma\alpha)^{n_{\lambda,i}(\sigma)}$.*

*Then for infinitely many $\lambda \in \mathcal{L}$ the representation $\rho_\lambda^{ss}$ is abelian.*

If $E = \mathbb{Q}$ this is theorem 1 of [**10**]. The same proof works in general. Replacing $\mathcal{L}$ by an infinite subset we may assume that the $\lambda$ have distinct residue characteristics and that the $n_{\lambda,i}(\sigma)$ are independent of $\lambda$. Again replacing $\mathcal{L}$ by an infinite subset and applying proposition 20 of [**10**] we may assume that there are characters $\psi_1, ..., \psi_d \in X(S_N)$ over $\overline{E}$ with $(\widetilde{\psi_i})_{l(\lambda)} = \overline{\theta_{\lambda,i}}$. Here and in the rest of this paragraph we use the notation of [**10**] and we suppose the prime of $\overline{E}$ above $l(\lambda)$ implicit in the notation is in fact chosen to be above $\lambda$. Let $\psi = \bigoplus_{i=1}^d \psi_i$. If $\wp$ is a prime of $K$ not dividing $N$ then as on page 292 of [**10**] one sees that $\psi(F_\wp)$ has characteristic polynomial $P_\wp(T)$. Let $\psi_\lambda$ be the corresponding $l(\lambda)$-adic representation of $\mathrm{Gal}\,(\overline{K}/K)$ as defined in section 3.4 of [**10**]. If $\wp \nmid Nl(\lambda)$ then $\psi_\lambda$ and $\rho_\lambda$ are both unramified at $\wp$ and $\psi_\lambda(\mathrm{Frob}\,_\wp)$ and $\rho_\lambda(\mathrm{Frob}\,_\wp)$ have the same characteristic polynomial $(P_\wp(T))$. The proposition follows.

## References

[1]  D.Blasius and J.Rogawski, *Motives for Hilbert modular forms*, Invent. Math. 114 (1993) 55-87.

[2]  H.Carayol, *Sur les représentations p-adiques associées aux formes modulaires de Hilbert*, Ann. Sci. ENS (4) 19 (1986) 409-468.

[3]  H.Carayol, *Formes modulaires et représentations Galoisiennes à valeurs dans un anneau local complet*, in "p-adic monodromy and the Birch-Swinnerton-Ryer conjecture" (eds. B.Mazur and G.Skerers), Contemporary Math. 165 (1994).

[4]  G.Faltings, *Crystalline cohomology and p-adic Galois representations*, Algebraic Analysis, Geometry and Number Theory, Proc. JAMI Inaugural Conference, Johns-Hopkins Univ. Press (1989), 25-79.

[5]  G.Henniart, *Représentations l-adiques abéliennes*, in Séminaires de Theorie des Nombres, Paris 1980-81 (ed. M.-J.Bertin), Birkhäuser (1982), 107-126.

[6]  H.Jacquet and R.P.Langlands, *Automorphic forms on $GL_2$*, LNM 114, Springer 1970.

[7]  H.Jacquet and J.Shalika, *On Euler products and the classification of automorphic representations I*, Am. J. Math. 103(2) (1981), 499-558.

[8]  M.Raynaud, *Schémas en groupes de type $(p, ..., p)$*, Bull. Soc. Math. France 102 (1974) 241-280.

[9]  K.Ribet, *Galois representations attached to eigenforms with nebentypus*, in modular Forms of One Variable V (ed. J.-P.Serre and D.Zagier), LNM 601 Springer 1977, 17-51.

[10] J.-P.Serre, *Propriétés galoisiennes des points d'ordre fini des courbes elliptiques*, Invent. Math. 15 (1972), 259-331.

[11] J.Tate, *Number theoretic background*, in Automorphic Forms, Representations and L-functions (ed. A.Borel and W.Casselman), Proc. Symp. Pure Math. XXXIII (2), AMS 1979, 3-26.

[12] R.Taylor, *On Galois representations associated to Hilbert modular forms*, Invent. Math. 98 (1989), 265-280.

QA 244 .E44 1995

ELLIPTIC CURVES, MODULAR
FORMS, & FERMAT'S LAST

### DATE DUE

| NOV 28 1998 | | | |
|---|---|---|---|
| FEB 2 5 2000 | | | |
| AUG 3 0 1999 | | | |
| DEC 1 5 2003 | | | |
| | | | |
| | | | |
| | | | |
| | | | |
| | | | |
| | | | |
| | | | |
| | | | |
| | | | |
| | | | |
| | | | |
| | | | |
| | | | |